Lecture Notes in Nanoscale Science and Technology

Volume 21

For further volumes:
http://www.springer.com/series/7544

Zhiming M. Wang
Editor

MoS_2

Materials, Physics, and Devices

Springer

Editor
Zhiming M. Wang
State Key Laboratory of Electronic
University of Electronic Science
and Technology
Chengdu
People's Republic of China

ISSN 2195-2159 ISSN 2195-2167 (electronic)
ISBN 978-3-319-02849-1 ISBN 978-3-319-02850-7 (eBook)
DOI 10.1007/978-3-319-02850-7
Springer Cham Heidelberg New York Dordrecht London

Library of Congress Control Number: 2013953630

Printed on acid-free paper

Springer is part of Springer Science+Business Media (www.springer.com)

Preface

Molybdenum Disulfide (MoS_2) is a family member of transition-metal dichalcogenides with a formula MX_2, where M and X correspond to a transition metal and chalcogen, respectively. Like graphene, MoS_2 and related dichalocogenides are two-dimensional (2D) layered materials, with a covalently bonded hexagonal network in layers stacked by weak Van der Waals forces between layers.

Graphene is a groundbreaking 2D material which has been extensively studied by the solid-state community. However, pristine graphene does not have a nonzero bandgap, a property which is mandatory for many electronic and optoelectronic applications. In contrast, it is demonstrated that MoS_2 undergoes a gradual transition from an indirect bandgap in bulk to a direct bandgap in a single layer material. Transport measurements further demonstrate high room-temperature mobility and large on/off ratio in single-layer MoS_2 field-effect transistors. These advantages led to a high expectation for the 2D MoS_2 materials system, which sparked a large amount of research involving both theory and experiment.

Chapters 1–5 of this book lay a theoretical foundation. In Chap. 1, progress in the theoretical study of 2D MoS_2 monolayers and nanoribbons is reviewed. Chapter 2 focuses on the electronic structure of exfoliated MoS_2. Chapter 3 reports on tunability of electronic and dielectric properties of MoS_2 by varying the layer thickness, applying mechanic strain, tuning the interlayer distance, and by applying an external electric field. Chapter 4 further shows the chemical trend, band alignment, alloying, and gap modulation on MoS_2 and its family by ab initio studies. Chapter 5 goes beyond the first principles of 2D-layered MoS_2 and includes the studies of the electronic states of one-dimensional and zero-dimensional MoS_2 nanostructures.

Experimental efforts are reviewed in Chaps. 6–8. Use of the micromechanical cleavage technique to obtain MoS_2 thin flakes down to a single layer is reviewed in Chap. 6. Several technologically relevant properties such as optical properties, elasticity, and electric field screening are also explored in this chapter. Chapter 7 studies the vibrational and electronic properties of MoS_2 using Raman, photoluminescence, and transport measurements. Chapter 8 investigates single-and few-layer MoS_2 by means of optical spectroscopy techniques, including photocarrier dynamics in time-resolved photoluminescence experiments.

Due to its important mechanical, electrical, and optical properties, MoS_2 has found enormous applications in the fields of electrochemical devices, hydrogen storage, catalysis, capacitors, solid lubricant, and as an intercalation host. Chapter 9 focuses on the application of MoS_2 nanomaterials to energy storage and conversion. Chapter 10 studies the valley polarization in transition-metal dichalcogenides by optical pumping, and reviews the emergence of valleytronics, a concept in electronics based on manipulating the valley index. The valley index, like charge and spin, can be one additional dimension of carrier freedom to control for in electronic and optoelectronic devices.

The study of 2D layered materials, especially transition-metal dichalcogenides represented by MoS_2, is an active and rapidly growing field of research. This field may lead to fabrication of nanodevices in a layer-by-layer fashion using 2D materials. The Editor, together with editorial assistant Mr. Xin Tong, wishes that this book will serve as a snapshot of the current research status including opportunities and challenges as well as a foundation for future efforts.

Chengdu, China Zhiming M. Wang

Contents

Contributors

Nicolás Agraït Universidad Autónoma de Madrid, Madrid, Spain, e-mail: nicolas.agrait@uam.es

P. K. Ahluwalia Department of Physics, Himachal Pradesh University, Shimla, Himachal Pradesh 171005, India, e-mail: pk_ahluwalia7@yahoo.com

Achintya Bera Department of Physics, Indian Institute of Science (IISc), Bangalore 560012, India, e-mail: achintya@physics.iisc.ernet.in

Andres Castellanos-Gomez Kavli Institute of Nanoscience, Delft University of Technology, NL-2628 CJ Delft, The Netherlands, e-mail: a.castellanosgomez@tudelft.nl

Yingchun Cheng Physical Sciences and Engineering Division, KAUST, Thuwal, Makkah 23955–6900, Saudi Arabia, e-mail: Yingchun.Cheng@kaust.edu.sa

Xiaodong Cui Department of Physics, The University of Hong Kong, Pokfulam road, Hong Kong, People's Republic of China, e-mail: xdcui@hku.hk

Junfeng Dai Department of Physics, South University of Science and Technology of China, Shenzhen 518055, People's Republic of China, e-mail: dai.jf@sustc.edu.cn

Shujiang Ding School of Chemical and Biomedical Engineering, Nanyang Technological University, 70 Nanyang Drive, Singapore 637457, Singapore, e-mail: dingsj@mail.xjtu.edu.cn

Jonathan Eroms Institut für Experimentelle und Angewandte Physik, University of Regensburg, D-93040 Regensburg, Germany, e-mail: Jonathan.Eroms@physik.uni-regensburg.de

Stefanie Heydrich Institut für Experimentelle und Angewandte Physik, University of Regensburg, D-93040 Regensburg, Germany, e-mail: Stefanie.Heydrich@physik.uni-regensburg.de

E. S. Kadantsev Department of Chemistry, University of Ottawa, Ottawa, ON K1N 6N5, Canada, e-mail: ekadants@uottawa.ca

Jun Kang State Key Laboratory for Superlattices and Microstructures, Institute of Semiconductors, Chinese Academy of Sciences, Beijing 100083, People's Republic of China, e-mail: jkang@semi.ac.cn

T. Korn Institut für Experimentelle und Angewandte Physik, University of Regensburg, D-93040 Regensburg, Germany, e-mail: tobias.korn@physik.uni-regensburg.de

A. Kumar Department of Physics, Himachal Pradesh University, Shimla, Himachal Pradesh 171005, India, e-mail: ashok.1777@yahoo.com

Jingbo Li State Key Laboratory for Superlattices and Microstructures, Institute of Semiconductors, Chinese Academy of Sciences, Beijing 100083, People's Republic of China, e-mail: jbli@semi.ac.cn

Jin Liang Xi'an Jiaotong University, No. 28, Xianning West Road, Xi'an 710049, Shaanxi, People's Republic of China, e-mail: liangjin1109@stu.xjtu.edu.cn

Hui Pan Institute of High Performance Computing, A*Star, Singapore, 138632, e-mail: panh@ihpc.a-star.edu.sg

Gerd Plechinger Institut für Experimentelle und Angewandte Physik, University of Regensburg, D-93040 Regensburg, Germany, e-mail: Gerd.Plechinger@physik.uni-regensburg.de

J. Quereda Universidad Autónoma de Madrid, Madrid, Spain, e-mail: jorge.quereda@uam.es

G. Rubio-Bollinger Universidad Autónoma de Madrid, Madrid, Spain, e-mail: gabino.rubio@uam.es

Franz Schrettenbrunner Institut für Experimentelle und Angewandte Physik, University of Regensburg, D-93040 Regensburg, Germany, e-mail: Franz-Xaver.Schrettenbrunner@physik.uni-regensburg.de

Christian Schüller Institut für Experimentelle und Angewandte Physik, University of Regensburg, D-93040 Regensburg, Germany, e-mail: Christian.Schueller@physik.uni-regensburg.de

Udo Schwingenschloegl Physical Sciences and Engineering Division, KAUST, Thuwal, Makkah 23955–6900, Saudi Arabia, e-mail: udo.Schwingenschlogl@kaust.edu.sa

A. K. Sood Department of Physics, Indian Institute of Science (IISc), Bangalore 560012, India, e-mail: asood@physics.iisc.ernet.in

Dieter Weiss Institut für Experimentelle und Angewandte Physik, University of Regensburg, D-93040 Regensburg, Germany, e-mail: Dieter.Weiss@physik.uni-regensburg.de

Xue Zhang Xi'an Jiaotong University, Xi'an, Shaanxi, People's Republic of China, e-mail: zhangxue0123@stu.xjtu.edu.cn

Chapter 1
Progress on the Theoretical Study of Two-Dimensional MoS_2 Monolayer and Nanoribbon

Hui Pan

Abstract Two-dimensional (2D) nanomaterials have attracted increasing attention because of their unusual physical and chemical properties. Among these 2D nanomaterials, the monolayers of layered transition metal dichalcogenides exhibit intriguing physical and chemical properties. In contrast to the graphene, they are direct gap semiconductors with tunable band structures by controlling the composition, functionalizing, and applying external fields. In this review, the recent progress on the first-principles studies of MoS_2 is presented. The electronic and optical properties of MoS_2 monolayer, spin Hall Effect, lattice dynamics, functionalization, and H adsorption and diffusion, are systematically reviewed to provide a broad overview on its applications in electronics, optoelectronics, spintronics, sensor, and membrane. The recent advances on the MoS_2 nanoribbon, including edge stability, edge-dependent magnetism and electronic properties, and its application in renewable energy storage, are also presented. The first principles studies, supported by experimental results, show that their morphologies and properties are useful for nanodevices, catalysts, and energy storage applications.

1.1 Introduction

The discovery of two-dimensional material, graphene, has triggered an extensive study on the two-dimensional materials for applications in next-generation nanodevices because of their easy fabrication [1–4], exceptional charge transport, thermal, optical, and mechanical properties [1–6], and potential diverse applications [4–12]. Graphene-analogous inorganic monolayer materials, including BN,

H. Pan (✉)
Faculty of Science and Technology, University of Macau, Macau SAR, China
e-mail: HuiPan@umac.mo; pansonh@yahoo.com

Z. M. Wang (ed.), *MoS_2*, Lecture Notes in Nanoscale Science and Technology 21,
DOI: 10.1007/978-3-319-02850-7_1,

SiC, GaN, ZnO, MnO_2, and even MOFs, have been reported [13–20]. In particular, the two-dimensional transition metal dichalcogenides show a wide range of electronic, optical, mechanical, chemical, and thermal properties [21]. In contrast to the zero-gap graphene, the two-dimensional transition metal dichalcogenides possess sizable bandgaps, very important to field-emission transistors (FET) and optoelectronic devices.

The transition metal dichalcogenides are a class of materials with the formula MX_2, where M is a transition metal element from group IV (Ti, Zr, Hf, and so on), group V (for instance V, Nb, or Ta) or group VI (Mo, W, and so on), and X is a chalcogen (S, Se, or Te). These materials have crystal structures consisting of weakly coupled sandwich layers X–M–X, where a M-atom layer is enclosed within two X layers and the atoms in layers are hexagonally packed [22] (Fig. 1.1a). Adjacent layers are weakly held together by van der Walls interaction to form the bulk crystal in a variety of polytypes, which vary in stacking orders and metal atom coordination (Fig. 1.1b) [21]. The overall symmetry of transition metal dichalcogenides is hexagonal or rhombohedral, and the metal atoms have octahedral or trigonal prismatic coordination.

Belonging to the family of layered transition metal dichalcogenides, molybdenum disulfide (MoS_2) has been widely used in numerous areas, such as hydrodesulfurization catalyst, photovoltaic cell, photocatalyst, nanotribology, lithium battery, and dry lubrication, due to their distinctive electronic, optical, and catalytic properties [22–34]. Bulk MoS_2 is a semiconductor with an indirect bandgap of 1.2 eV [35]. The report on synthesis of transition metal dichalcogenide nanotubes has triggered an extensive research of the inorganic nanostructures, including nanotubes, quantum dots, nanowires, and single layer, due to the outstanding physical and chemical properties [36–45]. The monolayer MoS_2 has recently attracted great interest because of its potential applications in two-dimensional nanodevices [44, 45], although it had been obtained and studied in the past several decades [46–49]. The monolayer MoS_2 is a direct gap semiconductor with a bandgap of 1.8 eV [44], and can be easily synthesized by using scotch tap or lithium-based intercalation [45–50]. The mobility of the monolayer MoS_2 can be at least 200 $cm^2V^{-1}s^{-1}$ at room temperature using hafnium oxide as a gate dielectric, and the monolayer transistor has the room temperature current on/off ratios of 1×10^8 and ultralow standby power dissipation [45]. Recently, the MoS_2 nanoribbons have been obtained by using electrochemical method [51]. The experimental achievements triggered the theoretical interests on the physical and chemical properties of two-dimensional (2D) MoS_2 nanostructures to reveal the origins of the observed electrical, optical, mechanical, and magnetic properties, and guide the design of novel MoS_2-based devices. First principles calculations based on the functional-density theory (DFT) can accurately predict their physical and chemical properties, provide useful guidance on the design of new materials for expected applications, and have been widely used in the study of bulk and nanomaterials. This review aims to introduce to the recent progress on the first-principles study of 2D MoS_2 nanostructures, and to suggest possibilities for future exciting developments.

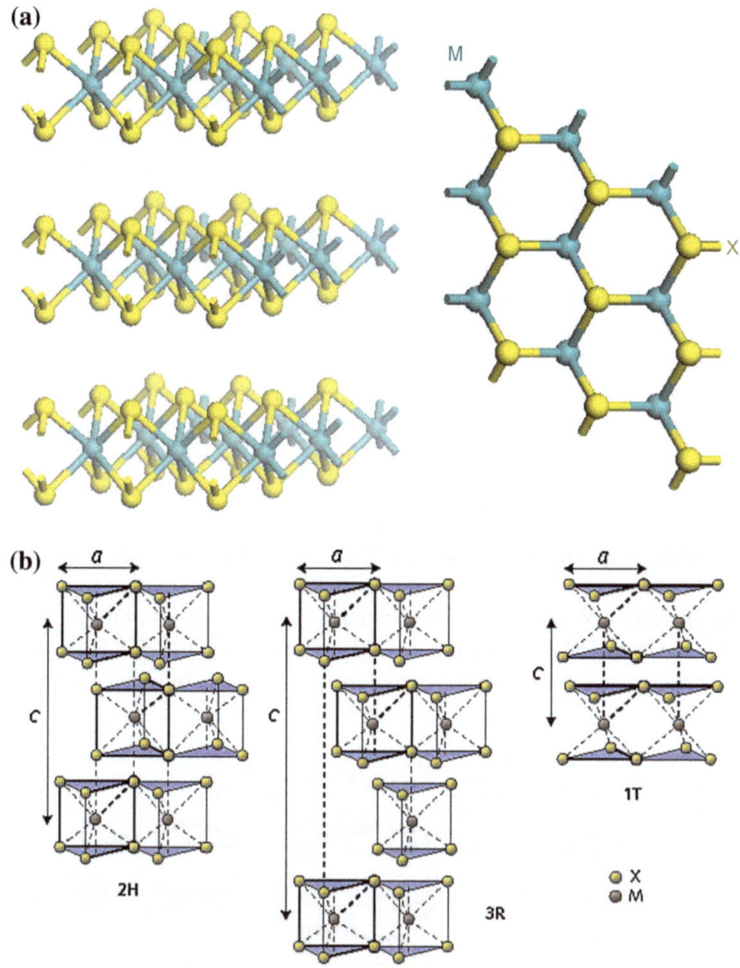

Fig. 1.1 **a** Three-dimensional schematic representation and top view of a typical MX_2 structure, with the chalcogen atoms (X) in *yellow* and the metal atoms (M) in *gray*. **b** Schematics of the structural polytypes: 2H (hexagonal symmetry, two layers per repeat unit, and trigonal prismatic coordination), 3R (rhombohedral symmetry, three layers per repeat unit, trigonal prismatic coordination), and 1T (tetragonal symmetry, one layer per repeat unit, octahedral coordination). **b** is reproduced with permission from Ref. [21]

1.2 MoS_2 Monolayer

1.2.1 Electronic Structure

The applications of MoS_2 monolayer in electronic and optoelectronic devices are directly dependent on its electronic properties, such as band structure and density of states. The band structure of bulk MoS_2 calculated from first-principles show indirect-semiconducting behavior with a bandgap of ~1.2 eV, which originates from transition from the top of valence band situated at Γ to the bottom of conduction band halfway between Γ and K high symmetry points (Fig. 1.2a) [52]. The optical direct bandgap is situated at K point. As the number of layers decreases, the indirect bandgap increases (Fig. 1.2b–e). In the monolayer, the material changes into a 2D direct bandgap semiconductor with a gap of 1.9 eV (Fig. 1.2f). At the same time, the optical direct gap (at the K point) stays almost unchanged (independent of the slab thickness) and close to the value of the optical direct bandgap (at the K point) of a bulk system. The calculated bandgap of MoS_2 monolayer varies from 1.6 to 1.9 eV from the literature due to different approximations for the exchange and correlation (XC) functionals [52–61]. The theoretical results based on the Perdew–Burke–Ernzerhof (PBE) functional form of the generalized gradient approximation (GGA) showed that the bandgap of MoS_2 monolayer is about 1.9 eV [52], which agrees with the experimental data (1.9 eV) observed from photoluminescence (PL) [44, 45]. It is known, however, that DFT always underestimates the bandgaps of materials due to the calculated unreliable exited states. Furthermore, the strong exciton binding due to the weak screening compared to bulk cases in low-dimensional systems may affect the bandgap. Therefore, the good bandgap agreement between theoretical and experimental results for MoS_2 monolayer may be a mere coincidence. The GW approach is expected to yield more accurate gaps and predicted that the gap of MoS_2 monolayer is 2.7–2.9 eV due to the effect of confinement and environment on the electronic structure and exciton binding energy [62–64]

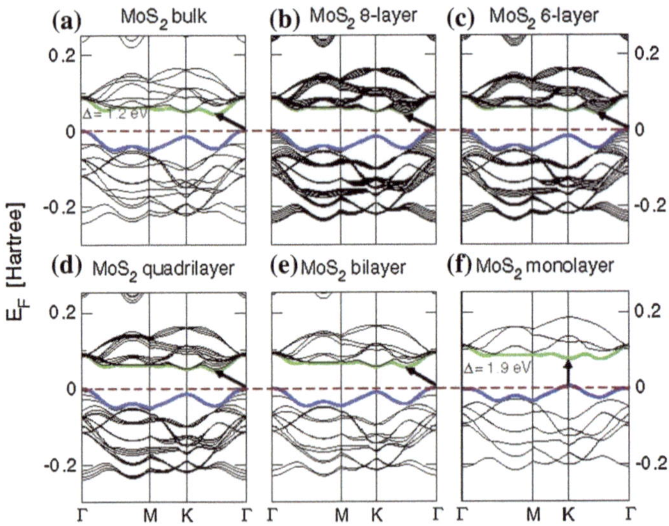

Fig. 1.2 Band structures of bulk MoS_2, its monolayer, as well as, polylayers calculated at the DFT/PBE level. The horizontal *dashed lines* indicate the Fermi level. The *arrows* indicate the fundamental bandgap (direct or indirect) for a given system. The top of valence band (*blue/dark gray*) and bottom of conduction band (*green/light gray*) are highlighted. Reproduced with permission from Ref. [52]

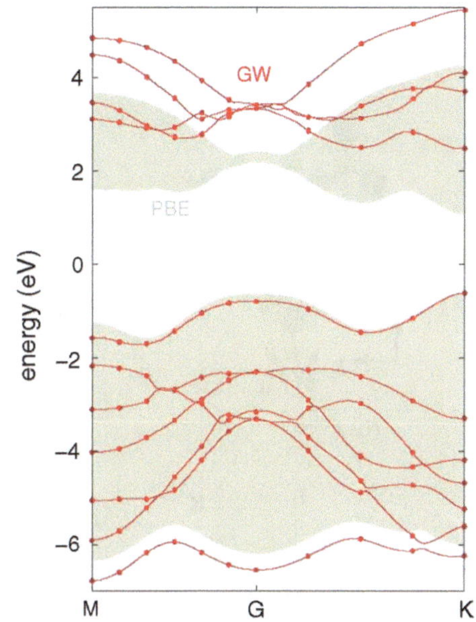

Fig. 1.3 The extrapolated ($1/d \rightarrow 0$) GW band structure (*red lines*) overlayed on top of the PBE band structure (*gray areas*). Reproduced with permission from Ref. [64]

(Fig. 1.3). It was argued that the experimentally observed gap was optical gap ($\sim$1.9 eV) and the exciton binding energy was about 0.8–1.0 eV [62–64]. The fundamental bandgap of MoS_2 monolayer, therefore, is about 2.8 eV. The direct experimental confirmation on the prediction is still not available, and the issue is open to question.

The band structures and bandgaps of MoS_2 monolayer are very sensitive to the external strain [65, 66]. Compared to that of graphene, a much smaller amount of strain is required to vary the bandgap of MoS_2 monolayer. The mechanical strains reduced the bandgap of semiconducting MoS_2 monolayer causing a direct-to-indirect bandgap and a semiconductor-to-metal transition (Fig. 1.4) [65]. These transitions, however, significantly depend on the type of applied. In addition, the results demonstrate that the homogeneous biaxial tensile strain of around 10 % leads to semiconductor-to-metal transition in all semiconducting MoS_2 monolayer [65]. By considering the accurate bandgap of MoS_2 monolayer, the required strain for the transition should be increased.

1.2.2 Optical Property

The optical properties of bulk MoS_2 had been thoroughly studied experimentally [67–69]. The calculated absorption spectrum of bulk MoS_2 showed that there are two distinct low-energy peaks at 1.88 and 2.06 eV [70], which were assumed to the direct transitions between a split valence band (VB) and the conduction band

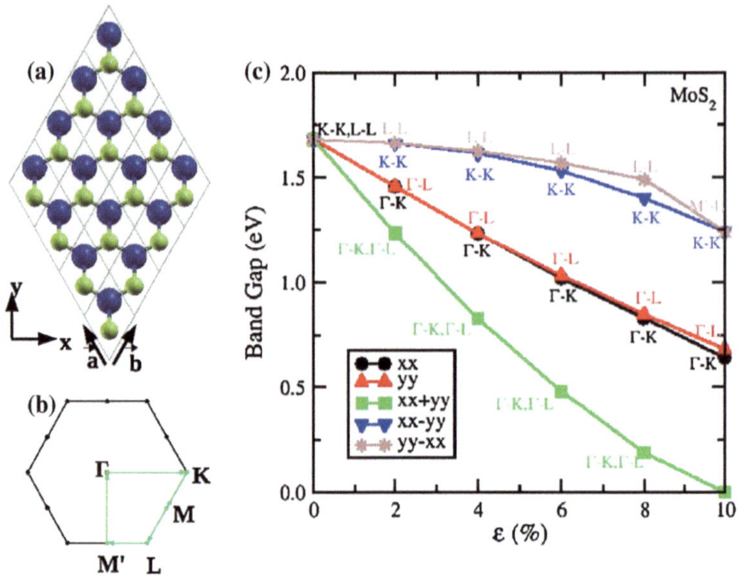

Fig. 1.4 **a** Top view of MoS_2 monolayer, where big and small spheres depict Mo and S atoms, respectively. The figure shows a 4 × 4 supercell. The strain along *x* or/and *y*-directions is applied by varying the *x* or/and *y* components of these lattice vectors, respectively. **b** Irreducible Brillouin zone of MoS_2 monolayer. **c** Bandgap of MoS_2 monolayer with respect to strain, ε, which varies from 0 to 10 %. Strain is applied to the optimized structures ($\varepsilon = 0$) through various approaches, such as uniaxial expansion in *x*-direction (*xx*), *y*-direction (*yy*), homogeneous expansion in both *x*- and *y*-directions (*xx* + *yy*), expansion in *x*-direction and compression in *y*-direction (xx−yy), and compression in *x*-direction and expansion in *y*-direction (*yy*−*xx*) with same magnitude of strain. The first three strain profiles correspond to tensile strain, while the latter two represent pure shear strain. Reproduced with permission from Ref. [65]

(CB) at the K high symmetry point of the Brillouin zone. Excitonic effects are important to understand the optical absorption spectrum of nanostructures and two-dimensional materials due to a reduced electron screening [71–74]. The first-principles calculations showed that electron energy loss spectroscopy (EELS) consists of two prominent resonance features for in-plane polarization (E ⊥ c) that lies (i) below 10 eV which is π plasmons peak and (ii) above 10 eV which is $\pi + \sigma$ plasmons peak [75]. The π plasmons feature in MoS_2 materials arises due to the collective π–π^* transitions, while $\pi + \sigma$ plasmons result from the π–π^* and σ–σ^* excitations [76, 77]. In MoS_2 with 18 valence electrons, 12 electrons per molecule with the mixture of s, p, and d orbitals form the σ valence band and the remaining six electrons are responsible for the π plasmon band [76]. The EELS for out-of-plane polarization (E ⊥ c) showed one prominent resonance feature above 10 eV due to $\pi + \sigma$ plasmons excitation. The $\pi + \sigma$ plasmons peak for the both perpendicular and parallel polarization are broad. Significant red shift in the energies of π and $\pi + \sigma$ plasmons was observed by reducing the layers from bulk to monolayer [75], which are consistent with the experimental results [78].

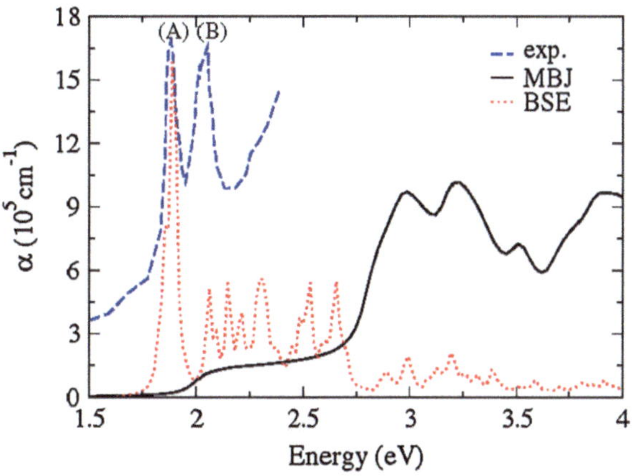

Fig. 1.5 Experimental as well as calculated (BSE and mBJ approximation) absorption spectrum of monolayer MoS_2. The experimental values are given in arbitrary units. Reproduced with permission from Ref. [79]

The interband transitions, however, remained unaffected by the reduction of layers. The detailed calculation on the absorption spectrum at low energy based on the modified Becke-Johnson (mBJ) approximation showed that the MoS_2 monolayer has two excitonic peaks at 1.88 eV (A exciton) and 2.06 eV (B exciton), respectively (Fig. 1.5) [79], which are similar to the bulky values [70] and consistent with the experimental data on monolayer [44, 80]. The high luminescence yield in MoS_2 monolayer can be attributed to the high binding energy of the A exciton, which strongly reduces the probability that excited electrons decay to the bottom of the CB before recombining [79]. The peaks labeled A and B agree with the energies of the A and B excitons, suggesting that they arise from the direct bandgap photoluminescence at the K point.

1.2.3 Spin Hall Effect

In crystalline semiconductors, it often happens that the conduction band bottom (CBB) and valence band top (VBT) are located at degenerate but inequivalent valleys in momentum space [81–83]. The valley index can be regarded as a discrete degree of freedom for low-energy carriers because of the large valley separation in momentum space. The valley polarization can be generated and manipulated by a general scheme based on inversion symmetry breaking, where the valley index can be associated with distinctive physical quantities such as the Berry curvature and orbital magnetic moment [84–86]. The inversion symmetry breaking leads to a valley Hall effect in 2D materials in which carriers in different valley flow to opposite edges with applied electric field [84]. Based on these observations, a growing interest has been focused on exploiting the valley index of 2D materials in electronic devices. The MoS_2 monolayer has attracted extensive attention in this because: (1) the inversion symmetry is explicitly broken in monolayer; (2) the conduction and valence bands harbors a multivalleyed structure; and (3) the spin-orbit coupling

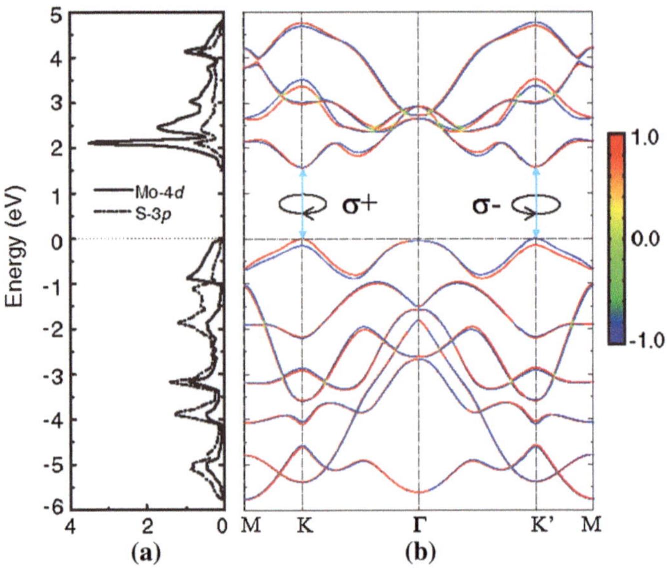

Fig. 1.6 The calculated electronic structure of MoS_2 with the spin-orbit coupling. **a** The partial density of states for Mo-4*d* and S-3*p* orbitals, respectively, in the unit of states/eV/cell. **b** The band structure with the projection of spin operator $\hat{}sz$ (color map). The *red* and *blue* colors indicate the spin-up and -down states, respectively. The optical transitions between the VBT and the CBB are coupled exclusively with $\sigma+$ ($\sigma-$) circular polarizations at the inequivalent valleys K (K′). Reproduced with permission from Ref. [91]

(SOC) is substantial due to the presence of heavy metal atoms [87–91]. Therefore, the MoS_2 monolayer may provide a perfect platform to investigate the interplay between spin and valley degrees of freedom. The calculated band structure of MoS_2 monolayer with the projection of spin operator clearly shows that there is a direct bandgap at the two inequivalent corners K and K′ of the Brillouin zone (Fig. 1.6b) [91]. The VBT states mainly attributed to the Mo-d and S-p electrons, while the CBB is dominated by Mo-d electrons (Fig. 1.6a). Furthermore, the inversion symmetry breaking results in a large spin splitting at the VBT with opposite spin moments at the two valleys [91, 92], indicating the valley index and carriers in the valence band can be distinguished by the spin index. The calculated Berry curvature shows strong peaks at both K and K′ with the opposite signs due to inversion symmetry breaking and strong SOC (Fig. 1.7a, b) [91]. The spin Berry curvature shows peaks at both K and K′ with the same signs because it is simply given by $s_z\Omega(K)$ (s_z: the spin operator; $\Omega(K)$: Berry curvature) (Fig. 1.7c, d). The spin Hall conductivity (σ^s) has one peak in *n*-doped MoS_2 monolayer, and two peaks with opposite values in *p*-doped MoS_2 monolayer (Fig. 1.8). Importantly, the spin Hall conductivity in monolayer is an order of magnitude larger than that in bulk. These findings may find applications for the integration of valleytronics and spintronics [87, 93].

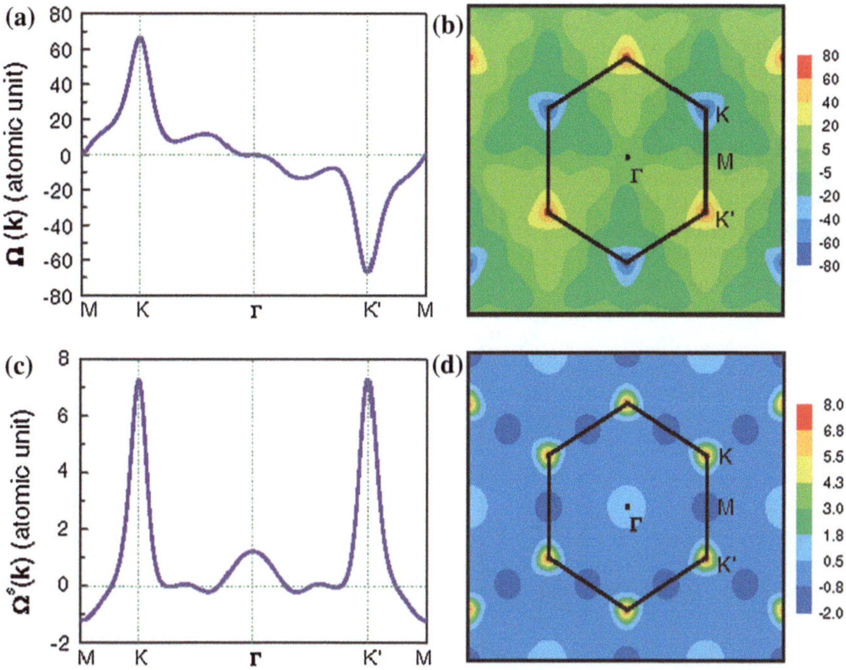

Fig. 1.7 The Berry curvatures of MoS_2 monolayer along the high symmetry lines (**a**) and in the 2D k plane (**b**). The spin Berry curvatures of MoS_2 monolayer along the high symmetry lines (**c**) and in the 2D k plane (**d**). Reproduced with permission from Ref. [91]

1.2.4 Lattice Dynamics

Raman infrared spectroscopy, as a valuable nondestructive and versatile method for structural characterization, has been widely used to study the electronic and vibrational properties in materials. The Raman spectrum is directly related to the lattice dynamics of materials, including phonon dispersion curves, phonon density of states, and infrared and Raman active modes. These properties can be predicted by first-principles calculations using the optimized atomic structures. The Mo atoms in bulk MoS_2 occupy sites of D_{3h} symmetry and S atoms obey C_{3v} symmetry. The symmetry of the bulk MoS_2 is D_{6h}, having 24 symmetry elements and 12 irreducible representations. Four second-order representations involve the lateral (in-plane) displacements of Mo and S atoms (Fig. 1.9) [94]. First-order representations are coupled with the displacements perpendicular to the layers of atoms or parallel to z axis. The MoS_2 monolayer has D_{3h} symmetry, 12 symmetry elements, and 6 irreducible representations. In order for an irreducible representation to be infrared-active mode, it must create a dipole moment in the system. For bulk MoS_2, E_{1u}, and A_{2u} modes are infrared active. E’ and A_2' modes are infrared-active modes of MoS_2 monolayer. Raman active modes induce polarization or quadruple moment

Fig. 1.8 **a** The intrinsic spin Hall conductivity as a function of the Fermi energy for monolayer MoS_2. **b** The *n*- and *p*-doping charge as a function of the Fermi energy. The low-doping regimes just above the CBM and below the VBM (indicated by *red circles*) are more relevant in experiments. Reproduced with permission from Ref. [91]

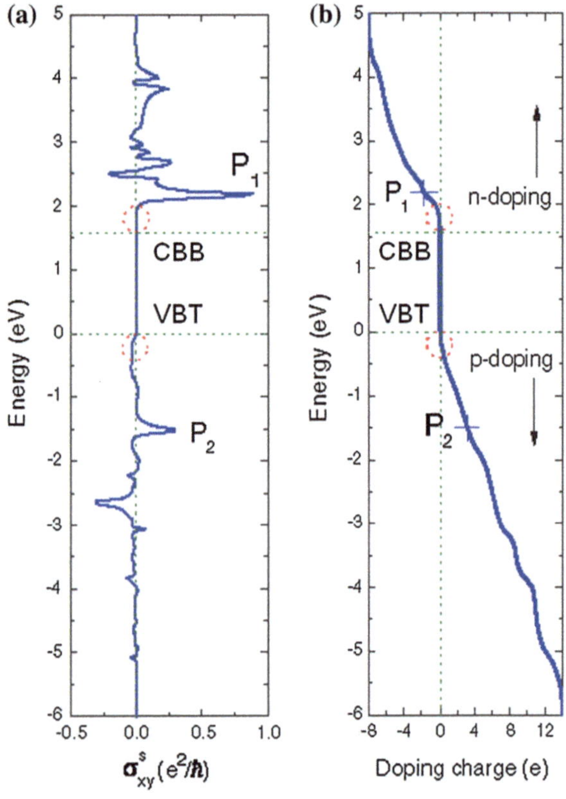

in the lattice. A_{1g}, E_{1g}, and E_{2g} modes are Raman active modes for bulk MoS_2, so as A_1', E', and E'' modes for MoS_2 monolayer (1H–MoS_2) [94]. The calculated acoustical and optical branches of 1H–MoS_2, Ω(k) are positive for any k in BZ, indicating that the suspended, single layer 1H–MoS_2 structure is stable (Fig. 1.9c). By using the experimental lattice parameters (a = 3.16 Å) and GGA + D method, Raman active modes are softened as the dimensionality reduced from 3D to 2D (Fig. 1.10), consistent with the experimental results [95]. It needs to point out that the calculated values of Raman and IF modes also depend on the calculation methods and software [62, 95].

Further study on the Raman scattering on multilayer and monolayer MoS_2 in the low-frequency range (<10 cm^{-1}) showed that there have been two Raman modes with a contrasting thickness dependence (Fig. 1.11) [96]. When increasing the number of MoS_2 layers, one mode shows a significant increase in frequency while the other decreases following a *1/N* (*N* denotes the number of unit layers) trend. With the aid of first-principles calculations, the former was assigned as the shear mode E_{2g}. The latter is distinguished as the compression vibrational mode, similar to the surface vibration of other epitaxial thin films [96]. The calculated in-plane stiffness (C) and Poisson's ratio (v) values for MoS_2 monolayer were 123–146 N/m and 0.25, respectively [66, 94].

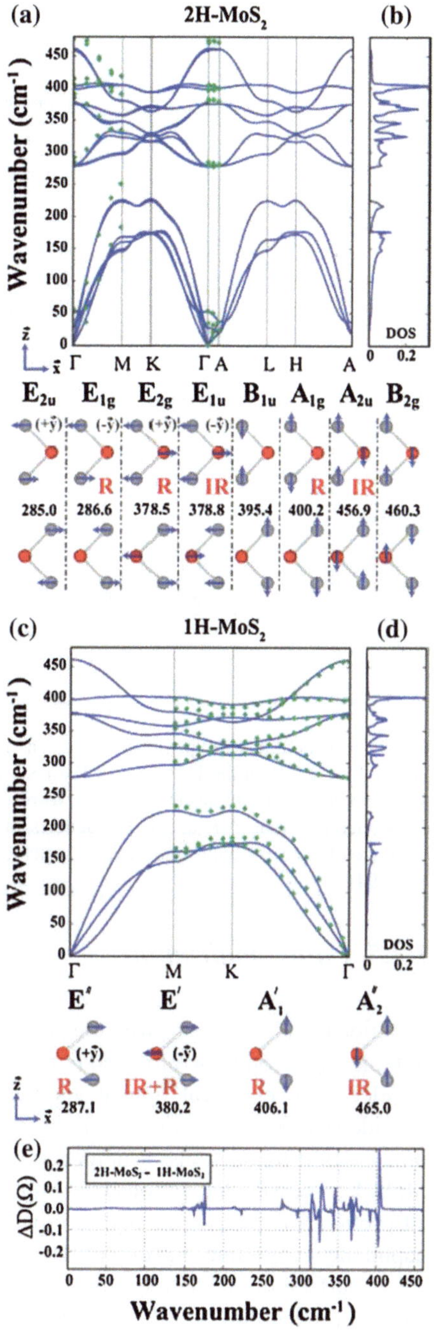

Fig. 1.9 **a** Calculated phonon dispersion curves of 2H–MoS_2, Ω(k) versus *k* along symmetry directions of BZ, and corresponding density of states (**b**). **c** and **d** are the same as (**a**) and (**b**) for 1H–MoS_2. **e** Difference of the densities of states of 2H–MoS_2 and 1H–MoS_2 (see text). Phonon branches derived from neutron scattering data6 and branches calculated by using a local basis set 46, 77 are indicated in (**a**) and (**c**) by *green* (light) *squares*, respectively. Infrared (IR) and Raman (R) active modes with symmetry representations and frequencies (cm^{-1}) at the Γ-point are indicated. Reproduced with permission from Ref. [94]

$2H\text{-}MoS_2$

Lattice Symmetry: $D_{6h} = D_6 \times i$

\# of Symmetry Elements: 24 — # of Irreducible Representations: 12

Atom	Wyckoff position	phonon modes
Mo	2c	$2\,A_{2u}+2\,B_{2g}+2\,E_{2g}+2\,E_{1u}$
S	4f	$A_{1g}+2\,A_{2u}+2\,B_{2g}+B_{1u}+2\,E_{1u}+2\,E_{2g}+E_{2u}+E_{1g}$

Raman $= A_{1g}+2\,E_{2g}+E_{1g}$
IR $= 2\,A_{2u}+2\,E_{1u}$

Lattice Constants	IR (cm^{-1})	Raman (cm^{-1})
$a = 3.220$ Å; $c = 12.411$ Å	$A_{2u}(456.9)$; E_{1u} (378.8)	A_{1g}(400.2), E_{2g}(378.5); E_{1g}(286.6)
(*) $a = 3.160$ Å; $c = 12.300$ Å	$A_{2u}(467.6)$; E_{1u} (381.6)	A_{1g}(407.7), E_{2g}(381.3); E_{1g}(277.8)

$1H\text{-}MoS_2$

Lattice Symmetry: $D_{3h} = D_3 \times i$

\# of Symmetry Elements: 12 — # of Irreducible Representations: 6

Atom	Wyckoff position	phonon modes
Mo	1a	$2\,A_2''+2\,E'$
S	2e	$A_1'+2\,A_2''+2\,E'+2\,E''$

Raman $= 2\,E'+E''+A_1'$
IR $= 2\,E'+2\,A_2''$

Lattice Constants	IR (cm^{-1})	Raman (cm^{-1})
$a = 3.221$ Å	A_2'' (465.0); E'(380.2)	A_1'(406.1); E'(380.2), E''(287.1)
(*) $a = 3.160$ Å	A_2'' (468.0); E'(381.2)	A_1'(397.8); E'(381.2), E''(274.5)
(**) $a = 3.140$ Å	A_2'' (467.5); E'(385.0)	A_1'(392.7); E'(385.0), E''(279.9)

Fig. 1.10 Calculated Frequencies of Raman (R) and Infrared (IR) Active Modes (in cm^{-1}) of 2H- and 1H–MoS_2 at the Γ-Point and Their Symmetry Analysis. The subscripts u and g represent antisymmetric and symmetric vibrations, respectively. The other subscript i ($i = 1, 2, 3$) indicates the stretching modes. IR and R frequencies of 2H- and 1H–MoS_2 are calculated for the fully optimized lattice constants and internal structural parameters. Reproduced with permission from Ref. [94]

1.2.5 Functionalization

Functionalization of nanostructures through chemical binding of atoms, molecules, or molecular groups has attracted much attention. It offers an efficient way to modify the electronic, chemical, optical, and mechanic properties of nanostructures, and to provide new functionalities to structures in nanoscale applications [97]. The functionalization by adatoms or defects on the surface of three-dimensional graphitic MoS_2 has showed that various interesting properties can be achieved [98–102]. The adsorption of thiophene, and aromatic (thiophene, benzothiophene, benzene, naphthalene, pyridine, and quinoline) and conjugated (butadiene) compounds on the active surface of crystalline MoS_2 can affect its catalytic activity in hydrodesulfurization process [98, 99]. It was found that neither S vacancy defects on the MoS_2 surface nor substitutional doping of Pd or Au induces a magnetic moment, whereas Fe and V induce magnetic moments when substituted for S atoms at the MoS_2 surface [101].

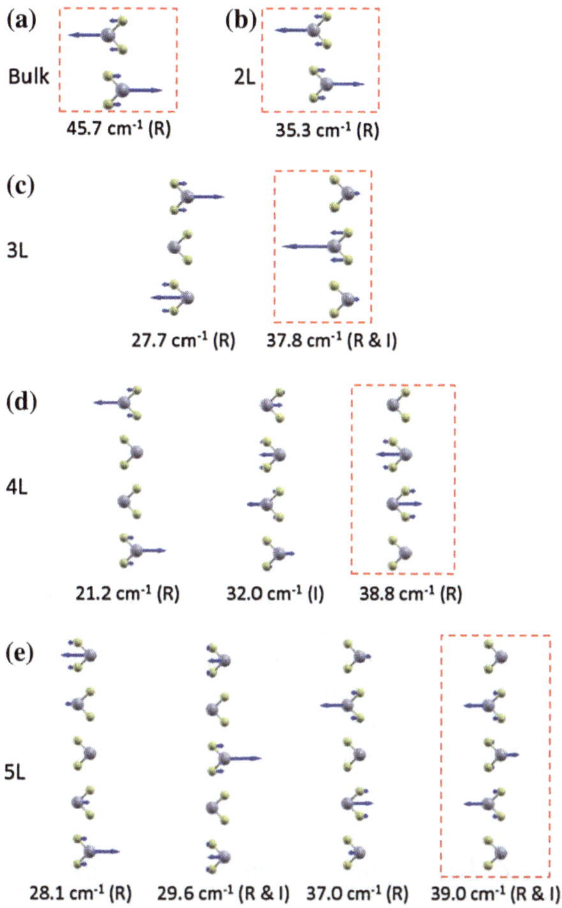

Fig. 1.11 Calculated phonon frequencies and displacement patterns for the shear modes of **a** bulk, **b** bilayer, **c** trilayer, **d** quadlayer, and **e** pentalayer MoS_2 at the Brillouin zone center obtained from DFT calculations. The Raman active (R) and infrared-active (I) modes are labeled. Reproduced with permission from Ref. [96]

Recently, Ataca and Ciraci reported a systematic study of the general effects of the adsorption of selected adatoms and vacancy defect on the electronic and magnetic properties of perfect MoS_2 monolayer [103]. Sixteen individual adatoms (i.e., C, Co, Cr, Fe, Ge, Mn, Mo, Ni, O, Pt, S, Sc, Si, Ti, V, and W) and five different types of vacancy defects (namely, Mo and S single vacancies, S_2 and MoS double vacancies, and MoS_2 triple vacancies) were investigated [103]. The possible adsorption positions for the adatoms are shown in Fig. 1.12. Six distinct adsorption sites depicted in Fig. 1.12, namely, Mo-1, Mo-2, S-1, S-2, S-3, and S-4, were distinguished after structure optimization. The data obtained from the first-principles calculations are summarized in Table 1.1. We see that several adatoms, such as O and W, can be adsorbed readily at diverse sites with significant binding energies, indicating that MoS_2 is suitable for functionalization. The MoS_2 monolayer can attain local magnetic moments through the adsorption of 3D transition metal atoms, as well as Si and Ge. The localized donor and acceptor

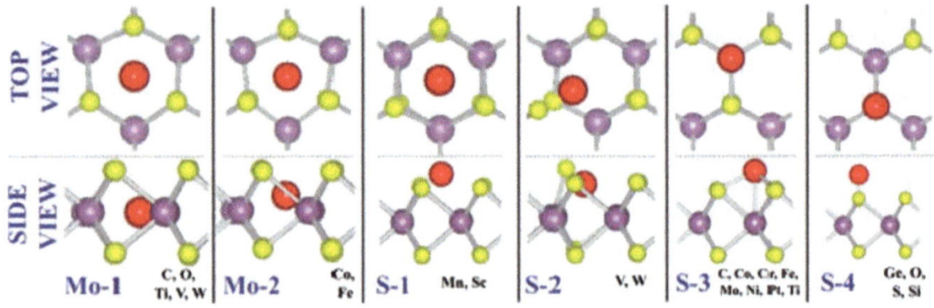

Fig. 1.12 Top- and side-view schematic representations of possible adsorption geometries of adatoms obtained after structure optimization. Adatoms and host Mo and S atoms are represented by *red*, *purple*, and *yellow balls*, respectively. Side views clarify the heights of adatoms from the Mo and S atomic planes. Different adsorption sites are specified below each entry as Mo-x or S-x, where Mo (S) indicates that the adatoms were placed initially (before structure optimization) on the Mo (S) plane. In the Mo-1 and Mo-2 geometries, the adatoms are in and slightly above the Mo layer, respectively. Reproduced with permission from Ref. [103]

states in the bandgap expand the utilization of MoS_2 in nanoelectronics and spintronics. Specific adatoms, such as C and O, attain significant excess charge upon adsorption onto the MoS_2 monolayer. Excess surface charge at higher coverage of adatoms can improve the tribological and catalytic properties of MoS_2 monolayer [103]. Additionally, although vacancy defects of S, S_2, Mo, and MoS created in MoS_2 monolayer do not induce any magnetic moment, the creation of MoS_2 triple vacancies results in a significant magnetic moment in the system.

1.2.6 Hydrogen Adsorption and Diffusion

Hydrogen is considered as one of the most promising and green energy carriers for the future, due to its abundance, easy synthesis, and nonpolluting nature [104–106]. The development of economical and environmentally sustainable hydrogen technologies, including production, storage, transportation, and separation, is a key factor for complete implementation of the hydrogen economy. However, the leakage of hydrogen at any of these stages can result in dangerous consequences. At the same time, special precautions in certain applications should be taken because H_2 acts as a strong reducing agent for many elements and has a high permeability through many materials [107]. As a result, we need to find ways to detect, quantify, and control the concentration of this colorless, odorless, and flammable gas. In order to meet the demands of the future hydrogen economy, a lot of research is ongoing to continuously improve sensitivity, selectivity, response time, and reliability in addition to reducing the size, cost, and power consumption.

Recently, Koh et al. [108] reported that the honeycomb structure of the MoS_2 monolayer may satisfy the requirements for gas sensing, more specifically H_2

Table 1.1 Calculated values for the properties of 16 adatoms adsorbed on MoS_2 monolayer. Reproduced with permission from Ref. [103]

Atom	Site[a]	h^b_{Mo} (Å)	h^c_S (Å)	d^d_{Mo} (Å)	d^e_S (Å)	E^f_b (eV)	$\mu^g_T(\mu_B)$	ρ^{*h} (e)	Φ^i(eV)	P^j (eÅ)	E^k_r (eV)
C	Mo-1	0.01	1.56	2.04	2.44	3.28	NM[l]	0.65	5.81	0.00	0.14, 0.19, 2.18, 4.85
	S-3	1.58	0.05	2.07	1.80	2.69	2.00	0.58	5.74	0.10	0.26↑, 0.30↓, 1.07↑, 1.13↑, 1.55↓, 1.60↓
Co	Mo-2	0.61	0.99	2.22	2.19	0.96	1.00	0.15	5.57	0.07	0.61↑, 0.66↑, 0.82↓, 0.87↓, 0.89↑, 1.13↓
	S-3	2.52	0.94	2.56	2.11	2.92	1.00	0.44	5.17	0.38	0.29↑, 0.51↑, 0.57↓, 0.62↑, 0.99↓, 1.00↓
Cr	S-3	2.79	1.21	2.77	2.26	1.08	4.00	0.91	4.25	0.75	0.19↑, 0.22↑, 0.59↑, 0.64↑, 0.68↑, 1.60↓, 1.73↓
Fe	Mo-2	0.31	1.29	2.21	2.26	0.39	2.00	0.34	5.07	0.04	0.32↑, 0.48↑, 0.87↑, 1.01↓, 1.15↓, 1.32↓
	S-3	2.52	0.93	2.53	2.14	2.42	2.00	0.59	4.99	0.46	0.03↑, 0.18↑, 0.21↑, 0.22↑, 0.91↓, 0.93↓, 1.31↓
Ge	S-4	3.83	2.26	4.28	2.30	1.18	2.00	0.39	4.78	0.27	1.10↑, 1.11↑, 1.72↓, 1.75↓
Mn	S-1	2.51	0.93	3.06	2.15	1.37	3.00	0.81	4.75	0.44	0.16↑, 0.17↑, 0.49↑, 0.50↑, 1.30↓, 1.98↓, 2.05↓
Mo	S-3	2.89	1.31	2.84	2.30	1.43	4.00	0.81	4.54	0.76	0.48↑, 0.79↑, 0.87↑, 0.91↑, 1.16↓, 1.42↓, 1.54↓
Ni	S-3	2.52	0.94	2.56	2.12	3.65	NM	0.36	5.58	0.31	0.12, 0.16, 0.21, 0.50, 0.51, 0.57
O	Mo-1	0.01	1.57	1.95	2.56	2.24	NM	1.11	5.72	0.00	6.36, 6.34, 5.66, 0.32
	S-4	2.94	1.38	3.51	1.49	3.99	NM	0.91	5.96	0.35	5.63, 1.16, 0.90
Pt	S-3	2.78	1.21	2.76	2.31	2.92	NM	0.08	5.66	0.26	0.26, 0.27, 0.37, 1.76
S	S-4	3.48	1.91	3.96	1.94	2.17	NM	0.11	5.96	0.23	0.00, 0.06, 0.11
Sc	S-1	2.84	1.25	3.30	2.29	2.63	1.00	1.45	4.31	1.18	1.28↑, 1.31↑, 1.35↓, 1.49↓, 2.25↓, 2.39↑
Si	S-4	3.73	2.16	4.18	2.17	1.39	2.00	0.58	4.94	0.14	0.98↑, 0.99↑, 1.65↓, 1.68↓
Ti	Mo-1	0.00	1.58	2.31	2.42	1.23	NM	1.14	5.71	0.00	0.26, 0.34, 0.36, 1.74, 1.77
	S-3	2.95	1.38	2.99	2.32	2.62	4.00	1.16	4.32	1.12	1.18↑, 1.24↑, 1.35↑, 1.37↑, 1.54↑, 1.77↓, 1.81↓

(continued)

Table 1.1 (continued)

Atom	Site[a]	h_{Mo}^{b} (Å)	h_S^c (Å)	d_{Mo}^d (Å)	d_S^e (Å)	E_b^f (eV)	$\mu_T^g(\mu_B)$	ρ^{*h} (e)	Φ^i(eV)	P^j (eÅ)	E_r^k (eV)
V	Mo-1	0.03	1.55	2.95	2.34	1.25	1.00	1.05	4.61	0.00	0.12↑, 0.16↓, 0.17↑, 0.21↓, 1.54↑, 1.67↑, 1.69↓
	S-2	1.87	0.19	2.56	2.07	2.76	1.00	1.05	5.18	0.04	0.07↑, 0.18↓, 0.84↑, 1.23↓, 1.45↑, 1.54↓
W	Mo-1	0.01	1.58	2.41	2.42	1.18	2.00	1.05	4.66	0.01	0.16↑, 0.21↓, 1.41↑, 1.49↑, 1.61↓, 1.81↓
	S-2	1.87	0.16	2.61	2.15	4.93	NM	0.85	5.58	0.12	0.13, 0.19, 0.58, 1.57, 1.75

[a] For specific adatoms, the first and second lines are associated with adsorption to the Mo layer and S layer sites, respectively. Other adatoms have only positive binding energies when adsorbed to the S layer site. Adsorption sites of adatoms are described in Fig. 1.13

[b] Height of the adatom from the Mo layer

[c] Height of the adatom from the nearest S layer

[d] Distance from the adatom to the nearest Mo atom

[e] Distance from the adatom to the nearest S atom

[f] Adatom binding energy

[g] Magnetic moment per supercell

[h] Excess charge on the adatom (where a negative sign indicates excess electrons)

[i] Photoelectric threshold (work function)

[j] Dipole moment calculated in the direction normal to MoS_2 surface

[k] Energies of localized states induced by adatoms, as measured from the top of the valence band. Occupied states are indicated in bold, and their spin alignments are denoted as either spin-up (↑) or spin-down (↓). States without the arrow sign indicating spin alignment are nonmagnetic

[l] NM indicates a nonmagnetic state with net $\mu = 0\ \mu_B$

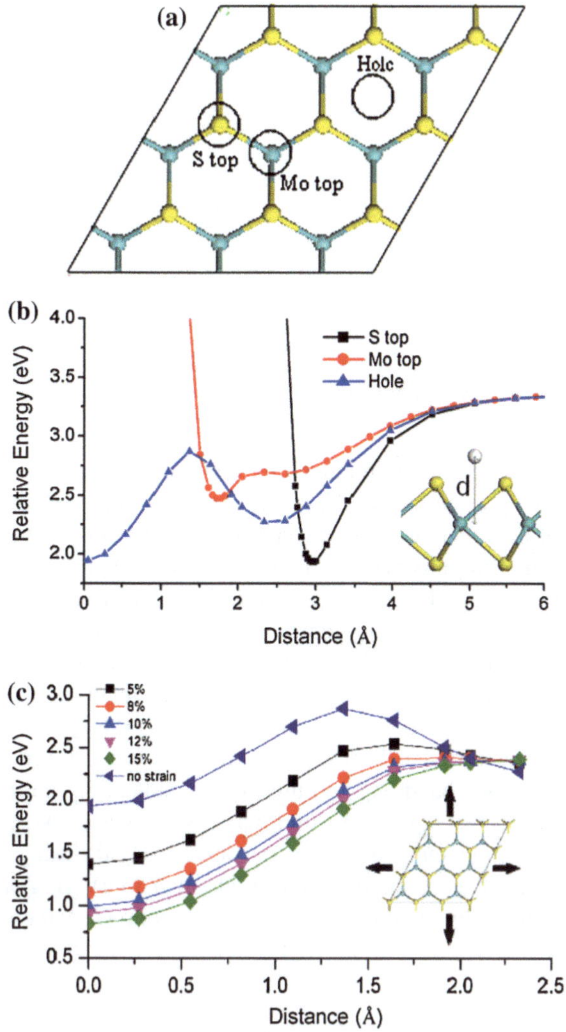

Fig. 1.13 **a** Top view of a monolayer of MoS_2 showing the S top, Mo top, and hole site for hydrogen adsorption. **b** Energy profile as a function of vertical distance of H atom along the three sites, on top of S, on top of Mo, and along the middle of hole. Inset shows the reference level for Mo atoms is set at 0, *X*-axis corresponds to distance of H atom from reference level. **c** Energy–distance profile along the hole site shows a decrease in energy barrier with increasing strain applied on the monolayer. Inset shows that the stretching is realized by applying strain along *x* and *y* directions of the monolayer. Reproduced with permission from Ref. [108]

sensing, or gas controller, or other novel applications in the H_2 infrastructure. There are three sites, Mo top, S top, and hole, for the adsorption of hydrogen atom on the surface of the monolayer (Fig. 1.13a). The calculated relative energy as a function of distance clearly shows that hydrogen atom prefers to bond to the S top with a length of 1.41 Å and a formation energy of 1.91 eV because the relative energy is lower at S top than those at other sites (Fig. 1.13b). If the hydrogen atom can overcome a barrier (0.57 eV) and diffuse into the inside, it will also be stable in the middle of the hexagonal ring (the "hole" site); this can be realized by applying strain on the monolayer. This strain is realized by stretching the monolayer, i.e., increasing the lattice constants (a and b) by x % (inset in Fig. 1.13c). The barrier is reduced under

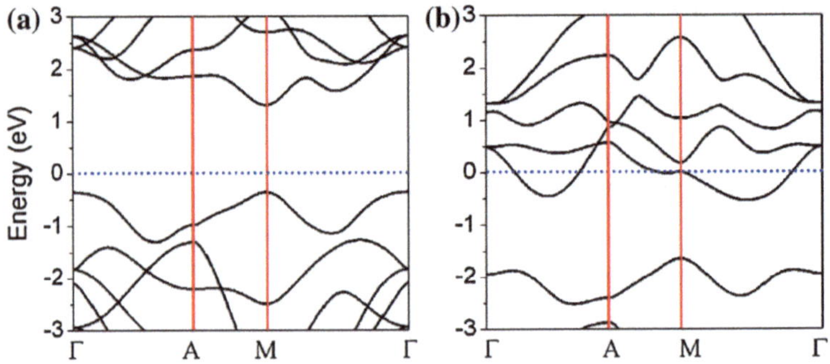

Fig. 1.14 Band structures of **a** pure and **b** H-covered MoS_2 monolayer. Reproduced with permission from Ref. [108]

strain and finally removed by applying strain up to 8 %, indicating that hydrogen atom can smoothly diffuse into the interior of the strained monolayer. The adsorption of hydrogen atom on the surface of the MoS_2 monolayer strongly affects its electronic property because of the formation of S–H bonds. The calculated band structure shows that the MoS_2 monolayer is an intrinsic direct semiconductor with a gap of 1.72 eV (Fig. 1.14a). For MoS_2 monolayer with S atoms at one side bonded with H atoms (H-covered MoS_2 monolayer), the lattice constants are extended by 1 %. The adsorption of hydrogen atoms results in the *n*-type conductivity because each hydrogen atom donates an electron into the system (Fig. 1.14b). Importantly, the bandgap of the H-covered monolayer is narrowed to 1.07 eV (Fig. 1.14b). The narrowed band gap and improved conductivity indicate that the MoS_2 monolayer may detect hydrogen with high sensitivity and fast response.

The calculated energy barrier for H_2 diffusing from one side of the MoS_2 monolayer to another side is about 6.56 eV (Fig. 1.15a). The energy barrier is reduced by applying strain to the monolayer. A reduced barrier of 1.38 eV can be

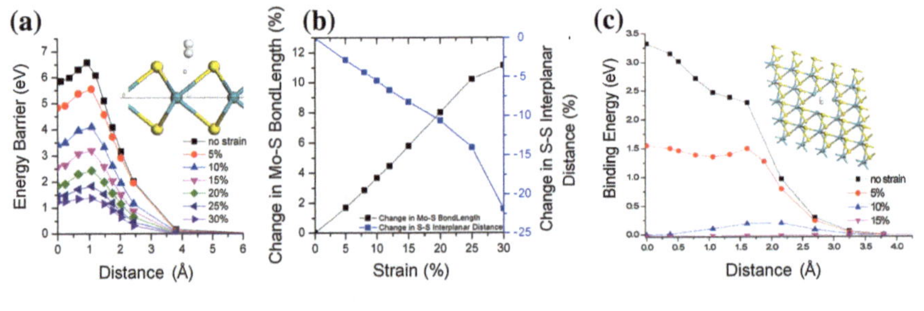

Fig. 1.15 **a** The H_2 diffusion energy barriers through the hole at various strains on the MoS_2 monolayer. **b** The change in Mo–S bond length and S–S interplanar distance as a function of strain on monolayer. **c** The H_2 diffusion energy barriers at various strains on the MoS_2 monolayer with S vacancies. Inset shows the position of the hydrogen molecule as it passes through the defect. Reproduced with permission from Ref. [108]

achieved with a strain of 30 %. The reduced barrier shows that the diffusivity of H_2 through the monolayer is greatly enhanced, indicating that the monolayer can control the H_2 concentration. Importantly, S–Mo bond length has only been extended by 10 % and the S–S distance within the interlayer is shortened by 20 % under a strain of 30 % (Fig. 1.15b). The changes of lattice parameters show that the monolayer keeps its structure under the strain and is reversible after releasing the strain, demonstrating that the MoS_2 monolayer can be a gas controller with high stability. The further calculations on the diffusion of H_2 through the MoS_2 monolayer with vacancies showed that the energy barrier for the diffusion of H_2 through the vacancy in the monolayer is about 3.3 eV, which is reduced to 0.2 eV at a strain of 10 % and totally removed at a strain of 15 %, indicating that the efficiency of H_2 diffusion through the monolayer can be controlled by creating S vacancies and applying strain (Fig. 1.15c). The diffusion possibility is enhanced by applying strain due to reduced energy barrier, indicating that the monolayer can be a mechanical valve to control the H_2 diffusion [108].

1.3 MoS_2 Nanoribbons

The electronic properties of the 2D materials can be efficiently tuned by their edge structures [109–115]. By controlling the edge structures, the 2D nanoribbons are able to show various functional features, including metallic, semi-metallic, semiconducting, and magnetic.

1.3.1 Edge Stability

The MoS_2 nanoribbons (MoS_2-NRs) can be directly obtained by cutting the MoS_2 monolayer (Fig. 1.16) [116]. According to the directions of termination, there exist two kinds of nanoribbons: armchair (ac-MoS_2-NR) and zigzag (zz-MoS_2-NR). Based on the symmetries of the S layer and Mo layer (Fig. 1.16), two ac-MoS_2-NRs can be constructed: symmetrical (ac-MoS_2-NR-s, Fig. 1.16a) and asymmetrical (ac-MoS_2-NR-u, Fig. 1.16b). Similarly, six zz-MoS_2-NRs can be formulated by considering the atoms at the edges and symmetry: one of the two edges can be terminated by Mo atoms and another by S atoms (zz-MoS_2-NR-s and zz-MoS_2-NR-u) (Fig. 1.16c, d), or both of the two edges can be Mo atoms (zz-MoS_2-NR-Mo-s and zz-MoS_2-NR-Mo-u) (Fig. 1.16e, f) or S atoms (zz-MoS_2-NR-S–s and zz-MoS_2-NR-S-u) (Fig. 1.16g, h).

From the calculated edge energies of the bare nanoribbons without spin-polarization (Fig. 1.17a), it can be noted that the edge energies of the S-terminated bare zigzag nanoribbons are negative (<-0.17 eV), especially the symmetrical one (-0.26 eV), and much less than those of others that have positive edge

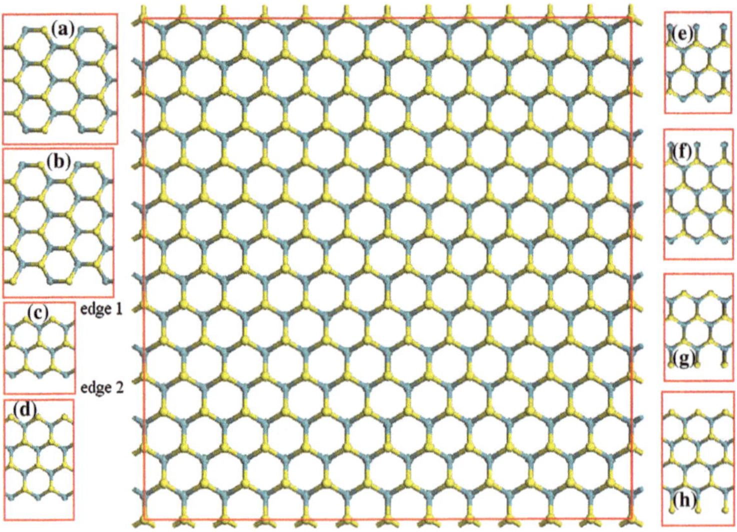

Fig. 1.16 The structure of MoS_2 monolayer (*middle* in the Figure) surrounded by nanoribbons. **a** armchair nanoribbon with symmetrical S/Mo layer (ac-MoS_2-NR-s), **b** armchair nanoribbon with asymmetrical S/Mo layer (ac-MoS_2-NR-a), **c** and **d** symmetrical and asymmetrical zigzag nanoribbon with two edges terminated by S and Mo atoms, respectively (zz-MoS_2-NR-s and zz-MoS_2-NR-a), **e** and **f** symmetrical and asymmetrical zigzag nanoribbon with two edges terminated by Mo atoms (zz-MoS_2-NR-Mo-s and zz-MoS_2-NR-Mo-a), respectively, and **g** and **h** symmetrical and asymmetrical zigzag nanoribbon with two edges terminated by S atoms (zz-MoS_2-NR-S–s and zz-MoS_2-NR-S-a), respectively. For the definition of symmetry in **a–h**, the dangling atoms (Mo and S) at one edge in the zigzag nanoribbons are not considered. Reproduced with permission from Ref. [116]

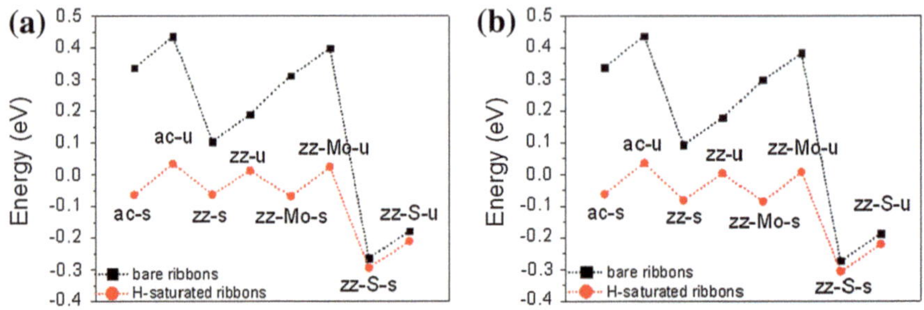

Fig. 1.17 The calculated edge energies of the nanoribbons. **a** without spin-polarization and **b** with spin-polarization. Reproduced with permission from Ref. [116]

energies, indicating that the nanoribbons can be stabilized by the S termination [116]. For other zigzag bare nanoribbons, the Mo-terminated zigzag bare nanoribbons are less stable than nanoribbons with two edges terminated by Mo and S atoms, respectively, because of their relatively higher edge energies (>0.3 eV).

For armchair nanoribbons, the edge energy of the symmetrical armchair nanoribbon (0.34 eV) is less than that of the symmetrical one (0.44 eV). Interestingly, the hydrogen saturation significantly lowers the edge energies of the nanoribbons by up to 0.4 eV, except the S-terminated zigzag nanoribbons that have the energies lowered by 0.03 eV (Fig. 1.17a), indicating that their edge stability can be improved by the hydrogen attachment. Although the edge energies of other nanoribbons are greatly reduced by the hydrogen saturation, the S-terminated zigzag nanoribbons are still most stable due to their lower edge energies (−0.29 and −0.21 eV for zz-MoS_2-NR-S–s and zz-MoS_2-NR-S-u, respectively). The calculated edge energies of the nanoribbons with spin-polarization give out the same trend (Fig. 1.17b). Generally, the symmetrical nanoribbons are more stable than the asymmetrical counterparts due to their relatively lower edge energies, regardless of the chirality and hydrogen saturation (Fig. 1.17). From the calculated edge energies and relaxed edge structures, we can see that the S-terminated zigzag nanoribbons are most stable and should be easily produced in experiments even without hydrogen saturation, while the hydrogen saturation should be necessary for the formation of other nanoribbons [116].

1.3.2 Magnetic Properties

The magnetic properties of the nanoribbons can be quantified by calculating the energy difference between the nonmagnetic and magnetic states ($E_{\mathrm{nonM}}-E_{\mathrm{M}}$) [116]. The armchair nanoribbons with and without hydrogen saturation are nonmagnetic because the energies of nonmagnetic states are equal to those of magnetic states (Fig. 1.18). The energies of magnetic states in zigzag nanoribbons are lower than those of nonmagnetic states (Fig. 1.18), regardless of the hydrogen saturation, indicating the zigzag nanoribbons are magnetic. The calculations showed that the armchair nanoribbons are nonmagnetic and zigzag nanoribbons are magnetic [116–119]. Interestingly, the hydrogen saturation at the edges may

Fig. 1.18 The calculated energy differences between the nonmagnetic and magnetic states of the nanoribbons ($E_{\mathrm{nonM}}-E_M$) divided by the number of the MoS_2 units at the edges. Reproduced with permission from Ref. [116]

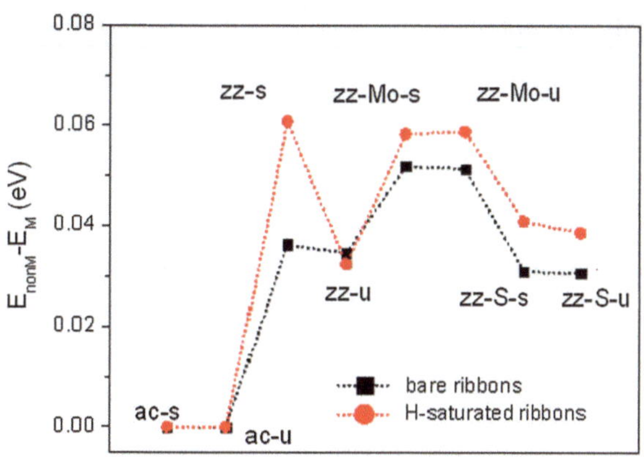

improve the transition temperature and stabilize the magnetic states because of the enhanced energy differences between the nonmagnetic and magnetic states of the zigzag ribbons (Fig. 1.18), except zz-MoS_2-NR-u (whose energy difference is slightly reduced [116–119]). The hydrogen saturated zz-MoS_2-NR-s and zz-MoS_2-NR-Mo-s/u may have higher transition temperature because of the relatively larger energy differences.

Theoretically, it has been predicted that zigzag-edged graphene nanoribbons (ZGNRs) have localized electrons at edge carbon atoms [120–125]. The coupling of the spin-polarized electrons at each edge is ferromagnetic, while that between two edges are antiferromagnetic. The calculated magnetic moments of the zigzag MoS_2 nanoribbons (Table 1.2) and energy differences (Fig. 1.18) show that the zigzag MoS_2 nanoribbons are ferromagnetic. The model proposed by Pan et al. [116] showed the origin of the ferromagnetism in zigzag MoS_2 nanoribbons (Fig. 1.19). Although, the spin orientations of S and Mo atoms at edge 1 in zz-MoS_2-NR-s (zz-MoS_2-NR-u, zz-MoS_2-NR-S–s, and zz-MoS_2-NR-S-u) are antiparallel (Fig. 1.19a), the moments of S atoms are larger than those of Mo atoms, leading to ferromagnetic coupling at this edge and parallel coupling between the two edges. For zz-MoS_2-NR-Mo-s and zz-MoS_2-NR-Mo-u, the spin polarizations at both edges and the coupling between two edges are parallel (Fig. 1.19b). The H-saturation at the edge states may adjust the magnetic moment and spin orientation, but the ferromagnetic couplings at each edge and between the two edges are stable, except H-saturated zz-MoS_2-NR-Mo-u. The spin polarizations at each edge of H-saturated zz-MoS_2-NR-Mo-u are parallel, but between the two edges are antiparallel (Fig. 1.19c). However, the magnetic moments at edge 1 are much larger than those at edge 2, resulting in its ferromagnetism (Fig. 1.19c). Therefore, the zigzag MoS_2 nanoribbons are ferromagnetic, regardless of edge structures and H-saturation. They should be more suitable for spintronic applications than graphene nanoribbons because ferromagnetic graphene nanoribbons can only be achieved by applying external electrical/magnetic field, and the magnetic

Table 1.2 Calculated magnetic moments of atoms at the edges of the MoS_2 nanoribbons. Reproduced with permission from Ref. [116]

MoS_2 nanoribbon		S (edge1/edge2) (μ_B)	Mo (edge1/edge2) (μ_B)
zz-MoS_2-NR-s	bare	0.38/0	−0.13/0.76
	with H	0/0	1.23/0.22
zz-MoS_2-NR-u	bare	0.37/0	−0.13/0.72
	with H	0/0.03	1.25/0.24
zz-MoS_2-NR-Mo-s	bare	0/0	3.39/0.74
	with H	0/0	1.79/−0.23
zz-MoS_2-NR-Mo-u	bare	0/0	3.39/0.72
	with H	0/0	1.81/0.24
zz-MoS_2-NR-S-s	bare	0.44/0.09	−0.12/0.19
	with H	−0.03/0.11	1.12/0.73
zz-MoS_2-NR-S-u	bare	0.44/0.09	−0.12/0.19
	with H	−0.03/0.11	1.13/0.73

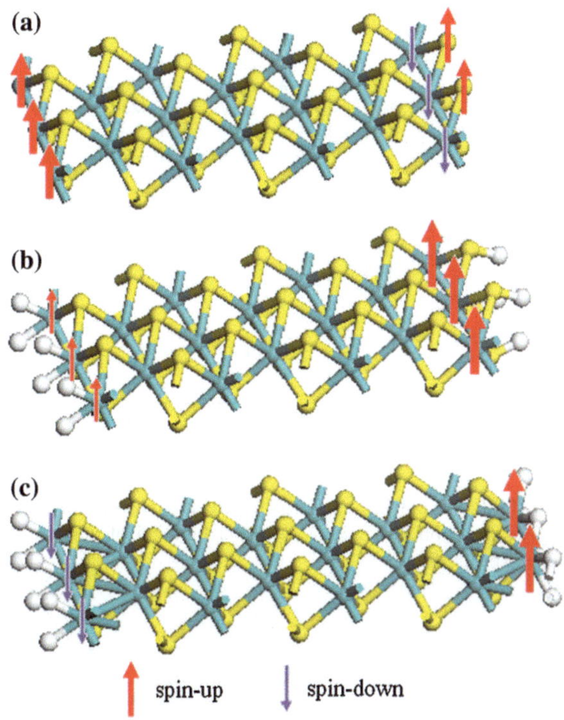

Fig. 1.19 The three ferromagnetic spin couplings in zigzag MoS_2 nanoribbons. **a** the alignments of magnetic moments between the two edges are parallel although the moments are antiparallel at one edge because the up-moments at the edge are much stronger than down-moments, **b** the alignments of magnetic moments between two edges and at each edge are parallel, **c** the alignments of magnetic moments at each edge are parallel, but between two edges are antiparallel. However, the moments at one of the edges are much stronger than those at another edge. Reproduced with permission from Ref. [116]

moments/states of zigzag MoS_2 nanoribbons are higher/stronger than those of graphene nanoribbons.

The magnetic properties of the zigzag nanoribbons were strongly affected by applied strain [126]. The magnetic states of zigzag MoS_2 nanoribbons were more stable and the transition temperature was enhanced under high strain because the energy differences between the nonmagnetic and magnetic states ($E_{nom}-E_{FM}$) were enhanced by applying external strain (Fig. 1.20). $E_{nom}-E_{FM}$ increased linearly with increasing the strain, regardless of the edge structures of the zigzag nanoribbons. For zz-MoS_2-NRs with one of the two edges terminated by Mo atoms and another by S atoms (zz-MoS_2-NRs and zz-MoS_2-NR-u), the $E_{nom}-E_{FM}$ is increased up to 4.9 times at the strain of 10 %. The magnetic moments are increased up to 4 times under a strain up to 10 % (Fig. 1.21).

1.3.3 Electronic Properties

The armchair nanoribbons are direct semiconductors with the bandgaps of 0.59 eV and 0.55 eV for ac-MoS_2-NR-s (Fig. 1.22a), and ac-MoS_2-NR-u (Fig. 1.22b), respectively [116, 117]. The armchair nanoribbons keep semiconducting after the

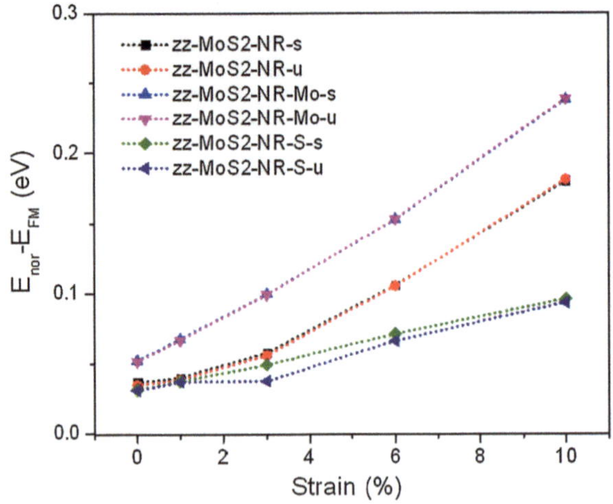

Fig. 1.20 The calculated energy differences between the nonmagnetic and magnetic states of the zigzag nanoribbons ($E_{\mathrm{nonM}}-E_{\mathrm{M}}$) as a function of applied strain. Reproduced with permission from Ref. [126]

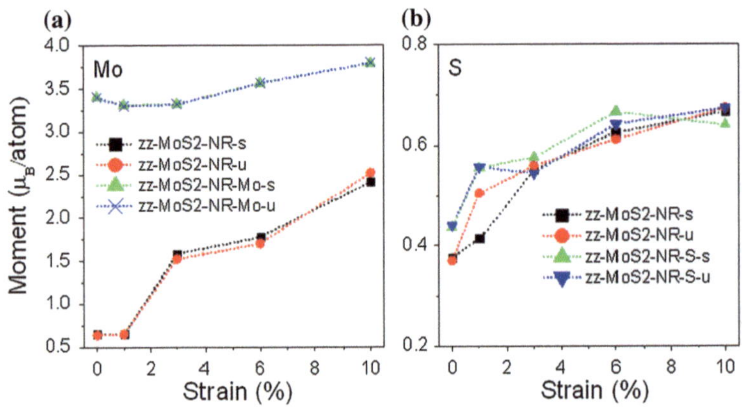

Fig. 1.21 The calculated magnetic moment per atom at the edge as a function of applied strain. **a** Mo, and **b** S atoms (unpublished)

hydrogen saturation (Fig. 1.23a, b). However, their conduction band bottoms are shifted up due to the saturation of the dangling states at the edges, resulting in the increase of their bandgaps (1.65 eV). Additional states appear in the gap region (Fig. 1.23a, b). The calculated PDOSs show that the localized states are dominated by the *d* electrons of Mo atoms at the edges because the dangling states cannot be exactly saturated by the hydrogen atoms (Fig. 1.23c, d) [116].

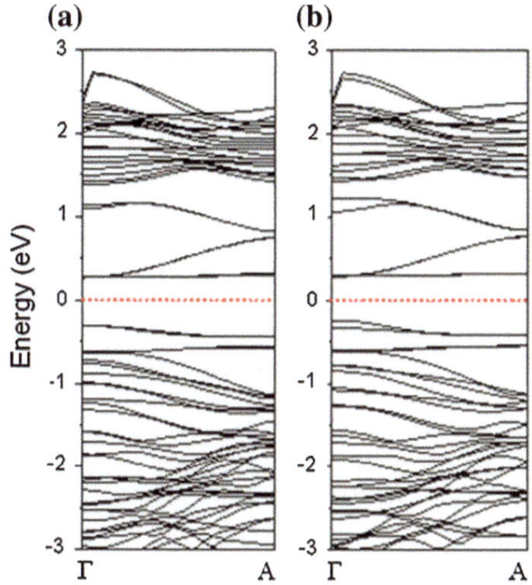

Fig. 1.22 The calculated band structures of **a** ac-MoS_2-NR-s and **b** ac-MoS_2-NR-u. Reproduced with permission from Ref. [116]

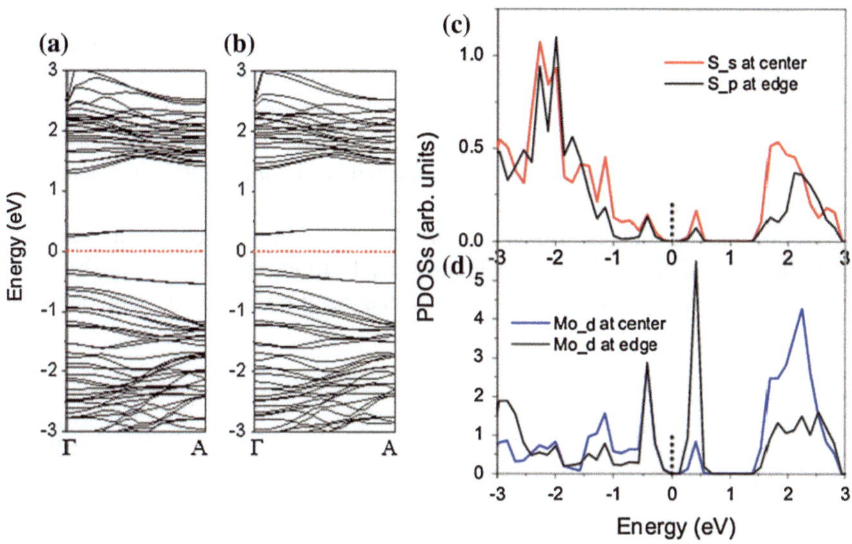

Fig. 1.23 The calculated band structures of H-saturated **a** ac-MoS_2-NR-s, and **b** ac-MoS_2-NR-u; and partial density of states (PDOSs) of H-saturated ac-MoS_2-NR-s. **c** S, and **d** Mo at the edges of ac-MoS_2-NR-s with H. Reproduced with permission from Ref. [116]

The external forces, including electrical field and strain, can efficiently adjust the electronic properties of armchair nanoribbons [126, 127]. The band structures of the nonmagnetic armchair MoS_2 nanoribbons changed from indirect character

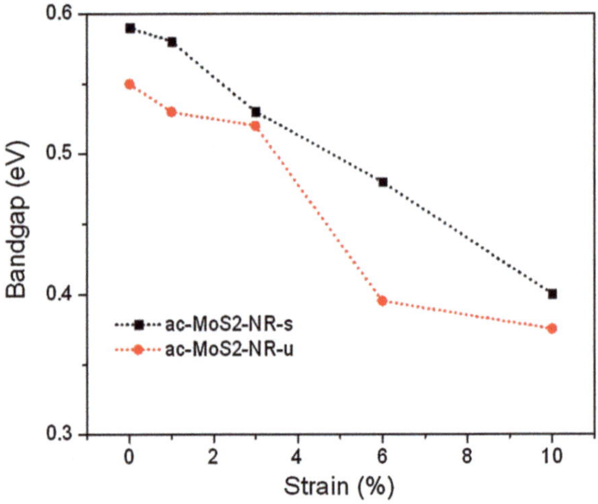

Fig. 1.24 The calculated bandgap as a function of applied strain. Reproduced with permission from Ref. [126]

to indirect, further to quasi-direct with increasing strain due to the shift of the energy states near the Fermi level [126]. The bandgaps of armchair MoS_2 nanoribbons are narrowed by applying the strain (Fig. 1.24). The bandgaps of ac-MoS_2-NR-s and ac-MoS_2-NR-u are 0.41 and 0.37 eV at the strain of 10 %, respectively, showing a reduction of $\sim$30 % in bandgaps [126]. By applying an external transverse electric field, the nanoribbon's bandgap can be significantly reduced, leading to a metal-insulator transition beyond a certain critical value (Fig. 1.25) [127]. The critical strength of transverse field for gap closure decreases as ribbon width increases. However, the perpendicular field did not modulate the bandgap of armchair nanoribbons [128].

All of the considered zigzag nanoribbons without hydrogen saturation are metallic, as indicated by the calculated band structures (Fig. 1.26a, b) [116]. Interestingly, the metallic zigzag ribbon can be transferred to semiconductors

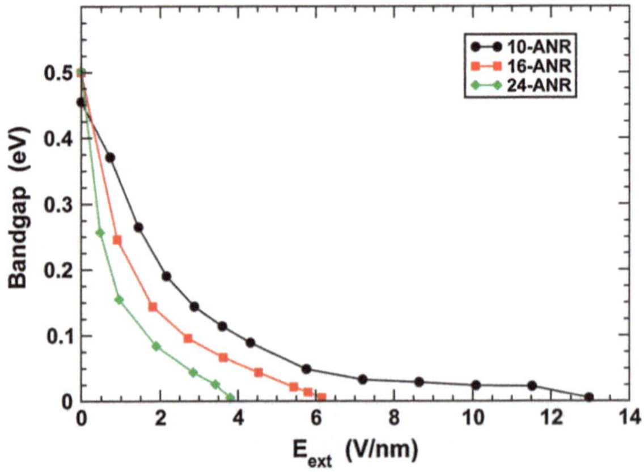

Fig. 1.25 Variation of the elementary LSDA bandgap with the applied transverse electric field, Eext, for 10-ANR (*black circles*), 16-ANR (*red squares*), and 24-ANR (*green diamonds*). Reproduced with permission from Ref. [127]

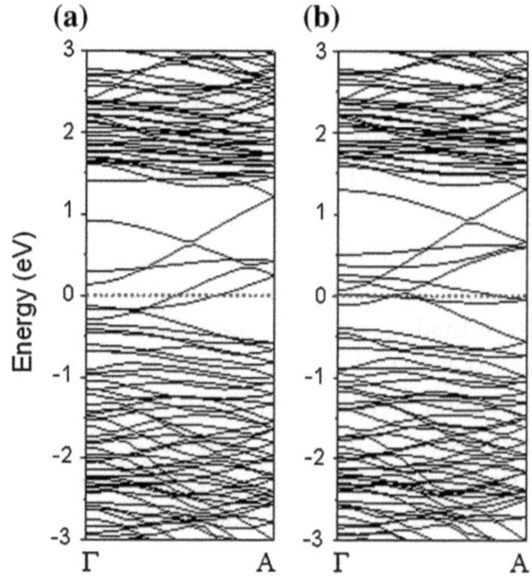

Fig. 1.26 The calculated band structures of zz-MoS_2-NR-s. **a** Spin-up and **b** spin-down. Reproduced with permission from Ref. [116]

(n- or p-type) or half metal by controlling the dangling states with hydrogen at the edges of the structure. Small bandgaps are observable in the band structures of H-passivated zz-MoS_2-NR-s and zz-MoS_2-NR-u, and the Fermi levels are in the conduction bands (Fig. 1.27a, b). The H-saturated Mo-terminated zigzag nanoribbons (zz-MoS_2-NR-Mo-s and zz-MoS_2-NR-Mo-u) become semimetal, where the spin-up band structure shows metallic (Fig. 1.27c), and spin-down band structure shows n-type semiconducting and has an indirect bandgap of 0.11 eV (Fig. 1.27d). Interestingly, the metallic S-terminated zigzag nanoribbons (zz-MoS_2-NR-S–s and zz-MoS_2-NR-S-u) transfer to p-type semiconductors with indirect bandgaps of ~0.3 eV after the hydrogen saturation (Fig. 1.27e, f) [116].

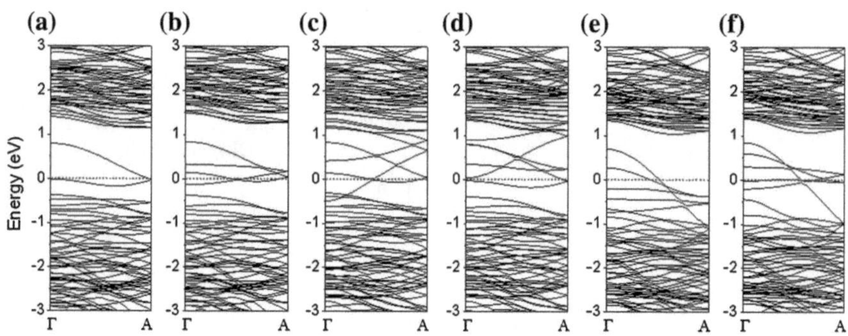

Fig. 1.27 The calculated band structures of H-passivated zigzag nanoribbon. **a** Spin-up and **b** spin-down of zz-MoS_2-NR-s, **c** spin-up and **d** spin-down of zz-MoS_2-NR-Mo-s, and **e** spin-up and **f** spin-down of zz-MoS_2-NR-S–s. Reproduced with permission from Ref. [116]

1.3.4 Application in Rechargeable Battery

As a layered material, MoS_2 is considered as a desirable intercalation host material since the guest atoms or molecules could reversibly intercalate and diffuse through its weak stacked layers and to be an ideal electrode material for advanced rechargeable batteries [129–136]. Li et al. recently reported that the Li adsorption and diffusion on MoS_2 zigzag nanoribbons were enhanced due to edge effects [137]. They found that the Li preferred to the edge sites with dangling states due to high binding energy. Interestingly, the Li atom only needs to overcome an energy barrier of 0.14 eV to migrate from one edge site to another, which is less than the diffusion barrier (0.21 eV) on the surface of monolayer [137]. However, the pathways along the transverse direction on the basal plane of nanoribbon (P10 = T5··· → ···T1, P11 = T5 → T6 → T7) had lower activation barriers than the axis direction considered above (Fig. 1.28). Thus, the Li atoms will prefer the migration along the transverse direction. Li atom will migrate to the edge by

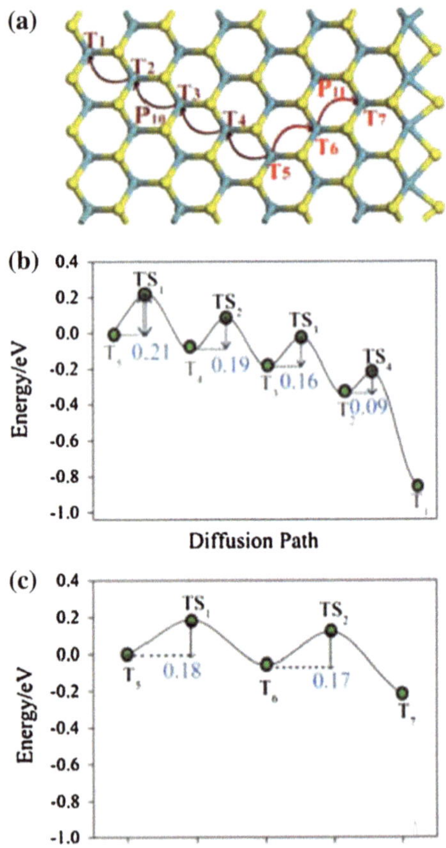

Fig. 1.28 **a** Schematic representation of two diffusion paths (P10, P11) on the basal plane of MoS_2 nanoribbon along the transverse direction. **b**, **c** Energy profiles for P10 and P11, respectively. Reproduced with permission from Ref. [137]

crossing the activation barriers, which decrease gradually toward the edge, regardless of its initial adsorption site on the nanoribbon. In addition to Li-battery, Yang et al. reported the application of MoS_2 nanoribbons in rechargeable Mg batteries [138]. It was found that the Mo top site at the edge of the nanoribbon was favorable for Mg adsorption. The maximum theoretical capacity of 223.2 mAh/g could be achieved by double-side Mg adsorptions. The calculated diffusion barrier on the surface of nanoribbon (0.48 eV) was much lower than that in bulk (2.61 eV). These theoretical studies show that MoS_2 nanoribbons may be used as electrodes in rechargeable battery with an excellent performance.

1.4 Conclusions

The MoS_2 monolayer and nanoribbon offer many opportunities for the investigation of fundamental phenomena and their practical applications. Their versatile and tunable properties make them applicable from energy storage and membrane to nanodevices (electronics, optoelectronics, and spintronics). This review provided a broad overview on the physical properties of MoS_2 monolayer and nanoribbon on the basis of first-principles calculations. Some key features of MoS_2 monolayer and nanoribbon, including bandgap, optical excitation, spin Hall Effect, strain effects, hydrogen adsorption and diffusion, and edge-dependent stability and magnetism, were highlighted. The theoretical studies showed that their applications in electronics, spintronics, and valleytronics could be achievable by controlling the doping, functionalization, and edge structures. At the same time, the 2D nanomaterials may be applicable in membrane for hydrogen diffusion and energy storage (battery). The predicted applications will require substantial collaborations from experimentalists.

References

1. Novoselov, K.S., Geim, A.K., Morozov, S.V., Jiang, D., Zhang, Y., Dubonos, S.V., Grigorieva, I.V., Firsov, A.A.: Electric field effect in atomically thin carbon films. Science **306**, 666–669 (2004)
2. Novoselov, K.S., Geim, A.K., Morozov, S.V., Jiang, D., Katsnelson, M.I., Grigorieva, I.V., Dubonos, S.V., Firsov, A.A.: Two-dimensional gas of massless Dirac fermions in graphene. Nature **438**, 197–200 (2005)
3. Du, X., Skachko, I., Duerr, F., Luican, A., Andrei, E.Y.: Fractional quantum hall effect and insulating phase of Dirac electrons in graphene. Nature **462**, 192–195 (2009)
4. Han, M.Y., Ozyilmaz, B., Zhang, Y., Kim, P.: Energy band-gap engineering of graphene nanoribbons. Phys. Rev. Lett. **98**, 206805 (2007)
5. Zhang, Y.B., Tan, Y.W., Stormer, H.L., Kim, P.: Experimental observation of the quantum hall effect and Berry's phase in graphene. Nature **2005**, 201–438 (2005)
6. Lee, C., Wei, X., Kysar, J.W., Hone, J.: Measurement of the elastic properties and intrinsic strength of monolayer graphene. Science **2008**, 321–385 (2008)

7. Castro Neto, A.H., Guinea, F., Peres, N.M.R., Novoselov, K.S., Geim, A.K.: The electronic properties of graphene. Rev. Mod. Phys. **81**, 106–162 (2009)
8. Das Sarma, S., Adam, S., Gwang, E.H., Rossi, E.: Electronic transport in two-dimensional graphene. Rev. Mod. Phys. **83**, 407–470 (2011)
9. Bao, X.L., Loh, K.P.: Graphene photonics, plasmonics, and broadband optoelectronics devices. ACS Nano. **6**, 3677–3694 (2012)
10. Wang, H.B., Maiyalagan, T., Wang, X.: Review on recent progress in nitrogen-doped graphene: synthesis, characterization, and its potential applications. ACS Catal. **2**, 781–794 (2012)
11. Stengl, V., Popelkova, D., Vlacil, P.: TiO_2-Graphene Nanocomposite as high performace photocatalysts. J. Phys. Chem. **C115**, 25209–25218 (2011)
12. Bellido, E.P., Seminario, J.M.: Graphene-based vibronic devices. J. Phys. Chem. C **116**, 8409–8416 (2012)
13. Eliseeva, N.S., Kuzubov, A.A., Ovchinnikov, S.G., Serzhantova, M.V., Tomilin, F.N., Fedorov, A.S.: Theoretical study of the magnetic properties of ordered vacancies in 2D hexagonal structures: Graphene, 2D-SiC, and h-BN. JETP Lett. **95**, 555–559 (2012)
14. Chen, Q., Hu, H., Chen, X., Wang, J.L.: Tailoring band gap in GaN sheet by chemical modification and electric field: Ab initio calculations. Appl. Phys. Lett. **98**, 053102 (2011)
15. Guo, H.Y., Zhao, Y., Lu, N., Kan, E.J., Zeng, X.C., Wu, X.J., Yang, J.L.: Tunable magnetism in a nonmetal-substituted ZnO monolayer: a first-principles study. J. Phys. Chem. C **116**, 11336–11342 (2012)
16. Tang, Q., Cui, Y., Li, Y.F., Zhou, Z., Chen, Z.F.: How do surface and edge effects alter the electronic properties of GaN nanoribbons? J. Phys. Chem. C **115**, 1724–1731 (2011)
17. Stock, N., Biswas, S.: Synthesis of metal-organic frameworks (MOFs): routes to various MOF topologies, morphologies, and composites. Chem. Rev. **112**, 933–969 (2012)
18. Tang, Q., Li, F.Y., Zhou, Z., Chen, Z.F.: Versatile electronic and magnetic properties of corrugated V_2O_5 two-dimensional crystal and its derived one-dimensional nanoribbons: a computational exploration. J. Phys. Chem. C **115**, 11983–11990 (2011)
19. Zhou, J., Sun, Q.: Magnetism of phthalocyanine-based organometallic single porous sheet. J. Am. Chem. Soc. **133**, 15113–15119 (2011)
20. Wang, L.Z., Takada, K., Kajiyama, A., Onoda, M., Michiue, Y., Zhang, L., Watanabe, M., Sasaki, T.: Synthesis of a Li-Mn-oxide with disordered layer stacking through flocculation of exfoliated MnO_2 nanosheets, and its electrochemical properties. Chem. Mater. **15**, 4508–4514 (2003)
21. Wang, Q.H., Kalantar-Zadeh, K., Kis, A., Coleman, J.N., Strano, M.S.: Electronics and optoelectronics of two-dimensional transition metal dichalcogenides. Nature Nanotechnol. **7**, 699–712 (2012)
22. Lee, P.A.: Physics and chemistry of materials with layered structures: optical and electrical properties. Reidel, Dordrecht (1976)
23. Aruchamy A (Ed.) Kluvert Photochemistry and Photovoltaics of Layered Semiconductors. Kluvert, Dordrecht (1992)
24. Lee, C., Li, Q., Kalb, W., Liu, X.Z., Berger, H., Carpick, R.W., Hone, J.: Frictional characteristics of atomically thin sheets. Science **328**, 76–80 (2010)
25. Hinnemann, B., Moses, P., Bonde, J., Jorgensen, K., Nielsen, J., Horch, S., Chorkendorff, I., Norskov, J.: Biornimetic hydrogen evolution: MoS_2 nanoparticles as catalyst for hydrogen evolution. J. Am. Chem. Soc. **127**, 5308–5309 (2005)
26. Jaramillo, T.F., Jorgensen, K.P., Bonde, J., Nielsen, J.H., Horch, S., Chorkendorff, I.: Identification of active edge sites for electrochemical H-2 evolution from MoS_2 nanocatalysts. Science **317**, 100–102 (2007)
27. Frame, F.A., Osterloh, F.E.: CdSe-MoS_2: A quantum size-confined photocatalyst for hydrogen evolution from water under visible light. J. Phys. Chem. C **114**, 10628–10633 (2010)
28. Min, S.X., Lu, G.X.: Sites for high efficient photocatalytic hydrogen evolution on a limited-layered MoS_2 cocatalyst confined on Graphene sheets-the role of Graphene. J. Phys. Chem. C **116**, 25415–25424 (2012)

29. Ionescu, A., Allouche, A., Ayeard, J.P., Rajzmann, M., Le Gall, R.: Study of gamma-alumina-supported hydrotreating catalyst: I. Adsorption of bare MoS_2 sheets on gamma-alumina surfaces. J. Phys. Chem. B **107**, 8490–8497 (2003)
30. Dominko, R., Arcon, D., Mrzel, A., Zorko, A., Cevc, P., Venturini, P., Gaberscek, M., Remskar, M., Mihailovic, D.: Dichalcogenide nanotube electrodes for Li-ion batteries. Adv. Mater. **14**, 1531–1534 (2002)
31. Hamilton, M.A., Alvarez, L.A., Mauntler, N.A., Argibay, N., Colbert, R., Burris, D.L., Muratore, C., Voevodin, A.A., Perry, S.S., Sawyer, W.G.: A possible link between macroscopic wear and temperature dependent friction behaviors of MoS_2 coatings. Tribology Lett. **32**, 91–98 (2008)
32. Kline, G., Kam, K., Ziegler, R., Parkinson, B.: Further-studies of the photo-electrochemical properties of the group-VI transition-metal dichalcogenides. Solar Energy Mater. **6**, 337–350 (1982)
33. Rapoport, L., Bilik, Y., Feldman, Y., Homyonfer, M., Cohen, S., Tenne, R.: Hollow nanoparticles of WS_2 as potential solid-state lubricants. Nature **387**, 791–793 (1997)
34. Voevodin, A.A., Zabinski, J.S.: Laser surface texturing for adaptive solid lubrication. Wear **261**, 1285–1292 (2006)
35. Kam, K.K., Parkinson, B.A.: Detailed photocurrent spectroscopy of the semiconducting group-VI transition-metal dichalcogenides. J. Phys. Chem. **86**, 463–467 (1982)
36. Tenne, R., Margulis, L., Genut, M., Hodes, G.: Phlyhedral and cylindrical structures of tungsten disulfide. Nature (London) **360**, 444–446 (1992)
37. Tenne, R., Homyonfer, M., Feldman, Y.: Nanoparticles of layered compounds with hollow cage structures (inorganic fullerene-like structures). Chem. Mater. **10**, 3225–3238 (1998)
38. Remškar, M.: Inorganic nanotubes. Adv. Mater. **16**, 1497–1504 (2004)
39. Li, W.J., Shi, E.W., Ko, J.M., Chen, Z.Z., Ogino, H., Fukuda, T.: Hydrothermal synthesis of MoS_2 nanowires. J. Crystal Growth **250**, 418–422 (2003)
40. Remskar, M., Mrzel, A., Skraba, Z., Jesih, A., Ceh, M., Demsar, J., Stadelmann, P., Levy, F., Mihailovic, D.: Self-assembly of subnanometer-diameter single-wall MoS_2 nanotubes. Science **292**, 479–481 (2001)
41. Albu-Yaron, A., Levy, M., Tenne, R., Popovitz-Biro, R., Weidenbach, M., Bar-Sadan, M., Houben, L., Enyashin, A.N., Seifert, G., Feuermann, D., Katz, E.A., Gordon, J.M.: MoS_2 hybrid nanostructures: from octahedral to quasi-spherical shells within individual nanoparticles. Angew. Chem. Inter. Edit. **50**, 1810–1814 (2011)
42. Hu, K.H., Hu, X.G., Sun, X.J.: Morphological effect of MoS_2 nanoparticles on catalytic oxidation and vacuum lubrication. Appl. Sur. Sci. **256**, 2517–2523 (2010)
43. Rosentsveig, R., Margolin, A., Gorodnev, A., Popovitz-Biro, R., Feldman, Y., Rapoport, L., Novema, Y., Naveh, G., Tenne, R.: Synthesis of fullerene-like MoS_2 nanoparticles and their tribological behavior. J. Mater. Chem. **19**, 4368–4374 (2009)
44. Mak, K.F., Lee, C., Hone, J., Shan, J., Heinz, T.F.: Atomically thin MoS_2: A new direct-gap semiconductor. Phys. Rev. Lett. **105**, 136805 (2010)
45. Radisavljevic, B., Radenovic, A., Brivio, J., Giacometti, V., Kis, A.: Single-layer MoS_2 transistors. Nature Nanotechnol. **6**, 147–150 (2011)
46. Banerjee, S., Richardson, W., Coleman, J., Chatterjee, A.: A new 3-terminal tunnel device. Electron Dev. Lett. **8**, 347–349 (1987)
47. Yang, D., Frindt, R.F.: Powder x-ray diffraction of two-dimensional materials. J. Appl. Phys. **79**, 2376–2385 (1996)
48. Frindt, R.F.: Single crystals of MoS_2 several molecular layers thick. J. Appl. Phys. **37**, 1928–1929 (1966)
49. Joensen, P., Frindt, R.F., Morrison, S.R.: Single layer of MoS_2. Mater. Res. Bull. **21**, 457–461 (1986)
50. Schumacher, A., Scandella, L., Kruse, N., Prins, R.: Single-layer MoS_2 on mica-studies by means of scanning force microscopy. Surf. Sci. **289**, L595–L598 (1993)

51. Li, Q., Newberg, J.T., Walter, E., Hemminger, J., Penner, R.: Polycrystalline molybdenum disulfide (2H-MoS_2) nano- and microribbons by electrochemical/chemical synthesis. Nano Lett. **4**, 277–281 (2004)
52. Kuc, A., Zibouche, N., Heine, T.: Influence of quantum confinement on the electronic structure of the transition metal sulfide TS_2. Phys. Rev. B **83**, 245213 (2011)
53. Liao, P.L., Carter, E.A.: New concepts and modeling strategies to design and evaluate photo-electro-catalysts based on transition metal oxides. Chem. Soc. Rev. **42**, 2401–2422 (2013)
54. Jiang, H.: Electronic band structures of molybdenum and tungsten dichalcogenides by the GW approach. J. Phys. Chem. C **116**, 7664–7671 (2012)
55. Ping, Y., Rocca, D., Galli, G.: Electronic excitations in light absorbers for photoelectrochemical energy conversion: first principles calculations based on many body perturbation theory. Chem. Soc. Rev. **42**, 2437–2469 (2013)
56. Matte, H.S.S., Gomathi, A., Manna, A.K., Late, D.J., Dutta, R., Pati, S.K., Rao, C.N.R.: MoS_2 and WS_2 Analogues of Graphene. Angew. Chem. Int. Ed. **49**, 4059–4062 (2010)
57. He, J., Wu, K., Sa, R., Li, Q., Wei, Y.: Magnetic properties of nonmetal atoms absorbed MoS_2 monolayers. Appl. Phys. Lett. **96**, 082504 (2010)
58. Enyashin, A., Gemming, S., Seifert, G.: Nanosized allotropes of molybdenum disulfide. Eur. Phys. J. **149**, 103–125 (2007)
59. Botello-Mendez, A.R., Lopez-Urias, F., Terrones, M., Terrones, H.: Metallic and ferromagnetic edges in molybdenum disulfide nanoribbons. Nanotechnology **20**, 325703 (2009)
60. Kumar, A., Ahluwalia, P.K.: Electronic structure of transition metal dichalcogenides monolayers 1H-MX_2 (M = Mo, W; X = S, Se, Te) from ab initio theory: new direct band gap semiconductors. Electronic structure of transition metal dichalcogenides monolayers 1H-MX2 (M = Mo, W; X = S, Se, Te) from ab initio theory: new direct band gap semiconductors. Eur. Phys. J. B **85**, 186 (2012)
61. Kadantsev, E.S., Hawrylak, P.: Electronic structure of a single MoS_2 monolayer. Solid State Commun. **152**, 909–913 (2012)
62. Ding, Y., Wang, Y.L., Ni, J., Shi, L., Shi, S.Q., Tang, W.H.: First principles study of structural, vibrational and electronic properties of graphene-like MX_2 (M = Mo, Nb, W, Ta; X = S, Se, Te) monolayers. Physica B **406**, 2254–2260 (2011)
63. Cheiwchanchamnangij, T., Lambrecht, W.R.L.: Quasiparticle band structure calculation of monolayer, bilayer, and bulk MoS_2. Phys. Rev. B **85**, 205302 (2012)
64. Komsa HP, Krasheninnikov AV (2012) Effects of confinement and environment on the electronic structure and exciton binding energy of MoS_2 from first principles. Phys. Rev. B **86**, 241201(R)
65. Johari, P., Shenoy, V.B.: Tuning the electronic properties of semiconducting transition metal dichalcogenides by applying mechanical strains. ACS Nano **6**, 5449–5456 (2012)
66. Yue, Q., Kang, J., Shao, Z.Z., Zhang, X.A., Chang, S.L., Wang, G., Qin, S.Q., Li, J.B.: Mechanical and electronic properties of monolayer MoS_2 under elastic strain. Phys. Lett. A **376**, 1166–1170 (2012)
67. Frindt RF, Yoffe AD.: Physical properties of layer structures—optical properties and photoconductivity of thin crystals of molybdenum disulfide. Proc. R. Soc. Lond. A **273**, 69 (1963)
68. Evans, B.L., Young, P.A.: Optical absorption and diffusion in molybdenum disulfide. Proc. R. Soc. Lond. A **284**, 402 (1965)
69. Wilson, J.A., Yoffe, A.D.: Transition metal dichalcogenides discussion and interpretation of observed optical, electrical and structural properties. Adv. Phys. **18**, 193 (1969)
70. Coehoorn, R., Haas, C., Dijkstra, J., Flipse, C.J.F., de Groot, R.A.: Electronic structure of $MoSe_2$, MoS_2, and WSe_2. 2. The nature of the optical band-gaps. Phys. Rev. B **35**, 6203–6206 (1987)
71. Hanke, W., Sham, L.J.: Many-particle effects in the optical-spectrum of a semiconductor. Phys. Rev. B **21**, 4656–4673 (1980)

72. Spataru, C.D., Ismail-Beigi, S., Benedict, L.X., Louie, S.G.: Excitonic effects and optical spectra of single-walled carbon nanotubes. Phys. Rev. Lett. **92**, 077402 (2004)
73. Wang, F., Dukovic, G., Brus, L.E., Heinz, T.F.: The optical resonances in carbon nanotubes arise from excitons. Science **308**, 838–841 (2005)
74. Arnaud, B., Lebegue, S., Rabiller, P., Alouani, M.: Huge excitonic effects in layered hexagonal boron nitride. Phys. Rev. Lett. **96**, 026402 (2006)
75. Kumar, A., Ahluwalia, P.K.: Tunable dielectric response of transition metals dichalcogenides MX2 (M = Mo, W; X = S, Se, Te): effect of quantum confinement. Physica B **407**, 4627–4634 (2012)
76. Liang, W.Y., Cundi, S.L.: Electron energy loss studies of transition metal dichalcogenides. Phis. Mag. **19**, 1031 (1969)
77. Mohan, B., Kumar, A., Ahluwalia, P.K.: A first principle study of interband transitions and electron energy loss in mono and bilayer graphene: effect of external electric field. Physica E **44**, 1670–1674 (2012)
78. Coleman, J.N., Lotya, M., ONeill, A., Bergin, S.D., King, P.J., Khan, U., Young, K., Gaucher, A., De, S., Smith, R.J., Shvets, I.V., Arora, S.K., Stanton, G., Kim, H.Y., Lee, K., et al.: Two-dimensional nanosheets produced by liquid exfoliation of layered materials. Science **331**, 568–571 (2011)
79. Singh, N., Jabbour, G., Schwingenschlogl, U.: Optical and photocatalytic properties of two-dimensional MoS_2. Eur. Phys. J. B **85**, 392 (2012)
80. Splendiani, A., Sun, L., Zhang, Y., Li, T., Kim, J., Chim, C.Y., Galli, G., Wang, F.: Emerging photoluminescence in monolayer MoS_2. Nano Lett. **10**, 1271–1275 (2010)
81. Rycerz, A., Tworzydlo, J., Beenakker, C.W.J.: Valley filter and valley valve in graphene. Nat. Phys. **3**, 172–175 (2007)
82. Zhu, Z., Collaudin, A., Fauque, B., Kang, W., Behnia, K.: Field-induced polarization of Dirac valleys in bismuth. Nat. Phys. **8**, 89–94 (2012)
83. Gunawan, O., Shkolnikov, Y.P., Vakili, K., Gokmen, T., De Poortere, E.P., Shayegan, M.: Valley susceptibility of an interacting two-dimensional electron system. Phys. Rev. Lett. **97**, 186404 (2006)
84. Xiao, D., Yao, W., Niu, Q.: Valley-contrasting physics in graphene: magnetic moment and topological transport. Phys. Rev. Lett. **99**, 236809 (2007)
85. Yao, W., Xiao, D., Niu, Q.: Valley-dependent optoelectronics from inversion symmetry breaking. Phys. Rev. B **77**, 235406 (2008)
86. Xiao, D., Chang, M.C., Niu, Q.: Berry phase effects on electronic properties. Rev. Mod. Phys. **82**, 1959–2007 (2010)
87. Xiao, D., Liu, G.B., Feng, W., Xu, X., Yao, W.: Coupled spin and valley physics in monolayers of MoS_2 and other group-VI dichalcogenides. Phys. Rev. Lett. **108**, 196802 (2012)
88. Li, T.S., Galli, G.L.: Electronic properties of MoS_2 nanoparticles. J. Phys. Chem. C **111**, 16192–16196 (2007)
89. Lebegue, S., Eriksson, O.: Phys. Rev. B **79**, 115409 (2009)
90. Zhu, Z.Y., Cheng, Y.C., Schwingenschlogj, U.: Giant spin-orbit-induced spin splitting in two-dimensional transition-metal dichalcogenide semiconductors. Phys. Rev. B **84**, 153402 (2011)
91. Feng, W.X., Yao, Y.G., Zhu, W.G., Zhou, J.J., Yao, W., Xiao, D.: Intrinsic spin Hall effect in monolayers of group-VI dichalcogenides: A first-principles study. Phys. Rev. B **86**, 165108 (2012)
92. Li, X., Zhang, F., Niu, Q.: Unconventional quantum hall effect and tunable spin hall effect in dirac materials: application to an isolated MoS_2 trilayer. Phys. Rev. Lett. **110**, 066803 (2013)
93. Zeng, H.L., Dai, J.F., Yao, W., Xiao, D., Cui, X.D.: Valley polarization in MoS_2 monolayers by optical pumping. Nature Nanotechnol. **7**, 490–493 (2012)
94. Ataca, C., Topsakal, M., Akturk, E., Ciraci, S.: A comparative study of lattice dynamics of three- and two-dimensional MoS_2. J. Phys. Chem. C **115**, 16354–16361 (2011)

95. Lee, C., Yan, H., Brus, L.E., Heinz, T.F., Hone, J., Ryu, S.: Anomalous lattice vibrations of single- and few-layer MoS_2. ACS Nano **4**, 2695–2700 (2010)
96. Zeng, H.L., Zhu, B.R., Liu, K., Fan, J.H., Cui, X.D., Zhang, Q.M.: Low-frequency Raman modes and electronic excitations in atomically thin MoS_2 films. Phys. Rev. B **86**, 241301(R) (2012)
97. Pan, H., Feng, Y.P., Lin, J.: *Ab initio* study of OH-functionalized single-wall carbon nanotubes. Phys. Rev. B. **70**, 245425 (2004)
98. Raybaud, P., Hafner, J., Kresse, G., Toulhoat, H.: Adsorption of thiophene on the catalytically active surface of MoS_2: An ab initio local-density-functional study. Phys. Rev. Lett. **80**, 1481–1484 (1998)
99. Moses, P.G., Mortensen, J.J., Lundqvist, B.I., Norskov, J.K.: Density functional study of the adsorption and van der Waals binding of aromatic and conjugated compounds on the basal plane of MoS_2. J. Chem. Phys. **130**, 104709 (2009)
100. Huang, M., Cho, K.: Density functional theory study of CO hydrogenation on a MoS_2 surface. J. Phys. Chem. C **113**, 5238–5243 (2009)
101. Fuhr, J., Saul, A., Sofo, J.: Scanning tunneling microscopy chemical signature of point defects on the MoS_2(0001) surface. Phys. Rev. Lett. **92**, 026802 (2004)
102. Ivanovskaya, V.V., Zobelli, A., Gloter, A., Brun, N., Serin, V., Colliex, C.: Ab initio study of bilateral doping within the MoS_2-NbS_2 system. Phys. Rev. B **78**, 134104 (2008)
103. Ataca, C., Ciraci, S.: Functionalization of single-Layer MoS_2 honeycomb structure. J. Phys. Chem. C **115**, 13303–13311 (2011)
104. Abbasi, T., Abbasi, S.A.: 'Renewable' hydrogen: prospects and challenges. Renew Sust. Energ. Rev. **15**, 3034–3040 (2011)
105. Reddy, R.G.: Fuel cell and hydrogen economy. J. Mater. Eng. Perform. **15**, 474–483 (2006)
106. Cheng, X., Shi, Z., Glass, N., Zhang, L., Zhang, J.J., Song, D.T., et al.: A review of PEM hydrogen fuel cell contamination: impacts, mechanisms, and mitigation. J. Power Sources **165**, 739–756 (2007)
107. Hubert, T., Boon-Brett, L., Black, G., Banach, U.: Hydrogen sensors—a review. Sens. Actuators, B **15**, 329–352 (2011)
108. Koh, E.W.K., Chiu, C.H., Lim, Y.K., Zhang, Y.W., Pan, H.: Hydrogen adsorption on and diffusion through MoS_2 monolayer: first-principles study. Int. J. Hydro. Energy **37**, 14323–14328 (2012)
109. Li, H.M., Dai, J., Li, J., Zhang, S., Zhou, J., Zhang, J.L., Chu, W.S., Chen, D.L., Zhao, H.F., Yang, J.L., Wu, Z.Y.: Electronic structures and magnetic properties of GaN sheets and nanoribbons. J. Phys. Chem. C **114**, 11390–11394 (2010)
110. Chai, G.L., Lin, C.S., Cheng, W.D.: Structure dependent electronic and magnetic properties of graphitic GaN-ZnO nanoribbons. J. Mater. Chem. **22**, 7708–7711 (2012)
111. Lin, Y., Connell, J.W.: Advances in 2D boron nitride nanostructures: nanosheets, nanoribbons, nanomeshes, and hybrids with graphene. Nanoscale **4**, 6908–6939 (2012)
112. Pan, L., Liu, H.J., Tan, X.J., Lv, H.Y., Shi, J., Tang, X.F., Zheng, G.: Thermoelectric properties of armchair and zigzag silicene nanoribbons. Phys. Chem. Chem. Phys. **14**, 13588–13593 (2012)
113. Lu, P., Zhang, Z.H., Guo, W.L.: Electronic structures of BC2 N nanoribbons. J. Phys. Chem. C **115**, 3572–3577 (2011)
114. Song, Y.L., Zhang, Y., Zhang, J.M.: Effects of the edge shape and the width on the structural and electronic properties of silicene nanoribbons. Appl. Sur. Sci. **256**, 6313–6317 (2010)
115. Kang, J., Wu, F.M., Li, J.B.: Symmetry-dependent transport properties and magnetoresistance in zigzag silicene nanoribbons. Appl. Phys. Lett. **100**, 233122 (2012)
116. Pan, H., Zhang, Y.W.: Edge-dependent structural, electronic and magnetic properties of MoS_2 nanoribbons. J. Mater. Chem. **22**, 7280–7290 (2012)
117. Li, Y.F., Zhou, Z., Zhang, S.B., Chen, Z.F.: MoS_2 Nanoribbons: high stability and unusual electronic and magnetic properties. J. Am. Chem. Soc. **130**, 16739–16744 (2008)

118. Ataca, C., Sahin, H., Akturk, E., Ciraci, S.: Mechanical and electronic properties of MoS_2 nanoribbons and their defects. J. Phys. Chem. C **115**, 3934–3941 (2011)
119. Wang, Z.Y., Li, H., Liu, Z., Shi, Z.J., Lu, J., Suenaga, K., Joung, S.K., Okazaki, T., Gu, Z.N., Zhou, J., Gao, Z.X., Li, G.P., Sanvito, S., Wang, E.G., Iijima, S.: Mixed low-dimensional nanomaterial: 2D ultranarrow MoS2 inorganic nanoribbons encapsulated in quasi-1D carbon nanotubes. J. Am. Chem. Soc. **132**, 13840–13847 (2010)
120. Son, Y.W., Cohen, M.L., Louie, S.D.: Energy gaps in graphene nanoribbons. Phys. Rev. Lett. **97**, 216803 (2006)
121. Hod, O., Barone, V., Peralta, J., Scuseria, G.: Enhanced half-metallicity in edge-oxidized zigzag graphene nanoribbons. Nano Lett. **7**, 2295–2299 (2007)
122. Zheng, H., Duley, W.: First-principles study of edge chemical modifications in graphene nanodots. Phys. Rev. B **78**, 045421 (2008)
123. Dutta, S., Manna, A., Pati, S.: Intrinsic half-metallicity in modified graphene nanoribbons. Phys. Rev. Lett. **102**, 096601 (2009)
124. Kan, E., Li, Z., Yang, J.L., Hou, J.G.: Half-metallicity in edge-modified zigzag graphene nanoribbons. J. Am. Chem. Soc. **130**, 4224 (2008)
125. Son, Y.W., Cohen, M.L., Louie, S.D.: Half-metallic graphene nanoribbons. Nature **444**, 347–349 (2006)
126. Pan, H., Zhang, Y.W.: Tuning the electronic and magnetic properties of MoS_2 nanoribbons by strain engineering. J Phys. Chem. **C116**, 11752–11757 (2012)
127. Dolui, K., Pemmaraju, C.D., Sanvito, S.: Electric field effects on armchair MoS_2 nanoribbons. ACS Nano **6**, 4823–4834 (2012)
128. Yue, Q., Chang, S.L., Kang, J., Zhang, X.A., Shao, Z.Z., Qin, S.Q., Li, J.B.: Bandgap tuning in armchair MoS_2 nanoribbon. J. Phys. Conden. Matter **24**, 335501 (2012)
129. Samaras, I., Saikh, S.I., Julien, C., Balkanski, M.: Lithium insertion in layered materials as battery cathodes. Mater. Sci. Eng. B **3**, 209–214 (1989)
130. Julien, C., Saikh, S.I., Nazri, G.A.: Electrochemical studies of disordered MoS_2 as cathode material in lithium batteries. Mater. Sci. Eng. B **15**, 73–77 (1992)
131. Miki, Y., Nakazato, D., Ikuta, H., Uchida, T., Wakihara, M.: Amorphous MoS_2 as the Cathode of Lithium secondary batteries. J. Power Sources **54**, 508–510 (1995)
132. Santiago, Y., Cabrera, C.R.: Surface analysis and electrochemistry of MoS_2 thin films prepared by intercalation-exfoliation techniques. J. Electrochem. Soc. **141**, 629–635 (1994)
133. Dominko, R., Arcon, D., Mrzel, A., Zorko, A., Cevc, P., Venturini, P., Gaberscek, M., Remskar, M., Mihailovic, D.: Dichalcogenide nanotube electrodes for Li-ion batteries. Adv. Mater. **14**, 1531–1534 (2002)
134. Wang, Q., Li, J.: Facilitated lithium storage in MoS_2 overlayers supported on coaxial carbon nanotubes. J. Phys. Chem. C **111**, 1675–1682 (2007)
135. Du, G., Guo, Z., Wang, S., Zeng, R., Chen, Z., Liu, H.: Superior stability and high capacity of restacked molybdenum disulfide as anode material for lithium ion batteries. Chem. Commun. **46**, 1106–1108 (2010)
136. Xiao, J., Choi, D., Cosimbescu, L., Koech, P., Liu, J., Lemmon, J.P.: Exfoliated MoS_2 nanocomposite as an anode material for Lithium ion batteries. Chem. Mater. **22**, 4522–4524 (2010)
137. Li, Y.F., Wu, D.H., Zhou, Z., Cabrera, C.R., Chen, Z.F.: Enhanced Li adsorption and diffusion on MoS_2 zigzag nanoribbons by edge effects: a computational study. J. Phys. Chem. Lett. **3**, 2221–2227 (2012)
138. Yang, S.Q., Li, D.X., Zhang, T.R., Tao, Z.L., Chen, J.: First-principles study of zigzag MoS_2 nanoribbon as a promising cathode material for rechargeable Mg batteries. J. Phys. Chem. C **116**, 1307–1312 (2012)

Chapter 2
Electronic Structure of Exfoliated MoS_2

Eugene Kadantsev

Abstract Molybdenum disulfide (MoS_2) is a transition metal dichalcogenide which has been a subject of intense research mainly due to its catalytic properties. MoS_2 is formed by two-dimensional (2D) graphene-like S–Mo–S layers held together by weak noncovalent interactions. Recently, MoS_2 has been exfoliated into individual S–M–S monolayers and the electronic and optical properties of a single MoS_2 layer have been investigated. It was demonstrated that a single layer MoS_2 undergoes indirect to direct band gap transition, which enables a wide range of optoelectronic applications. I will review the electronic structure of exfoliated MoS_2 and discuss several ways in which it can be manipulated. I will also review potential applications of exfoliated MoS_2 in the exciting new field of valleytronics.

2.1 Introduction

The discovery of exfoliated graphene and its unique electronic, optical, and magnetic properties [1] culminated in the award of the 2010 Nobel Prize in physics to Konstantin Novoselov and Andre Geim. Much of the enthusiasm which surrounds graphene can be attributed to its 2D nature. Graphene and other 2D materials have the advantage that 2D geometry is compatible with the established manufacturing techniques used in the semiconductor industry. The pristine graphene, however, does not have a nonzero band gap, a property which is mandatory for many electronic applications. As such, graphene cannot be used as silicon replacement in the electronic devices. Several attempts have been made to overcome this difficulty and open up the gap, for example, by applying the electric field [2–4] or by quantum confinement effect in graphene-based nanostructures

E. Kadantsev (✉)
Department of Chemistry, Centre for Catalysis Research and Innovation,
University of Ottawa, 10 Marie Curie, Ottawa K1N 6N5, Canada
e-mail: ekadants@yahoo.ca

Z. M. Wang (ed.), *MoS_2*, Lecture Notes in Nanoscale Science and Technology 21,
DOI: 10.1007/978-3-319-02850-7_2,

(nanoribbons [5–8] and quantum dots [9–12]). Some progress has been made along these lines. For example, it was experimentally demonstrated [13] that the electric field can be used to drive up the gap in two-sheet graphene (from 0 to 250 meV). However, graphene is made up of carbon atoms only and one can expect that the possibilities for engineering the desired structure of energy levels in graphene-based materials are, to a certain degree, limited.

The realization of graphene's intrinsic limitations prompted research in materials which are both similar to graphene in that they possess 2D translational invariance and, hence, can be potentially used as building blocks of novel optoelectronic devices and yet distinct in their chemical composition and electronic properties. The broad class of 2D materials of this type [14, 15] encompasses transition metal dichalcogenides [16], transition metal oxides [17], topological insulators [18], silicene [19, 20] germanene [21], and hexagonal boron nitrade [22].

In this review, I will focus on the electronic structure properties of exfoliated molybdenum disulfide (MoS_2). Bulk MoS_2 is a prototypical metal dichalcogenide material which has been the subject of research for almost 40 years [23–26]. Exfoliated MoS_2 is an example of 2D material with electronic properties that are different from both graphene and bulk MoS_2. One important distinction from graphene is that the conduction and valence band manifolds in MoS_2 are made up of molybdenum *d* and sulfur *p* atomic orbitals. The chemical diversity and the presence of *d* electrons in the conduction and valence bands signal a possibility of interactions that are not encountered in the *p* bonded graphene. The second important distinction is the strong spin-orbital coupling experienced by molybdenum 4*d* electrons. The spin-orbital coupling manifests itself by removing Kramer's degeneracy in systems without inversion symmetry as it happens in zinc-blende semiconductors [27] and exfoliated odd layer MoS_2. The inversion symmetry can also be destroyed, for example, by application of the transverse (out-of-plane) electric field. The coupling of spin and momentum space (or "valley") degrees of freedom at the valence band edges suppresses spin and valley relaxation and leads to a coexistence of spin- and valley-Hall effect [28].

This review is organized as follows. After brief description of geometry, the role of noncovalent interactions in bonding, and methods that are used to obtain single (multi-) layer MoS_2, I will review electronic structure properties of exfoliated MoS_2 highlighting the differences with the bulk electronic structure. I will discuss several ways in which the electronic structure can be manipulated. Finally, I will review the theory behind and potential applications of exfoliated MoS_2 in the exciting field of valleytronics. This will be followed by a concluding summary and some thoughts on the future directions.

2.2 Structure of Bulk MoS_2 and Preparation of Single (Multi-) Layer MoS_2

Molybdenum disulfide is a "layered" transition metal dichalcogenide which has been widely used as a solid lubricant, a catalyst for the hydrodesulfurization reaction [29–32], and a sulfur-tolerant anode catalyst in high-temperature solid oxide fuel cells [33]. The "layered" structure of MoS_2 is formed by graphene-like hexagonal arrangement of Mo and S atoms bonded together to give S–Mo–S sandwiches (Fig. 2.1). The S–Mo–S units are stacked on top of each other and are held together by weak noncovalent interactions. In this arrangement, in each layer Mo atom is covalently bonded to six sulfur atoms, whereas each sulfur atom is connected to three Mo atoms. MoS_2 bulk unit cell belongs to space group $P6_3/mmc$ and contains six atoms (two Mo and four S). The structure is uniquely determined by the hexagonal lattice constant a, the out-of-plane lattice constant c, and the internal displacement parameter z. The experimental lattice constants and the internal displacement parameter were determined as $a = 3.16$ Å, $c = 12.58$ Å, and $z = 0.12$, respectively.

The noncovalent interactions play a crucial role in the bonding of MoS_2 sheets in bulk MoS_2 [34]. Figure 2.2a shows the binding energy of bulk (two-sheet) MoS_2 with respect to the decomposition into two MoS_2 layers as a function of the out-of-plane lattice constant c. Figure 2.2a shows two types of ab initio density functional [35, 36] calculations: one with (black filled circles) and other without (red squares) Grimme's dispersion corrections [37]. Perdew-Burke-Ernzerhof (PBE) exchange correlation functional [38] was used in both of these calculations. Figure 2.2a shows that the calculations that do not involve corrections for noncovalent interactions (red solid squares) do not give binding (the binding energy is positive) and favor unphysically large out-of-plane lattice constant c or, in other words, dissociation of MoS_2 into individual sheets. On the other hand, the calculations which include the corrections for noncovalent interactions (black-filled circles on

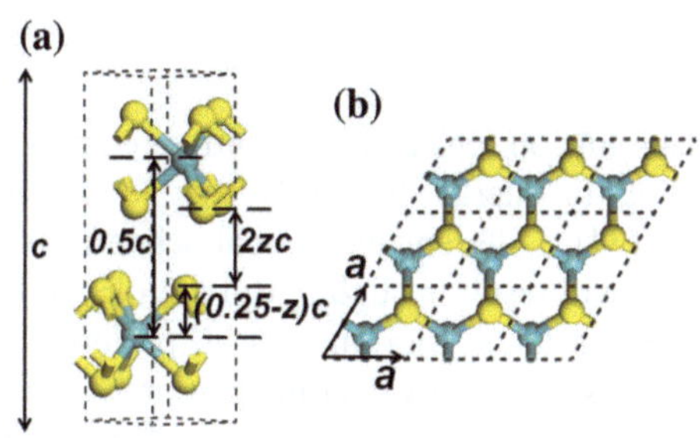

Fig. 2.1 Geometry of bulk MoS_2 (space group $P6_3/mmc$). **a** Side view of the bulk unit cell (*dashed line*). The bulk unit cell is made up of two Mo atoms (*blue*) and four S atoms (*yellow*). Each of the Mo atoms is coordinated to six S atoms in a triangular prismatic fashion. The vertical separation between intra- and interlayer sulfurs is (0.5–2z)c and 2zc, respectively. **b** Top view of the graphene-like single S–Mo–S monolayer (3 × 3 surface unit cell)

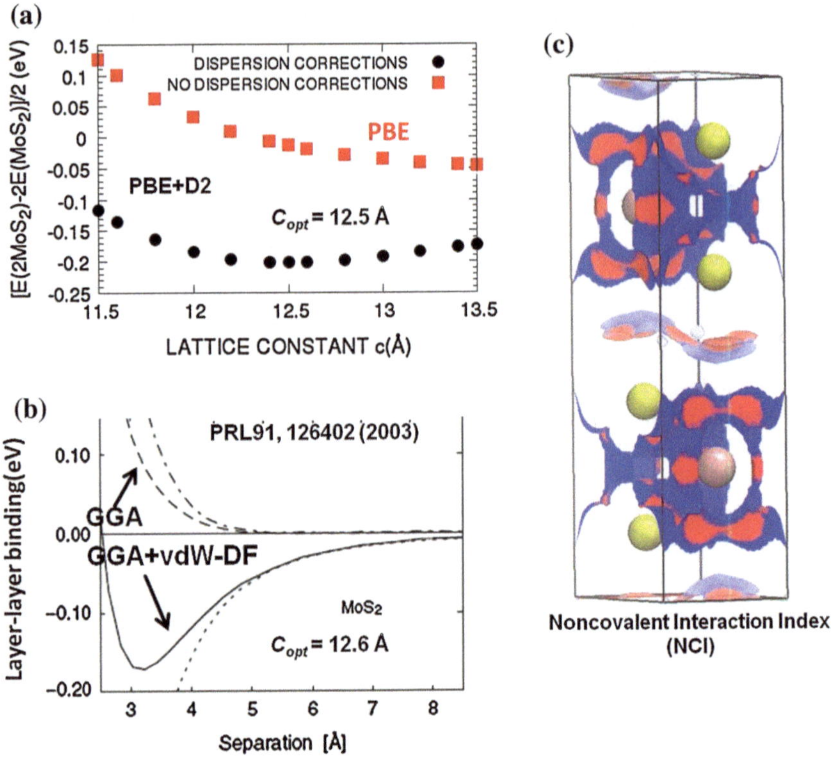

Fig. 2.2 Noncovalent interactions in bulk MoS_2. **a** Binding energy of bulk MoS_2 with respect to the decomposition into two MoS_2 layers with (*black solid circles*) and without (*red squares*) Grimme's D2 dispersion corrections. **b** Interlayer binding energy as a function of separation of two MoS_2 layers (*black solid line*) using vdW−DF approach (from Ref. [34]). **c** Noncovalent interaction index—reduced density gradient isosurface $s(\vec{r}) = 0.5$ with density cut-off of $\rho(\vec{r}) < 0.05$ a.u. The isosurface is colored according to the type of the noncovalent interactions with *red* regions corresponding to bonding interactions and *blue* regions corresponding to the nonbonding interactions. The MoS_2 unit cells consists of 2 [Mo(*pink*)2S(*yellow*)] units stacked on top of each other in the *c*- (out-of-plane) direction

Fig. 2.2a and solid black line on Fig. 2.2b) correctly predict binding in bulk MoS_2. The layer–layer binding in energy in MoS_2 was found to be approximately 200 meV per unit cell [34]. Different theoretical schemes for the description of noncovalent interactions will result in slightly different layer–layer binding energy and out-of-plane lattice constants. Björkman et al. performed a systematic study [39] of different corrections for noncovalent interactions using a large test set of 74 layered solids and recommended the usage of nonempirical nonlocal functionals [40–42]. Nevertheless, the problem of capturing noncovalent interactions with Density Functional Theory is complicated and is a subject of ongoing research.

Noncovalent interactions reveal themselves in the spatial regions characterized by low density and reduced density gradient [43]. Figure 2.2c shows reduced gradient isosurface $s(\vec{r}) = 0.5$ for spatial regions of electron density with $\rho(\vec{r}) < 0.05$ a. u. in bulk MoS_2 unit cell. The isosurface of the reduced density gradient is colored according to the interaction type. The red and blue regions correspond to bonding and nonbonding interactions, respectively. Similar to graphene (Fig. 3h of [43]), the noncovalent interactions between MoS_2 sheets manifest themselves in the interlayer space. A distinctive feature of the interlayer isosurface of the reduced density gradient is the bonding (red) region between two sulfur atoms located in two different MoS_2 sheets (top S from the bottom sheet interacts with bottom S from the top sheet).

The weak (as compared to the covalent bond strength of 2 eV) interlayer binding in MoS_2 enables the synthesis of single (multi-) layer MoS_2 using exfoliation. The mechanical exfoliation [44, 45] is the most reliable method when it comes to the preparation of single (multi)- layer MoS_2. The photoluminescence (PL) spectra which confirm the existence of a direct band gap in a single MoS_2 layer was obtained for mechanically exfoliated MoS_2 flakes. The typical area of these flakes was reported in the proximity of 25–200 μm^2 [46, 47]. The mechanically exfoliated MoS_2 was also used in the *n*- [48] and *p*-type [49] transistors. Other synthesis methods include exfoliation with organic solvents and lithium atom intercalation. One of the disadvantages of exfoliation in organic solvents is the difficulty of preparation of technologically important monolayer sheets. In principle, lithium atom intercalation can produce monolayer sheets; however, this method is prone to the structural deformations. These structural deformations may affect the electronic and optical properties of the samples in an uncontrollable way. Large-area monolayer MoS_2 can also be synthesized using chemical vapor deposition with size of the flakes reaching up to 80 μm [50]. Once the mono (multi-) layer MoS_2 is synthesized, it can be characterized using a variety of experimental techniques including photoluminescence and Raman spectroscopy.

2.3 Photoluminescence and Electronic Structure Properties of Exfoliated MoS_2

The bulk MoS_2 is a semiconductor with an indirect band gap of 1.29 eV. Several experimental studies [46, 47] have now conclusively demonstrated that MoS_2 becomes a direct band gap semiconductor when thinned down to a single layer (one S–Mo–S sandwich). This experimental observation was supported by numerous ab initio calculations [46, 51–55].

The band structure of a single MoS_2 monolayer along the lines connecting high-symmetry points of the Brillouin zone is shown in Fig. 2.3b. The red (blue) lines correspond to the valence (conduction) bands. Also, Fig. 2.3b shows the top valence and conduction bands in bulk MoS_2 (green lines on Fig. 2.3b). In the case of the bulk (green lines), the maximum of the valence (conduction) band is located

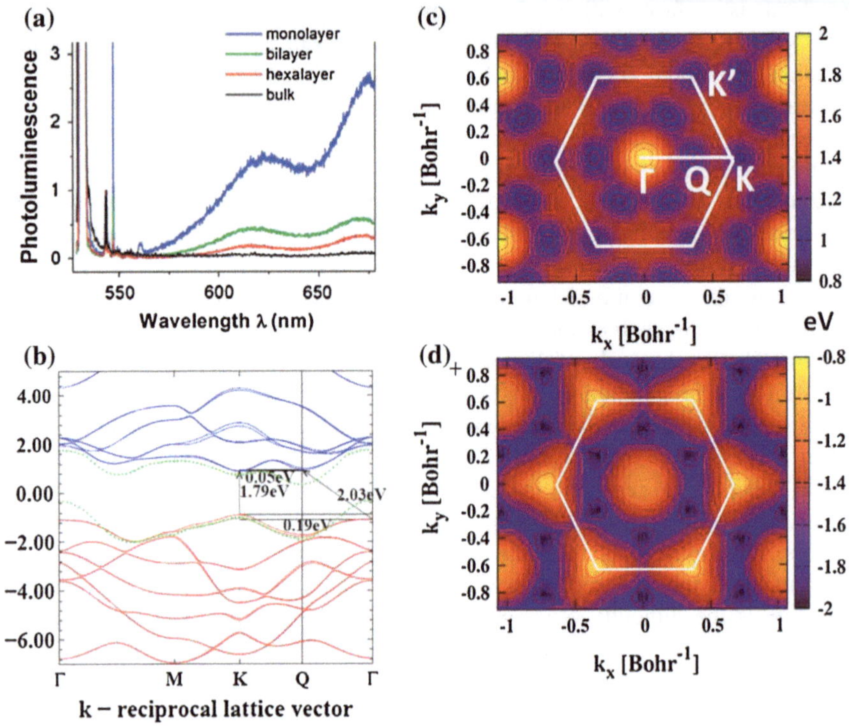

Fig. 2.3 Electronic structure of a single MoS_2 monolayer. **a** Layer dependence of PL in MoS_2 for monolayer, bilayer, hexalayer, and bulk samples taken from Splendiani et al. [46]. **b** Band structure of MoS_2 monolayer and bulk MoS_2 (Ref. [52]). The *red* (*blue*) *lines* represent the valence (conduction) band manifold in MoS_2 monolayer. The *green lines* show the top of the valence and bottom of the conduction bands of bulk MoS_2. Bulk MoS_2 is an indirect $\Gamma \rightarrow$ Q band gap semiconductor, whereas MoS_2 monolayer is a direct K $\rightarrow$ K band gap semiconductor. **c**, **d** Topology of the bottom of the conduction and top of the valence bands in a MoS_2 monolayer over the whole Brillouin zone. The topologies are characterized by relatively weak dispersion. The effective masses at K and Γ points are isotropic, whereas the effective masses at Q point are somewhat anisotropic

at Γ (Q) **k**-point of the Brillouin zone, where the Q point is located along the $\Gamma \rightarrow$ K line and, approximately, $0.53 \mid \Gamma K \mid$ away from the Γ-point. The valence band at Γ point of the Brillouin zone is diffused and is composed of Mo *d* and sulfur *p* orbitals of antibonding character [51]. Therefore, the band energy at Γ **k**-point is strongly affected by the interlayer separation. On the other hand, the valence and conduction bands at K-point of the Brillouin zone are made up of *d*-type orbitals only and are strongly localized on Mo atom. Hence, the band energies at K **k**-point are somewhat insensitive to the layer separation. As the number of MoS_2 layers decreases from the infinitely large number (bulk) to only a few, the valence (conduction) band energy at Γ (Q) **k**-point progressively stabilizes (destabilizes), which results in the increase of the magnitude of the indirect band gap. The direct band gap at K **k**-point practically does not change with the number of layers. For a MoS_2 monolayer, the direct band gap at K-point of the Brillouin zone is smaller than the indirect band gap which experimentally manifests itself (Fig. 2.3a) through strong photoluminescence (PL). This PL is only present in a monolayer and decreases as the number of MoS_2 sheets increases. All electron calculations [52] with variational inclusion of spin-orbital coupling give the direct (indirect) band gap in monolayer as 1.79 eV (2.03 eV). This computed band gap is in reasonable agreement with experimentally measured emission energy at 1.83–1.98 eV. The valence band maximum (VBMAX) and conduction band minimum (CBMIN) at K-point are located 0.19 above and 0.05 eV below the energies at Γ and Q points, respectively (Fig. 2.3b).

Figure 2.3 also shows the topology of conduction band minimum (CBMIN—Fig. 2.3c) and valence band maximum (VBMAX—Fig. 2.3d) of a MoS_2 monolayer over the whole Brillouin zone. The top valence band attains its maximum values at Γ and K-points of the Brillouin zone. The minimum is attained along the line connecting Γ and M points. For the bottom of the conduction band, the minimum values occur at K and Q points and the maximum values are attained at Γ and M points, respectively.

The electron and hole effective masses which allow for the calculation of exciton spectra and mobilities were also determined in exfoliated MoS_2 [52, 54, 55]. For a monolayer, the electron (Fig. 2.3c) and hole (Fig. 2.3d) effective masses at K-point are expected to be isotropic, whereas the conduction band effective mass at the Q point is expected to be somewhat anisotropic. The hole (electron) effective masses at K-point were found to be 0.44 (0.54), 0.64 (0.48), and 0.43–0.46 (0.34–0.35) in Refs. [52, 54, 55], respectively. The electron effective mass at the bottom of the conduction band (Q point) was found to be in the 0.54–0.98 and 0.55–1.18 ranges. The smaller effective mass at Q point of the Brillouin zone characterizes stronger dispersion along the ΓK direction (longitudinal), whereas the larger effective mass characterizes the weaker dispersion direction which is transverse to ΓK. The dispersion of the valence band maximum around Γ point of the Brillouin zone is very weak (Fig. 2.3d). The hole effective mass at Γ point was reported as 3.52 and 5.2 in Refs. [52, 55], respectively.

One of the ways in which MoS_2 differs from graphene is the presence of strong spin–orbit interactions. The coupling between the spin and angular momentum of electrons creates an internal magnetic field that can break down Kramer's degeneracy in systems without inversion symmetry, such as, for example, zinc-blende semiconductors [27, 56–58] as the twofold degeneracy throughout the Brillouin zone is no longer required [56]. The bulk MoS_2 with two MoS_2 monolayers per unit cell possesses the inversion symmetry. In other words, if one uses one of the Mo atoms as the inversion center, the "inverted" sulfur atoms of the first MoS_2 monolayer will transform (with an application of a non-primitive translation) into the sulfur atoms of the second MoS_2 monolayer. However, in the case of a single MoS_2 sheet, the sulfur atoms will be transformed into an empty site. Therefore, the transition from bulk MoS_2 to a monolayer MoS_2 removes inversion symmetry and breaks down the Kramer's degeneracy in most of the Brillouin zone.

The intrinsic spin-splittings for CBMIN and VBMAX bands in one MoS_2 sheet were explored in detail in Ref. [52]. Figure 2.4 shows the absolute values of spin-splittings for VBMAX (Fig. 2.4a) and CBMIN (Fig. 2.4b) over the whole Brillouin zone. VBMAX spin-splittings are very small in the vicinity of the Γ point (dark region in the center in Fig. 2.4a) and are exactly zero at the Γ point. The maximum spin-splitting for VBMAX is attained at K-point of the Brillouin zone and equals 145 meV (six bright regions in Fig. 2.4a arranged in a graphene-like lattice).

The spin-splittings pattern for CBMIN band is more complex with several local minima and maxima (Fig. 2.4b). The CBMIN splittings are by a factor of 2 smaller than those for VBMAX. The maximum value of about 80 meV is attained in the neighborhood of Q point which is located, approximately, halfway between K and Γ points of the Brillouin zone. There is also a smaller double maxima along KK line which are separated by minima at K and M points. The CBMIN intrinsic spin-splitting is exactly zero at the Γ point and is very small at K-point (3 meV).

2.4 Manipulation of Electronic Structure Through Strain and Electric Field

Electronic devices such as transistors with favorable characteristics [48] have been already realized on a single sheet MoS_2. It is, however, highly desirable that we are able to further manipulate the electronic structure of 2D MoS_2 as it increases a number of potential applications of MoS_2. Two very common approaches to

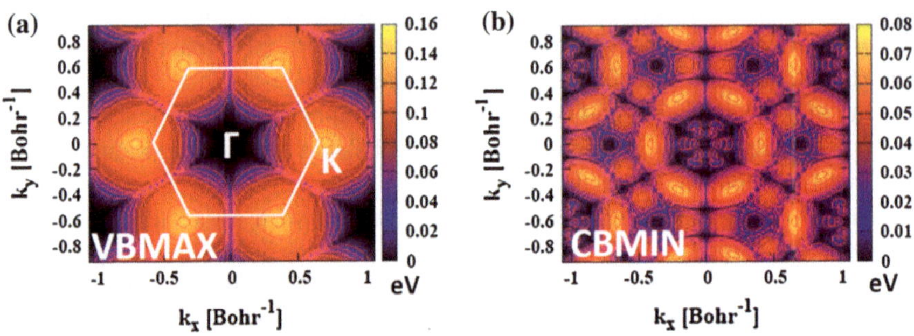

Fig. 2.4 Absolute magnitude of the intrinsic (Dresselhaus) spin-splitting of the valence band maximum (VBMAX) and conduction band minimum (CBMIN) over the whole Brillouin zone. The color scale bar units are in eVs. The global maximum for VBMAX (CBMIN) is located at K (at Q point halfway along ΓK line). The regions with small (zero) spin-splittings are shown as *dark* (*black*) *colors*. The regions of the Brillouin zone with larger spin-splittings are shown with progressively *lighter colors*. The results are taken from Ref. [52]

manipulation of the electronic structure involve application of strain or application of external (electric and magnetic) fields. In the latter case, the application of electric fields is preferred as the electric fields are easier to work with.

The electronic band structure of a semiconductor is modified by strain. For example, strain plays the most crucial role in the determination of properties of multiexciton complexes in semiconductor epitaxial quantum dots and core shell nanocrystals. Several calculations of the strain-modified electronic band structure of 2D MoS_2 were reported. Johari and Shenoy [59] performed Density Functional calculations using several different types of strains which included elongation of the hexagonal lattice constant, uniaxial, and shear deformations (Fig. 2.5a). It was found that the fundamental gap generally decreases with tensile strain. The magnitude of the change in band energies depends on the type of strain used. Among the three types of strain considered in Ref. [59], the largest change in the band gap was given by the tensile deformation of the hexagonal lattice constant.

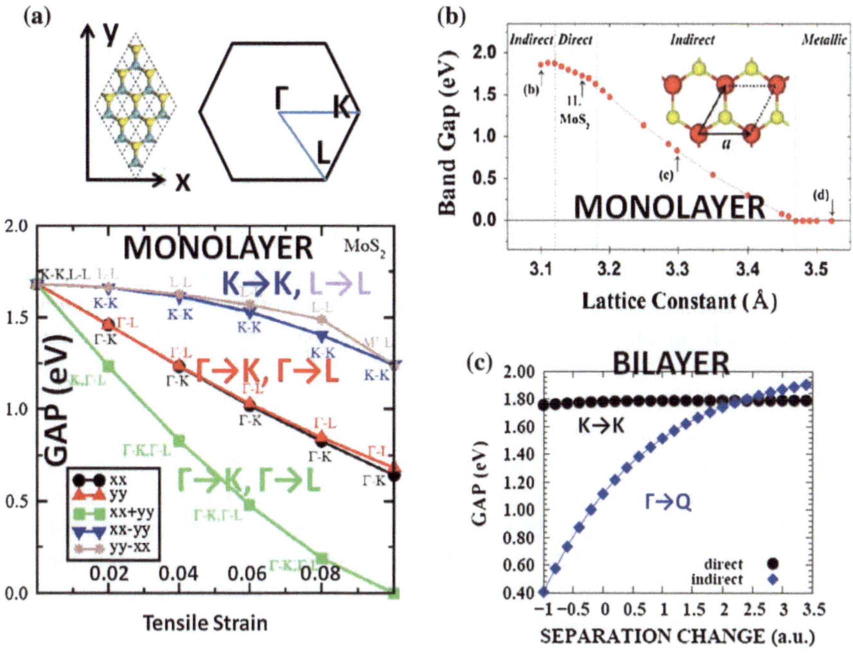

Fig. 2.5 The manipulation of electronic band structure in exfoliated MoS_2 with strain. **a** Band gap change as a function of tensile, uniaxial, and shear deformations (Density functional calculations of Johari and Shenoy from Ref. [59]). **b** Band gap change due to compressive and tensile deformation of the hexagonal lattice constant (Density functional calculations of Yun et al. from Ref. [55]). **c** The direct (K → K) and indirect (Γ → Q) band gap (eV) in two sheet MoS_2 film as a function of change in the separation (Bohr). The "zero change in separation" corresponds to the bulk set-up. The direct K → K band gap is insensitive to the separation, whereas the indirect band gap strongly depends on it

Also, in this case, a MoS_2 layer undergoes a direct to indirect band gap transition due to the shift of VBMAX from K to Γ point.

In another recent study, Yun et al. [55] investigated both compressive and tensile deformation of the hexagonal lattice constant. Figure 2.5b shows the results obtained by Yun et al. [55]. A slightly reduced hexagonal lattice constant $a = 3.10$ Å (−1.9 % compressive stress) slightly increases the band gap from 1.73 to 1.86 eV. However, this slight increase is accompanied by the direct to indirect band gap transition—the shift in CBMIN from K to Q points. Further increase in the compressive strain leads to metallicity [60]. In the case of tensile stress, the band gap decreases as the hexagonal constant increases and attains value of 0.83 at the hexagonal lattice of $a = 3.30$ Å (4 % tensile strain). Further increase in the tensile strain leads to metallicity. The position of VBMAX also undergoes change from K-point to Γ point while the nature of CBMIN remains the same (CBMIN is located at the K-point). Overall, the results obtained by Yun et al. and Johari et al. indicate that the direct band gap nature of a single MoS_2 sheet is very sensitive to strain effects.

In Ref. [52], a computational experiment in which two sheets of MoS_2 where pulled apart (or brought closer together) in the vertical (out-of-plane or c) direction. This experiment can be thought of as an application of the tensile and compressive strains in the out-of-plane direction. The separation between the two sheets was changed and the evolution of direct K → K and indirect Γ → Q band gaps was monitored (Fig. 2.5c). The "zero change" in the separation corresponds to the bulk geometry. Figure 2.5c shows that the direct K → K gap is insensitive to the separation between the two MoS_2 sheets and is very close in value to the direct band gap in single MoS_2 monolayer. As the two sheets are brought closer together (negative change in the separation on Fig. 2.5c), the indirect band gap decreases dramatically. Figure 2.5c indicates that, at some point, the compressive uniaxial strain in the vertical direction induces the indirect to direct band gap transition.

Electric fields have been successfully used [13] to open up band gaps in bilayer graphene. Ramasubramaniam et al. [61] performed a theoretical investigation of the band gap tuning in bilayer MoS_2 with an external electric field applied in the out-of-plane direction. In the case of bilayer MoS_2, the nature of the fundamental band gap at zero electric field is the same as in the bulk—the gap is between VBMAX at Γ and CBMIN at Q. Ramasubramaniam et al. reported that the application of an external electric field shifts CBMIN from Q to K-point of the Brillouin zone. As the magnitude of the electric field increases, the fundamental band gap in bilayer MoS_2 (and several other metal dichalcogenides) smoothly decreases to zero. The magnitude of the external field required to induce semiconductor to metal transition for bilayer MoS_2 was reported to be as 0.26 V/Å [61].

The mechanism of this transition was attributed to a giant Stark effect. Upon the application of an external field, the valence and conduction bands undergo subband mixing which leads to a field-induced splitting of the electronic levels. The field-induced repulsion among the electronic levels leads to a destabilization of

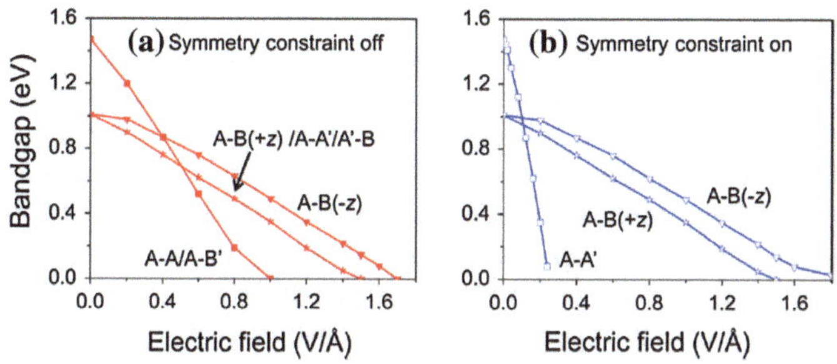

Fig. 2.6 The manipulation of energy levels in exfoliated two-sheet MoS_2 with electric field (the results are taken from Ref. [62]). The calculations are performed using different stackings of individual S–M–S layers. The bulk structure corresponds to A–A′ stacking. The band gap smoothly decreases as the magnitude of the electric field increases. *Left* Electronic structure calculations are performed without symmetrization. *Right* Electronic structure calculations are performed employing symmetrization. The symmetrization results in the artificial underestimation of the magnitude of the "transition" field (0.20 vs. 1.5 V/Å)

VBMAX and stabilization of CBMIN and to a decrease in the band gap. The band edges, which were initially distributed over both layers in MoS_2, localize on either top or bottom layer. As the electric field increases, the gap decreases and eventually closes.

Another investigation of the effects of the out-of-plane electric field in bilayer MoS_2 was carried by Liu et al. [62]. Liu et al. performed their calculations for five allotropes of bilayer MoS_2. The allotropes were distinct in the way the individual layers were stacked. Liu et al. confirmed that as the electric field increases, the band gap in the bilayer MoS_2 slowly decreases and closes. The value for the "transition" electric field at which this closing occurs obtained by Liu et al. was substantially larger (1.6 V/Å) than the "transition" field value reported by Ramasubramaniam et al. (0.26 V/Å). This discrepancy was attributed to the symmetrization used in calculations from Ref. [61]. I have also performed out-of-plane electric field calculations in bilayer MoS_2 with no symmetrization and confirmed the magnitude of the "transition" electric field reported by Liu et al. in Ref. [62] (Fig. 2.6).

2.5 Application of Exfoliated MoS_2 in Valleytronics

The electronic and spintronic devices are based on the transport of electric charge and electron spins, respectively. Valleytronics attempts to employ yet another so-called electron's "valley" degree of freedom. The term "valley" refers to two or more degenerate minima in the conduction band or maxima in the valence band. For example, in graphene (and graphene-like dichalcogenides including MoS_2) the valleys occur at the nonequivalent K and K′ points of the Brillouin zone

(Fig. 2.3c and d). A first crucial step in the realization of the valleytronic device is to be able to achieve the so-called "valley polarization" or, in other words, to selectively populate charge carriers in one of the momentum valleys. A second step is to be able to observe "valley-polarized" transport. Due to technical reasons, the valley polarization is hard to achieve. The valley polarization has been demonstrated through the application of uniaxial strain and magnetic fields but in both cases, the experimental set-up is complicated. Also, neither strain nor magnetic field valley polarization allows for an "on demand" control.

Recently, it has been experimentally demonstrated that the valley degree of freedom can be controlled in a single layer MoS_2 [63–65]. The valley polarization was achieved by optically exciting electrons with a circularly polarized light. Due to the optical selection rules, a given circular polarization will only excite electrons with either spin up or down. In a MoS_2 monolayer, the inversion symmetry is lost and the Kramer's degeneracy is broken. VBMAX at K and K′ points of the Brillouin zone in a MoS_2 monolayer is occupied with electrons of opposite spins (Fig. 2.7a). Therefore, the excitation of one of the spins (up or down) will also mean excitation of one of the valleys (K or K′). The valley polarization in a single layer MoS_2 was demonstrated in Refs. [63–65]. The single layers of MoS_2 were optically excited with right- or left-hand circularly polarized light. The optical excitation with circularly polarized light was followed with measurements of the helicity of the resulting PL. These measurements confirmed valley polarization in a single MoS_2 sheet. Zheng et al. [64] and Mak et al. [63] have also probed two-layer MoS_2. Due to the presence of the inversion symmetry in two-layer MoS_2 and the Kramer's degeneracy at K and K′ points (Fig. 2.7b), no valley polarization was observed. The coexistence of valley and spin Hall effect in a MoS_2 monolayer has also been theoretically demonstrated [28]. This important property can be used to provide accumulations of spin- and valley-polarized carriers with long relaxation times.

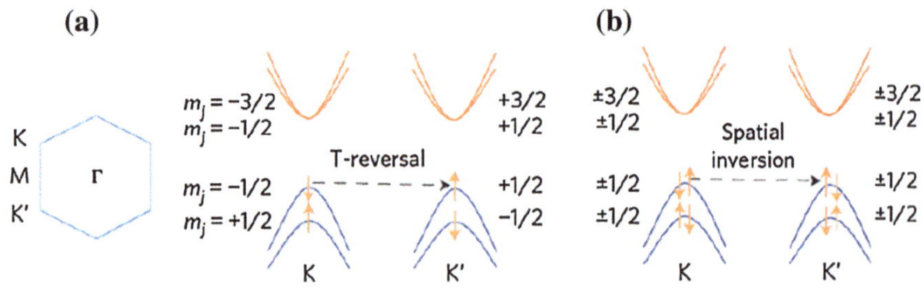

Fig. 2.7 Diagram of the electronic band structure of the CBMIN and VBMAX at K and K′ points of the Brillouin zone in **a** MoS_2 monolayer and **b** MoS_2 bilayer. **a** The loss of the inversion symmetry in a single MoS_2 layer leads to the VBMAX at K (K′) with spin *down* (*up*) electrons. Excitation of a given spin (*down* or *up*) which can be accomplished using a circular polarized light will also lead to a selective excitation of a given valley K or K′. **b** In bilayer MoS_2, the excitation with a circular polarized light will populate both valleys. The band structure diagram is taken from Ref. [63]

2.6 Conclusions and Future Directions

The basic electronic structure properties of exfoliated MoS_2 layers are now well understood. MoS_2 undergoes an indirect to direct band gap transition when is thinned down to one layer. This transition stems from the fact that the confinement in the vertical direction stabilizes the top of the valence band at Γ and, hence, increases the indirect band gap. At the same time, this vertical confinement does not affect localized states at K-point of the Brillouin zone. Furthermore, the inversion symmetry is not present for odd MoS_2 layers (including a MoS_2 monolayer). The absence of the inversion symmetry leads to a loss of the Kramer's degeneracy of electronic bands in most of the Brillouin zone. The spin-orbital splitting is strong and the intrinsic spin-splittings can be of order 100 meV. The loss of the inversion symmetry in conjunction with strong spin-orbital coupling leads to a set of unique optical selection rules that couple spin and valley degrees of freedom. The fundamental band gap can be tuned with strain and application of the external electric field. It was found that the electronic structure of MoS_2 is somewhat sensitive to strain and that large electric fields induce metallicity. From the theoretical perspective, there are several largely unexplored areas: phonons, the impact of doping, and functionalization. It would be interesting to consider hybrid systems, for example, an assembly of two different metal dichalcogenide sheets. For these heterosystems, the band lineups and deformation potentials will need to be established. MoS_2-derived quantum dots and nanoribbons could be interesting from a theoretical point of view. The multiexcitonic complexes in nanostructures based on MoS_2 might be very interesting as an example of a system in which many-body Coulomb interactions are largely unscreened. From experimental point of view, the major problem seems to be the controllable and scalable synthesis of MoS_2 sheets. Without a robust synthesis, the replacement of silicon as a major building block of electronic devices is not possible. It is my hope that recent discovery of many unique properties of exfoliated MoS_2 will inspire further research on this and other 2D materials.

References

1. Neto, A.H.C., Guinea, F., Peres, N.M.R., Novoselov, K.S., Geim, A.K.: Rev. Mod. Phys. **81**, 109 (2009)
2. Ohta, T., Bostwick, A., Seyller, T., Horn, K., Rotenberg, E.: Science **313**, 951 (2006)
3. Castro, E.V., Novoselov, K.S., Morozov, S.V., Peres, N.M.R., Lopes dos Santos, J.M.B., Nilsson, J., Guinea, F., Geim, A.K., Castro Neto, A.H.: Phys. Rev. Lett. **99**, 216802 (2007)
4. Oostinga, J. B., Heersche, H. B., Liu, X., Morpurgo, A. F., Vandersypen, L. M. K.: Nature Mater. **7**, 151 (2007)
5. Nakada, K., Fujita, M., Dresselhaus, G., Dresselhaus, M.S.: Phys. Rev. B **54**, 17954 (1996)
6. Wakabayashi, K., Fujita, M., Ajiki, H., Sigrist, M.: Phys. Rev. B **59**, 8271 (1999)
7. Wunsch, B., Stauber, T., Sols, F., Guinea, F.: Phys. Rev. Lett. **101**, 036803 (2008)
8. Yang, L., Cohen, M.L., Louie, S.G.: Phys. Rev. Lett. **101**, 186401 (2008)

9. Wunsch, B., Stauber, T., Guinea, F.: Phys. Rev. B **77**, 035316 (2008)
10. Wurm, J., Rycerz, A., Adagideli, I., Wimmer, M., Richter, K., Baranger, H. U.: Phys. Rev. Lett. **102**, 056806 (2009)
11. Libisch, F., Stampfer, C., Burgdorfer, J.: Phys. Rev. B **79**, 115423 (2009)
12. Sheng, W.-D., Korkusinski, M., Güçlü, A.D., Zielinski, M., Potasz, P., Kadantsev, E.S., Voznyy, O., Hawrylak, P.: Front. Phys. **7**, 328 (2012)
13. Zhang, Y., Tang, T.-T., Girit, C., Hao, Z., Martin, M. C., Zettl, A., Crommie, M. F., Shen, Y. R., Wang, F.: Nature **459**, 820 (2009)
14. Xu, M., Liang, T., Shi, M., Chen, H.: Chem. Rev. **113**, 3766 (2013)
15. Butler, S. Z., Hollen, S. M., Cao, L., Cui, Y., Gupta, J. A., Gutierrez, H. R., Heinz, T. F., Hong, S. S., Huang, J., Ismach, A. F., Johnston-Halperin, E., Kuno, M., Plashnitsa, V. V., Robinson, R. D., Ruoff, R. S., Salahuddin, S., Shan J., et al.: ACS Nano. **7**, 2898 (2013)
16. Wang, Q. H., Kalantar-Zadeh, K., Kis, A., Coleman, J. N., Strano, M. S.: Nature Nanotech. **7**, 699 (2012)
17. Osada, M., Sasaki, T.: Adv. Mater. **24**, 210 (2012)
18. Zhang, H., Liu, C.-X., Qi, X.-L., Dai, X., Fang, Z., Zhang, S.-C.: Nature Phys. **5**, 438 (2009)
19. Aufray, B., Kara, A., Vizzini, S., Oughaddou, H., Léandri, C., Ealet, B., Le Lay, G.: Appl. Phys. Lett. **96**, 183102 (2010)
20. Lalmi, B., Oughaddou, H., Enriquez, H., Kara, A., Vizzini, S., Ealet, B., Aufray, B.: Appl. Phys. Lett. **97**, 223109 (2010)
21. Takeda, K., Shiraishi, K.: Phys. Rev. B **50**, 14916 (1994)
22. Watanabe, K., Taniguchi, T., Kanda, H.: Nature Mater. **3**, 404 (2004)
23. Yoffe, A.D.: Chem. Soc. Rev. **5**, 51 (1976)
24. Bromley, R.A.: Phys. Lett. A **33**, 1970 (242)
25. Harper P. G., Edmondson, D. R.: Phys. Stat. Sol. (b) **44**, 59 (1971)
26. Mattheiss, L.F.: Phys. Rev. B **8**, 3719 (1973)
27. Chantis, A.N., Christensen, N.E., Svane, A., Cardona, M.: Phys. Rev. B **81**, 205205 (2010)
28. Xiao, D., Liu, G.-B., Feng, W., Xu, X., Yao, W.: Phys. Rev. Lett. **108**, 196802 (2012)
29. Raybaud, P., Hafner, J., Kresse, G., Kasztelan, S., Toulhoat, H.: J. Catal. **189**, 129 (2000)
30. Schweiger, H., Raybaud, P., Kresse, G., Toulhoat, H.: J. Catal. **207**, 76 (2002)
31. Byskov, L.S., Nørskov, J.K., Clausen, B.S., Topsøe, H.: J. Catal. **187**, 109 (1999)
32. Sun, M., Adjaye, J., Nelson, A.E.: Appl. Catal. A **263**, 131 (2004)
33. Galea, N.M., Kadantsev, E.S., Ziegler, T., Phys, J.: Chem. C **113**, 193 (2009)
34. Rydberg, H., Dion, M., Jacobson, N., Schröder, E., Hyldgaard, P., Simak, S.I., Langreth, D.C., Lundqvist, B.I.: Phys. Rev. Lett. **91**, 126402 (2003)
35. Hohenberg, P., Kohn, W.: Phys. Rev. **136**, B864 (1964)
36. Kohn, W., Sham, L.J.: Phys. Rev. **140**, A1133 (1965)
37. Grimme, S.: J. Comput. Chem. **27**, 1787 (2006)
38. Perdew, J.P., Burke, K., Ernzerhof, M.: Phys. Rev. Lett. **77**, 3865 (1996)
39. Bjorkman, T., Gulans, A., Krasheninnikov, A.V., Nieminen, R.M.: J. Phys.: Condens. Matter **24**, 2012 (424218)
40. Dion, M., Rydberg, H., Schroder, E., Langreth, D.C., Lundqvist, B.I.: Phys. Rev. Lett. **92**, 246401 (2004)
41. Lee, K., Murray, E.D., Kong, L., Lundqvist, B.I., Langreth, D.C.: Phys. Rev. B **82**, 081101 (2010)
42. Vydrov, O.A., Van Voorhis, T.: J. Chem. Phys. **133**, 244103 (2010)
43. Johnson, E. R., Keinan, S., Mori-Sánchez, P., Contreras-García, J., Cohen, A. J., Yang, W.: JACS **132**, 6498 (2010)
44. Novoselov, K. S. et al.: Proc. Natl. Acad. Sci. USA **102**, 10451 (2005)
45. Frindt, R.F.: J. Appl. Phys. **37**, 1928 (1966)
46. Splendiani, A., Sun, L., Zhang, Y., Li, T., Kim, J., Chim, C.-Y., Galli, G., Wang, F.: Nano Lett. **10**, 2010 (1271)
47. Mak, K.F., Lee, C., Hone, J., Shan, J., Heinz, T.F.: Phys. Rev. Lett. **105**, 136805 (2010)

48. Radisavljevic, B., Radenovic, A., Brivio, J., Giacometti, V., Kis, A.: Nature Nanotech. **6**, 147 (2011)
49. Fontana, M., Deppe, T., Boyd, A. K., Rinzan, M., Liu, A. Y., Paranjape, M., Barbara, P.: Scientific Reports **3**, 1634 (2013)
50. van der Zande, A. M., Huang, P. Y., Chenet, D. A., Berkelbach, T. C., You, Y., Lee, G.-H., Heinz, T. F., Reichman, D. R., Muller, D. A., Hone, J. C.: Nature Mater. **12**, 554 (2013)
51. Li, T., Galli, G.: J. Phys. Chem. C **111**, 16192 (2007)
52. Kadantsev, E.S., Hawrylak, P.: Solid State Commun. **152**, 2012 (909)
53. Lebègue, S., Eriksson, O.: Phys. Rev. B **79**, 115409 (2009)
54. Cheiwchanchamnangij, T., Lambrecht, W.R.L.: Phys. Rev. B **85**, 205302 (2012)
55. Yun, W.S., Han, S.W., Hong, S.C., Kim, I.G., Lee, J.D.: Phys. Rev. B **85**, 033305 (2012)
56. Dresselhaus, G.: Phys. Rev. **100**, 580 (1955)
57. Cardona, M., Christensen, N.E., Fasol, G.: Phys. Rev. B **38**, 1806 (1988)
58. Luo, J.-W., Bester, G., Zunger, A.: Phys. Rev. Lett. **102**, 056405 (2009)
59. Johari, P., Shenoy, V.B.: ACS Nano **6**, 2012 (5449)
60. Scalise, E., Houssa, M., Pourtois, G., Afanas'ev, V., Stesmans, A.: Nano Res. **5**, 43 (2012)
61. Ramasubramaniam, A., Naveh, D., Towe, E.: Phys. Rev. B **84**, 205325 (2011)
62. Liu, Q., Li, L., Li, Y., Gao, Z., Chen, Z., Lu, J., Phys, J.: Chem. C **116**, 21556 (2012)
63. Mak, K. F., He, K., Shan, J., Heinz, T. F.: Nature Nanotech. **7**, 494 (2012)
64. Zeng, H., Dai, J., Yao, W., Xiao, D., Cui, X.: Nature Nanotech. **7**, 490 (2012)
65. Cao, T., Wang, G., Han, W., Ye, H., Zhu, C., Shi, J., Niu, Q., Tan, P., Wang, E., Liu, B., Feng, J.: Nature Commun. **3**, 887 (2012)

Chapter 3
Tunable Electronic and Dielectric Properties of Molybdenum Disulfide

Ashok Kumar and P. K. Ahluwalia

Abstract We report tunability in electronic and dielectric properties of a technologically promising nanomaterial MoS_2. The properties of MoS_2 can be tuned by varying the layer thickness, by applying mechanic strain, by tuning the interlayer distance, and by applying external electric field. Reducing the slab thickness systematically from bulk to monolayers causes blue shift in the band gap energies, thereby, resulting in tunability of the electronic band gap. By reducing the number of layers from bulk to monolayer limit, electron energy loss spectra (EELS) shows red shift in the energies of both π and $\pi + \sigma$ plasmons. Mechanical strains reduce the band gap of monolayer MoS_2 by causing a direct-to-indirect band gap transitions and finally rendering it into metal at critical values depending on the types of applied strain. Dielectric properties of monolayer MoS_2 too get influenced by the type of applied strain. Imaginary part of dielectric function (ε_2) shows redshift in the structure peak energy on the application of strains with significant dependence on the types of applied strain. In-plane strains also cause semiconductor-metal transitions (e_T) in bilayer sheets of MoS_2. The energy gap of semiconducting bilayer MoS_2 gets reduced continuously by reducing the bilayer separation, eventually rendering it metallic at critical value of interlayer distance. Electrically gated semiconducting bilayer MoS_2 is also found to show reduction in the band gap on increasing the magnitude of electric field and results in band gap closure at a critical value of the field.

A. Kumar (✉) · P. K. Ahluwalia
Physics Department, Himachal Pradesh University, Shimla 171005, India
e-mail: ashok.1777@yahoo.com

P. K. Ahluwalia
e-mail: pk_ahluwalia7@yahoo.com

Z. M. Wang (ed.), *MoS_2*, Lecture Notes in Nanoscale Science and Technology 21,
DOI: 10.1007/978-3-319-02850-7_3,

3.1 Introduction

It had been believed for several years that free-standing two dimensional (2D) crystals were unlikely to exist, because they would ultimately return to three-dimensional (3D) structures. This fact was supported in the theoretical investigations by Peierls [1], Landau [2], and Lifshitz and Pitaevskii [3] which reveal that pure 2D lattice could not exist at nonzero temperatures. Also, the Mermin–Wagner theorem [4] rigorously proved that 2D materials would not exhibit long-range order, because thermal fluctuations would cause atomic displacements of the same magnitude as that of the distance between the atoms and thereby make the 2D materials unstable. Also it is known experimentally that thin films cannot be synthesized below a certain thickness due to the occurrence of decompositions. However, lately it appeared that isolated sheets of graphene could be obtained by mechanical exfoliation of a graphite crystal [5], which therefore contradicted the theoretical contention that the 2D crystals could not exist.

The experimental synthesis of two-dimensional crystals has lead to the emergence of a novel physics such as linear energy dispersion with momentum near the K point of the Brillouin zone in case of graphene [6] and, therefore, the corresponding quasiparticles could be described by a Dirac-type Hamiltonian. Although graphene was the groundbreaking 2D material, later on with rapid developments in search of such 2D materials caused wide interest among the researchers [7–12], leading to investigation of a large number of two-dimensional crystals such as graphene [13], BN [14], transition-metal dichalcogenide [15], silicene [16], germanane [17] etc. Lately, low-dimensional systems have emerged as an area of much interest due to their entirely different behavior than corresponding bulk counterparts under different physical conditions such as under mechanical strain and application of electric field.

Nanoscience and consequent nanotechnologies have been dominated by low-dimensional honeycomb-structured carbon-based materials in the last two decades. Graphene, a two-dimensional (2D) honeycomb structure of carbon, has been studied in great detail by researchers [6, 18–20], but it has no band gap. However, many important applications in optics and transistor technology require a band gap, leading to vigorous research efforts to functionalize it to achieve band gap manipulation. In contrast to pristine graphene, monolayer MoS_2 which also comes under a new classification "graphene analogs" [21] with the similar layered structure as of graphene, has a direct band gap of approximately 1.90 eV [22]. 2D-MoS_2 has, therefore, come up as a competing material for nanoelectronics and is expected to show potential in wide range of practical applications [9, 11, 23–29].

The band gap in monolayer MoS_2 has attracted interest for use in transistor technology and successful implementation of a monolayer MoS_2-based field effect transistor (FET) [30], with HfO_2 as a gate insulator has been achieved. The few layered MoS_2 film is also expected to be a perfect transistor material [31–33]. The key features of MoS_2 transistor include (i) high mobility (200 $cm^2V^{-1}s^{-1}$) at room temperature, (ii) a large on/off current ratio (1×10^8), (iii) a ultralow standby power dissipation, (iv) thermal stability up to 1,100 °C, (v) the absence of dangling bonds, (vi) a high degree of immunity to short channel effects, i.e., drain-induced barrier lowering of approximately 10 mV/V, and (vii) abrupt electrical switching with a subthreshold swing that can be as low as 60 mV/decade [34].

The application of mechanically exfoliated monolayer, bilayer, trilayer, and quadrilayer MoS_2 FETs for NO gas sensing has also been exploited with the detection limits of 0.8 ppm [35]. Results showed that single layer MoS_2 exhibits a rapid and dramatic response upon exposer to NO, but the signal was found to be unstable. In contrast, the MoS_2 FETs comprised of higher number of layers exhibit both stability and sensitivity to the gas. Because of the energy-gap modulation by the number of MoS_2 layers, it is possible to exploit such characteristics for selective light detection. It was shown that photo-transistors consisting of trilayer MoS_2 exhibit excellent photo detection capabilities for red light, while those with monolayer and bilayer turned out to be useful for green light section [33].

MoS_2 in its bulk form has been previously used in many applications including photocatalysis, optoelectronics, and photovoltaics [36–41]. The weak forces between the layers and the anisotropic character lead to MoS_2 being put into lubricant applications [42]. MoS_2 has also been found to show many interesting optical properties including excitonic effects in it [36, 43–46] which are important for device applications. An extensive survey of the properties of bulk MoS_2 has been made by Wilson and Yoffe [36, 47]. Beal et al. [43] measured the transmission spectra in the energy range of 0–4.0 eV, and Beal and Hughes [44] measured the reflectivity spectrum for $\overleftarrow{E} \perp c$. Hughes and Liang [45] have measured the vacuum ultraviolet reflectivity spectra in the range of 4.5–14 eV for $\overleftarrow{E} \perp c$. Zeppenfeld [48] has measured the electronic energy losses and optical anisotropy of single MoS_2 crystal. Liang and Cundy [49] have made detailed study of the EELS of layered transition-metal dichalcogenides (LTMDs) including MoS_2.

3.1.1 Crystal Structure of MoS_2

Naturally occurring MoS_2 has two polytypes, $2H-MoS_2$ and $3R-MoS_2$, both having the trigonal prismatic coordination [36]. $2H-MoS_2$ has two layers per unit cell which are stacked in the hexagonal (H) symmetry having AB type of stacking as shown in Fig. 3.1a, b, while $3R-MoS_2$ has three layers per unit cell (Fig. 3 of [36]) in the rhombohedral (R) symmetry. The $2H-MoS_2$ is dominant and more stable in nature and $3R-MoS_2$ transforms to $2H-MoS_2$ upon heating [50]. In all the layered materials including MoS_2, intralayer bonding is very strong as compared to interlayer bonding which is weak. The bulk crystal of $2H-MoS_2$ is built up of van der Walls bonded S–Mo–S units which does not have inversion symmetry as shown in Fig. 3.1a. The bulk crystal of $2H-MoS_2$ crystallizes in a hexagonal structure with space group $P6_3/mmc$ corresponding to space group number 194 [36]. The each stable S–Mo–S units is referred to as a monolayer, $1H-MoS_2$, as shown in Fig. 3.1c. The monolayer consists of two hexagonal

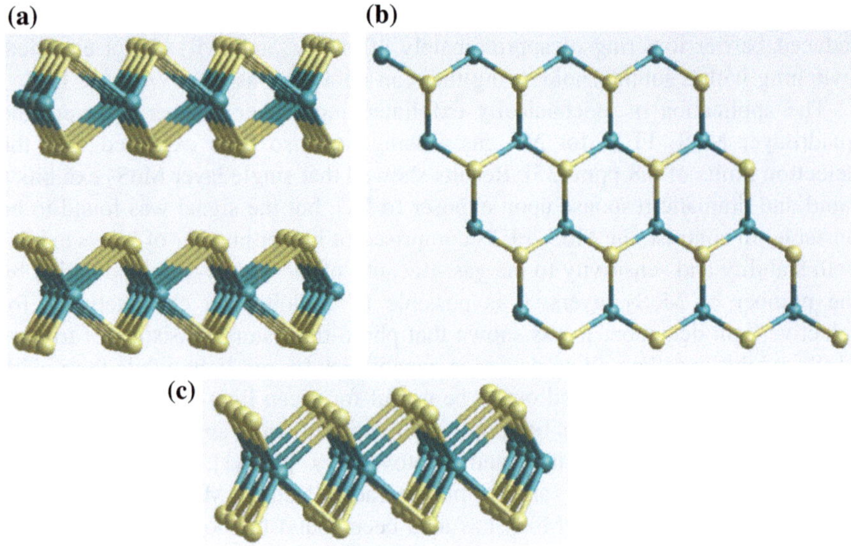

Fig. 3.1 **a** *Side view* of bulk MoS_2 (van der Walls bonded MoS_2 units), **b** *top view* of both bulk MoS_2 and monolayer MoS_2, **c** *side view* of monolayer MoS_2 [59]. Reproduced with permission from Ref. [59]. Copyright 2012 Elsevier Publishing

planes of chalcogen atoms and an intermediate sandwiched hexagonal plane of metal atoms. Because of the weak interactions between the layers and the strong interactions within the layers, shearing takes place more easily and the formation of ultrathin crystals of $1H - MoS_2$ has become possible. Transmission electron microscope (TEM) is one of the most direct methods to study the morphology and crystal structure of 2D materials. The difference between the diffraction patterns of single and multilayer sheets of MoS_2 which is caused by the reduced dimension and loss of sixfold symmetry [51], is used to identify the single layer nanosheets by using TEM.

3.1.2 Synthesis of Ultrathin MoS_2

The formation of ultrathin crystals of MoS_2 has been achieved by Novoselov et al [52], by the same micromechanical cleavage technique, as was applied to graphene. Mechanical exfoliation remains the best method for separating layered crystals with a view to investigating their physical properties. They extracted 6.5 Å thick monolayer with honeycomb structure. Single layer MoS_2 nanocrystal of 30 Å width has also been synthesized on Au (111) surface [53]. Liquid exfoliation of nanosheet of MoS_2 and other transition metal dichalcogenides has also been reported by Coleman et al. [15] which is a very promising technique to obtain large quantity of nanosheets. In the most experimental studies, the bulk MoS_2 bulk crystal is used as a precursor to prepare the isolated MoS_2 nanosheets [9, 11, 54]. The single layer MoS_2 was reported to be obtained by the method of using *n*-butyl lithium in hexane as the intercalation agent to insert lithium ions into the layered structures followed by exfoliation in water with ultrasonication [21]. Very recently, MoS_2 and other 2D materials have been obtained by directly exfoliating their powder materials using supercritical CO_2 assisted with ultrasound [55]. Chemical vapor deposition (CVD) method for growing atomically thin films of MoS_2 on insulating substrates have also been recently reported [56–58]. These methods use different solid precursors (sulfur powder and MoO_3 powder) heated at high temperature which results into a thin layer of Mo metal deposited onto a wafer heated with solid sulfur [57]. Substrate is then dip-coated in a solution of $(NH_4)_2MoS_4$ and heated in the presence of sulfur gas [58]. Although, in CVD, the final MoS_2 film thickness is dependent on the initial precursor and precise control of number of layers over a large area has not yet been achieved, but it is hoped that further work will lead to production of uniform, large area sheets with controllable number of layers. Also, various single layer and few layers of LTMDs including MoS_2 have been predicted stable in first principles structure optimization and phonon calculations based on DFT [10].

3.2 Thickness Dependent Properties of Layered MoS_2

Interestingly, structural parameters of monolayer MoS_2 remain nearly same that of corresponding structure parameter of bulk counterpart [59]. The comparison of the structural parameters of monolayer MoS_2 and bulk MoS_2 is shown in Table 3.1.

Bulk MoS_2 is a semiconductor with an indirect band gap of about 1.23 eV [37]. Conduction band minimum is found between the Γ and K high symmetry point, while valence band maximum is found at Γ point revealing the indirect band gap of bulk MoS_2 [59, 60]. Conduction band minimum and valence band maximum for

Table 3.1 Structural parameters of bulk MoS_2 and monolayer MoS_2 using both local density approximation (LDA) and generalized gradient approximation (GGA)

Property	Monolayer MoS_2		Bulk MoS_2		
	LDA[a]	GGA[a]	LDA[a]	GGA[a]	Experimental
c/a ratio (in Å)	–	–	3.84	4.01	3.89[b]
Lattice constant (in Å)	3.13	3.23	3.13	3.23	3.16[b]
Mo–S bond length (in Å)	2.39	2.45	2.39	2.45	2.41[b]
S–Mo–S bond angle (in degree)	81.73	80.88	81.74	80.84	81.30[b]
S–S distance (in Å)	3.13	3.18	3.12	3.18	3.19[b]
Band gap (in eV)	1.89	1.55	0.75	1.05	1.23[c]

Distance S–S is the vertical distance between two sulfur atoms in S–Mo–S layer. Experimental values for bulk MoS_2 are also given for the sake of comparison

[a] Ref. [59], [b] Ref. [36], [c] Ref. [37]

monolayer MoS_2 is found at K high symmetry point revealing monolayer as a direct band gap semiconductor [59, 60]. The bands around the energy gap are relatively flat which are expected from the '*d*' character of the electron states at these energies. The states around Fermi energy are mainly due to the Mo-*d* orbitals, while strong hybridization between Mo-*d* and *S-p* orbitals below the Fermi energy has been observed [59, 60]. The bulk band gap calculated with both local density approximation (LDA) and generalized gradient approximation (GGA), as shown in Table 3.1, underestimates the experimental band gap. The calculated direct band gap in monolayer MoS_2 is reasonably good agreement with the band gap 1.80 eV measured using complementary techniques of optical absorption, photoluminescence, and photoconductivity of monolayer thick MoS_2 [22].

It is interesting to note that the measured optical band gap is always smaller than fundamental band gap due to large excitonic binding energy. The calculated electronic direct band gap for monolayer MoS_2 is almost double the calculated indirect band gap of corresponding bulk MoS_2 [59, 60]. The underestimation of DFT band gap is due to the well known discrepancies in semi-local functionals and inherent intractability of DFT approach. The band gap problem can be addressed more accurately by using alternative methods such as GW approximation [61, 62] or using hybrid functionals [63].

However, the recent quasiparticle self-consistent GW calculations [61, 62] of monolayer MoS_2 reveals the large overestimation of band gap from the experimental value, which may be attributed to large excitonic effects possessed by monolayer MoS_2.

3.2.1 Thickness Dependent Electronic Band Structure of MoS_2

Theoretical calculation has predicted an indirect to direct band gap transition when MoS_2 is thinned to a single layer [60, 64, 65]. It is found that, as the number of layers decreases from bulk to the monolayer limit, the indirect band gap increases and becomes so large in the monolayers that the materials change to direct band gap semiconductors (Fig. 3.2). Decreasing the number of layers from bulk causes a gradual blue shift in the indirect band gap energy up to the bilayer limit, and after that the indirect band gap changes to a direct band gap in the monolayer limit [60]. The magnitudes of the blue shift in the band gap energies is found to be 1.14 eV as we go from the bulk phase (indirect band gap) to monolayer limit (direct band gap). At the same time, magnitudes of the blue shift in the direct band gap energies at the "K" point is found to be 0.19 eV. Therefore, the change in the indirect band gap energy is significantly larger than that of the direct band gap energy. The blue

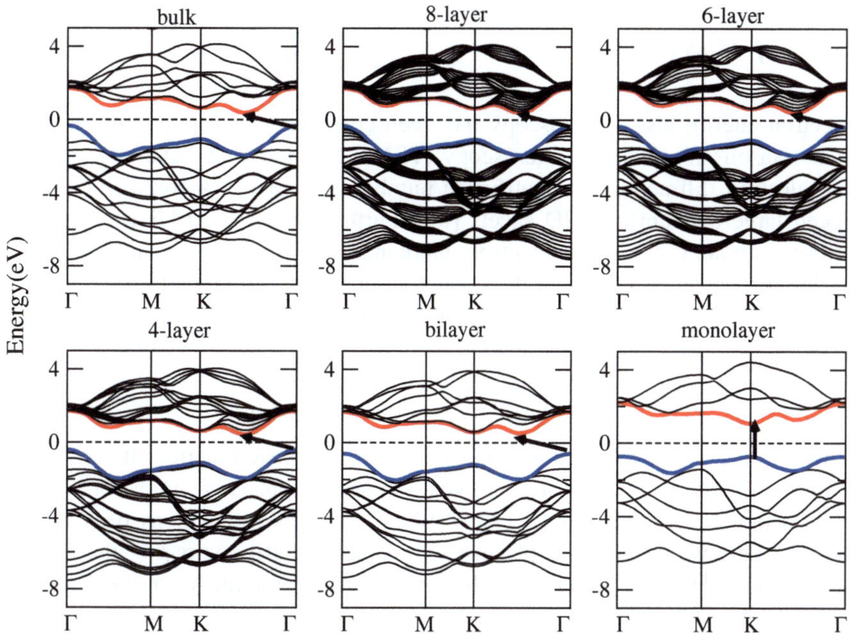

Fig. 3.2 Electronic band structure of bulk MoS_2, its monolayer, as well as polylayers. *Top* of the valence band (*blue*) and *bottom* of the conduction band (*red*) are highlighted. The *arrow* indicates the smallest value of the indirect or direct band gap. The Fermi level is set at 0 eV. Reproduced with permission from Ref. [60]. Copyright 2012 Springer Publishing

shift in the band gap energies is maximum when one goes from bilayer to the monolayer of MoS_2 [59].

The conduction band states at K point are mainly due to the localized d orbitals of the Mo atoms, located in the middle of the S–Mo–S layer sandwiches and relatively unaffected by interlayer coupling. However, the states near the Γ point are due to the combinations of antibonding p_z orbitals of S atoms and the d orbitals of Mo atoms and have a strong interlayer coupling effect [66]. Therefore, as the layer number changes, the direct excitonic states near the *k*-point are relatively unchanged, but the transition at the Γ point shifts significantly from indirect one to a larger, direct one. Therefore, change in the band structure with layer number is due to quantum confinement and the resulting change in hybridization between the p_z orbitals on S atoms and d orbitals on Mo atoms [22, 66]. The band gap of bulk MoS_2 is in the infrared region, while reducing the size of the material to the monolayer limit shifts the band gap toward the visible region. Thus, MoS_2 change its electronic properties from indirect semiconductors in the bulk phase to direct semiconductors in the monolayer limit, opening up an interesting transition to potential applications for photonics, optoelectronics, and sensing.

3.2.2 Thickness Dependent Electron Energy Loss Spectra of MoS_2

Electron energy loss spectroscopy has been considered as a powerful technique over transmission electron microscopy (TEM), coupled with flake edge analysis and electron diffraction to identify the number of layers in a commercially produced large and uniform 2D sheets [67]. Furthermore, the EELS can also give information about the single electron interband transitions which can be identified in the lower energy side from the collective plasma oscillations [49].

The EELS can be calculated from dielectric function using the expression:

$$\mathrm{Im}\left\{\frac{-1}{\varepsilon(\omega)}\right\} = \frac{\varepsilon_2(\omega)}{\varepsilon_1^2(\omega) + \varepsilon_2^2(\omega)} \tag{3.1}$$

where ε_1 and ε_2 are real and imaginary parts of dielectric function. It can be seen from Fig. 3.3 that EELS consists of two prominent resonance features for in-plane-polarization ($E \perp c$) that lies (i) below 10 eV which is π plasmons peak and (ii) above 10 eV which is $\pi + \sigma$ plasmons peak. It is worth comparing the bonding scheme and plasmons features of MoS_2 with that of graphite since these are well

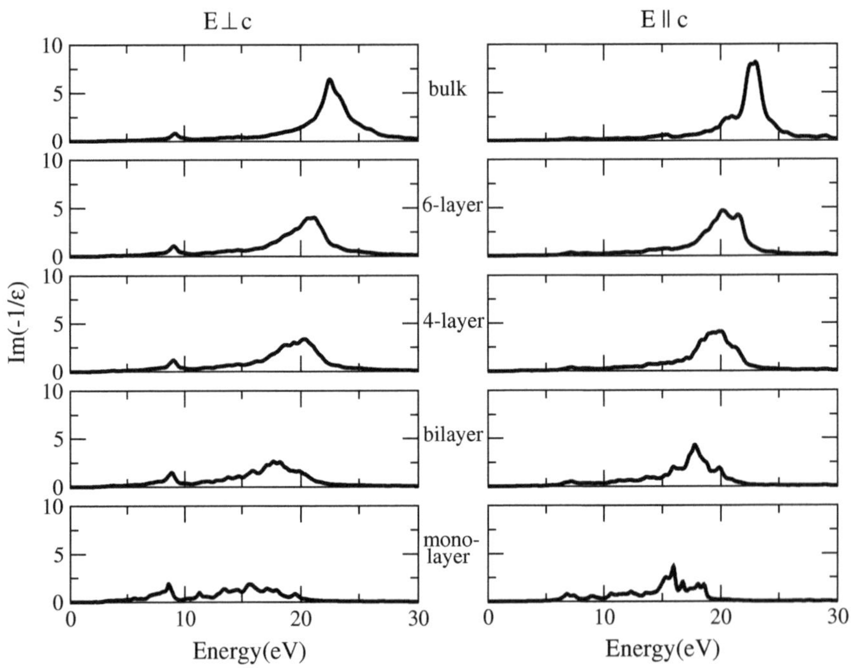

Fig. 3.3 Electron energy loss spectra (EELS) of MoS_2 for electric vector perpendicular to c-axis ($E \perp c$) and electric vector parallel to c-axis ($E \parallel c$), with number of layers. Reproduced with permission from Ref. [70]. Copyright 2012 Elsevier Publishing

understood in case of graphite. It was suggested that despite of less flattened surface of MoS_2 materials as compared to graphite, the bonding scheme in both types of materials remains similar except the narrow d_{z^2} bands in MoS_2 [49]. Moreover, it was found recently that plasmons featuring in rippled graphene [68] remain similar to that of flat graphene [6, 68, 69] except the shift in the π and σ plasmons peaks due to strong hybridization between π and σ orbitals in rippled graphene. The π plasmons feature in MoS_2 arises due to the collective $\pi - \pi^*$ transitions, while $\pi + \sigma$ plasmons results from the $\pi - \sigma^*$ and $\sigma - \sigma^*$ excitations [48, 49]. In MoS_2 with 18 valence electrons, the strong covalent bonding between the atoms of Mo and S is made up of 12 electrons per molecule having mixed s, p and d orbitals which results into the σ valence band and the remaining six electrons are responsible for the π plasmon band [49]. Among the six electrons in π plasmon band, two occupy the d_{z^2} orbitals of the metal atom.

Electron energy loss spectra for out-of-plane polarization ($E \parallel c$) of MoS_2 (Fig. 3.3) consists of one prominent resonance feature above 10 eV [70] due to $\pi + \sigma$ plasmons excitation. The common feature between EELS of two polarizations is that $\pi + \sigma$ plasmons peak for the both perpendicular and parallel polarization remains broad. The loss spectra shifts toward lower energies even by the variation of slab thickness from bilayer to monolayer [70, 71]. π plasmon peaks shift slightly but prominent red shift can be seen in $\pi + \sigma$ plasmon peaks for both perpendicular and parallel polarization as one goes from bulk to monolayer limit. It emerges that as we go from bulk to monolayer, the average concentration of electrons decreases which results in considerable enhancement in the effective mass of electrons and hence a lower value of ω_p for monolayers, which is consistent with the equation:

$$\omega_p^2 = \frac{Ne^2}{\varepsilon_0 m_{\text{eff}}} \tag{3.2}$$

where N, ε_0, e, and m_{eff} are number density of electrons, permittivity of the free space, charge on the electrons and effective mass of the electrons, respectively. It is also observed that intensity of the plasmon peaks varies with the change in the number of layers as can be seen in Fig. 3.3. The intensity of the π plasmon peak increases while intensity of $\pi + \sigma$ plasmon peak decreases by reducing the number of layers from bulk to monolayer [70, 71]. The experimental EELS of MoS_2 [15] also reveal strong sensitivity to the number of layers. Therefore, EELS can also be used as a technique to identify the number of layers in 2D sheets of MoS_2.

3.2.3 Thickness Dependent Photoluminescence of MoS_2

The evolution of electronic structure of MoS_2 nanosheets with various thicknesses gets reflected in the photoluminescence (PL) spectra. The bulk MoS_2 showed negligible PL, while thinner MoS_2 nanosheets exhibit pronounced emissions at

~670 and ~627 nm arising from the direct excitonic transitions at the K point of brillouin zone [66]. The PL spectrum of monolayer MoS_2 is centered at 1.90 eV, which is attributed to a direct-gap luminescence. Few layer samples display multiple emission peaks. The peak coinciding with that of the monolayer emission peak shifts to the red and broadens slightly with increasing thickness. It has been shown that the PL intensity depends inversely on the layer number of MoS_2 nanosheets. The monolayer exhibits the strongest PL intensity and highest incident photon to converted electron ratio [22, 66] as compared to the thickest layered MoS_2 which exhibited weakest PL intensity. Furthermore, with the reduction of thickness of MoS_2 films, the absorption intensities decrease [72]. The absorption spectrum of MoS_2 showed two absorption bands centered at 600 and 700 nm, while the emission spectrum consists of one major peak at 660 nm and one minor peak at 610 nm [72]. The emission intensity gradually decreases with increasing film thickness, which is due to the weak interlayer coupling between the restacked MoS_2 sheets [22]. A slight red shift in both the absorption resonance and photoluminescence energy with increasing film thickness was found [22, 72]. The small shift is consistent with the fact that the direct gap is only slightly sensitive to layer thickness at the K point due to the quantum confinement effect [60]. The photoluminescence intensity of layered MoS_2 can be enhanced by Ag@SiO_2 core/shell composites due to a metal-enhanced fluorescence [73]. These features of MoS_2 make it attractive for novel electronic and optoelectronic devices such as solar cells and light-emitting diodes.

3.2.4 Thickness Dependent Raman Spectra of MoS_2

Furthermore, thickness-dependent electronic structure of MoS_2 nanosheets can be reflected in the Raman scattering of MoS_2 nanosheets. The Raman characteristics, namely, the frequency, intensity, and width of the two peaks (due to strong in-plane E^1_{2g} and out-of-plane A_{1g} vibrations) influence strongly by changing the layer number [74]. The blue shift of A_{1g} and the red shift of E_1 has been observed with increasing MoS_2 layer number. This opposite direction of the frequency shift has been partly attributed to the Columbic interaction and possible stacking-induced change of the intra-layer bonding [74].

3.3 Mechanical Strain Mediated Properties of Monolayer MoS_2

Pristine graphene has been shown as one of the strongest materials ever measured, with a breaking strength of 42 N m^{-1} and the Youngs modulus of 1.0 TPa [75]. Theoretical study of monolayer MoS_2 report an in-plane stiffness and Poisson's ratio calculated in the harmonic elastic strain range to be 123 N m^{-1} and 0.25, respectively, indicating that monolayer MoS_2 is much softer than graphene [76]. Elastic properties of the freely suspended MoS_2 nanosheets, with thicknesses ranging from 5 to 25 layers have been measured by using an AFM tip which applied load in the center of the suspended region of a MoS_2 nanosheet [77]. The sheet deformation (δ) was determined by the equation $\delta = \Delta Z_{\text{piezo}} - \Delta Z_c$, where ΔZ_{piezo} is the displacement of the AFM scanning piezotube and ΔZ_c is the deflection of an AFM cantilever. The average Youngs modulus of the MoS_2 sheet is found exceptionally high, i.e., $E = 0.33 \pm 0.07$ TPa, which is comparable to that of graphene oxide. This value is also higher than the Youngs modulus of bulk MoS_2 (i.e., 0.24 TPa). In another work [78], the measured Youngs modulus of a monolayer MoS_2 has been reported to be ~0.27 TPa, higher than that of the bulk form. This discrepancy is attributed to the presence of stacking faults in the low density thinner flakes compared to the bulk crystal. It has been also found that the measured strength of individual single layer MoS_2 lies in the range of 6–11 % of its Youngs modulus [78], which is the theoretical upper limit of a materials breaking strength (i.e., the intrinsic strength of inter atomic bonds in the crystal). The low pre-tension and high elasticity of MoS_2 [76–78] makes it attractive semiconductors for flexible electronic and optoelectronic devices, as well as for composite films.

The controlled introduction of strain into semiconductors offers an important degree of flexibility in studies involving basic materials as well as device engineering. Strain in a crystalline solid modifies the lattice constants and reduces the crystal symmetry, leading to significant shifts in the energy band edges [79, 80]. When investigated under an applied strain, electronic structure of 2D-MoS_2 can be tuned [79–82] and can potentially offer mechanism for tuning electronic transport properties ranging from semiconducting to metallic. The nanomaterials, whose electronic properties can be controlled by the mechanical strain are greatly desired for the applications in nano electromechanical systems (NEMS) as well as nano optomechanical systems (NOMs). Strain engineering has been successfully used in the semiconductor industry to impressively improve the performance levels of the conventional metal-oxide-semiconductor field-effect transistors (MOSFETs) [83, 84]. Interestingly, lattice mismatch between deposited thick layer of HfO_2 and monolayer MoS_2 in recent MoS_2 -based transistor has been regarded analogous to tensile strain which results in an increase in the carrier mobilities [65].

3.3.1 Mechanical Strain-Mediated Structural Properties of Monolayer MoS_2

The theoretical mechanical strains ($e = \Delta a/a_0$, where a_0 is unstrained lattice constant and Δa is change in the lattice constant after deforming the lattice) can be applied to the supercell of monolayer MoS_2, with respect to the calculated lattice constant (a_0) [79]. For achieving this, the strained cell has been modeled by applying tensile as well as compression strains by varying the lattice value 'a' with strain 'e' as $a \rightarrow ae$. The tensile strain was applied as an uniaxial expansion in x direction (+a) and a biaxial expansion in x and y directions (+a+b). Similarly, compression strains $-a$ and $-a-b$ were applied by uniaxial compression in x direction and biaxial compression in x and y directions, respectively. All the above four configurations are symmetric with same magnitudes of the strain. Asymmetric biaxial strain (+$a-b$) can also be applied by expanding the lattice constant in x direction and compressing it in y direction by the same magnitude. It was found that the in-plane strains not only influence the Mo–S bond length in MoS_2 sheet, but also the vertical S–S distance and S–Mo–S bond angle varies [79, 81]. The unstrained structural parameters and 10 % strained structural parameters are listed in Table 3.2.

Tensile strains elongate the Mo–S bond lengths, while these are shortened by compression strains. Vertical S–S distances are decreased by the in-plane tensile strains, while compression strains increased the S–S distances. Also, S–Mo–S bond angles show decrease with the in-plane-applied tensile strains, while they are found to increase with compression strains. The influence of biaxial strains is found to be more pronounced than uniaxial strains. The structural parameters in asymmetric strain are found to be in between the tensile and compression strains [79]. These changes in the structural properties can be understood easily, since we fix the in-plane lattice vectors, the coordinates would relax perpendicular to the sheet, which results into the change in the S–S distance as well as S–Mo–S bond angle. These changes have significant impact on the electronic structure as semiconducting direct band gap of unstrained monolayer MoS_2 have been found to first change into indirect band gap, then finally to metallic, with the increased magnitude of applied strain [79–81].

Table 3.2 Calculated structural parameters (Mo–S bond length, $d_{Mo–S}$ vertical distance between S atoms, $d_{S–S}$, S–Mo–S bond angle, $\Theta_{S–Mo–S}$) for MoS_2 with DFT/GGA

Parameter[a]	Unstrained	+a	+a+b	$-a$	$-a-b$	+$a-b$
$d_{Mo–S}$	2.450	2.486	2.543	2.416	2.397	2.437
$d_{S–S}$	3.178	3.082	3.001	3.285	3.422	3.155
$\Theta_{S–Mo–S}$	80.878	76.570	72.364	85.692	91.101	80.721

The calculated values of structural parameters at the applied strain (+a, +a+b, $-a$, $-a-b$, +$a-b$) are at $e = 0.1$, i.e., 10 %. All the tensile and compression strains are symmetric and applied homogeneously. +a, +a+b, $-a$, $a-b$ and +$a-b$ indicates, respectively, uniaxial tensile, biaxial tensile, uniaxial compression, biaxial compression, and asymmetric biaxial strain

[a] Ref. [79]

3.3.2 Mechanical Strain-Mediated Electronic Properties of Monolayer MoS_2

The systematic change in the electronic band structure of strained MoS_2 sheet on the application of five different types of strain is shown in Fig. 3.4. It is found that uniaxial tensile strain (+*a*) results in raising (lowering) of VB (CB) at Γ (K) high symmetry point. CB at Γ and M point, and VB at K point are found to remain unchanged while slow upward shift has been found in VB at M point with uniaxial tensile strain (+*a*). At the applied uniaxial tensile strain (+*a*) greater than 12 %, CBM and VBM at K and Γ point, respectively, remain unchanged, while VB at M point and CB at Γ point are found to shift toward Fermi level. Ultimately at 35 % applied strain (Table 3.3), VB at M point crosses the Fermi level resulting in strain-induced semiconductor-metal transition (e_T) [79].

Tensile biaxial strain (+*a*+*b*) also shows similar behavior, however, shift in the bands are found more pronounced than uniaxial tensile strain (+*a*). e_T for tensile biaxial (+*a*+*b*) strain is found to occur at 10 % of applied strain [79–81]. Smaller compression strain up to 3–4 % shows exactly opposite trends than tensile strain [79, 82]. At relatively higher magnitude of strain, gradual raising of VB at M point and lowering in K-Γ direction can be seen in Fig. 3.4 for both uniaxial (−*a*) and

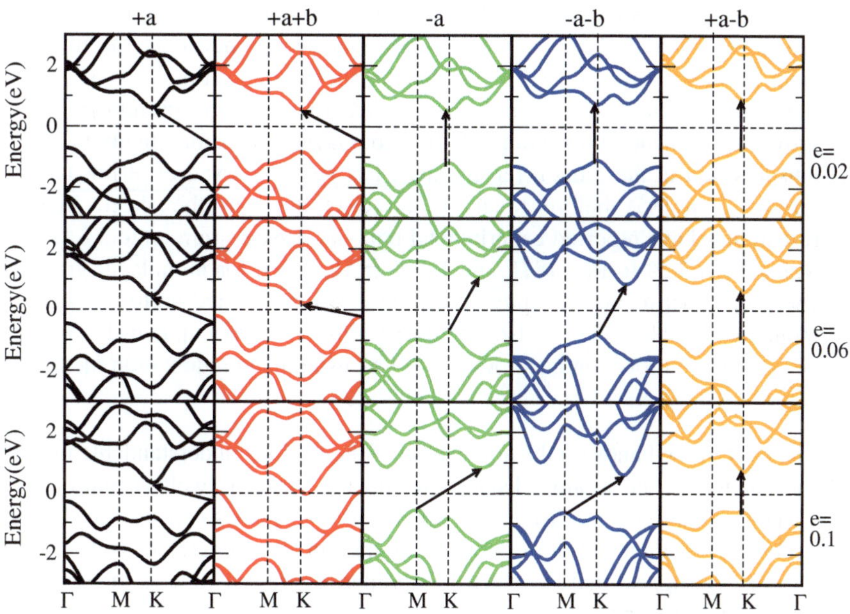

Fig. 3.4 Electronic band structure of strained monolayer MoS_2 corresponding to the applied strain ($e = \Delta a/a_0$) 0.02, 0.06, and 0.1, when different type of strains (+*a*, +*a*+*b*, −*a*, *a*−b, and +*a*−*b*) are applied to unstrained structure. +*a*, +*a*+*b*, −*a*, *a*−*b*, and +*a*−*b* indicate, respectively, uniaxial tensile, biaxial tensile, uniaxial compression, biaxial compression, and asymmetric biaxial strain. Reproduced with permission from Ref. [79]. Copyright 2013 Elsevier Publishing

Table 3.3 The calculated % value of applied strains ($\frac{\Delta a}{a_0} * 100$) at which semiconductor-metal transitions (e_T) occur and the % value of applied strains up to which direct band gap (E_D) remains preserved in monolayer MoS_2

Property	+a	+a+b	−a	−a−b	+a−b
e_T (in %)	35.0[a]	10.0[a], 9.8[b], 10.0[c], 10.0[d]	17.0[a]	15.0[a], 15.0[c]	19.0
E_D (in %)	0.5[a]	0.3[a], 0.3[b]	5.0[a]	2.5[a], 1.3[b]	10.0[a], 10.0[d]
DP (in eV)[a]	−12.0	−22.5	7.5	14.5	−1.0

The band gap deformation potential (DP) in linear regime ($e < 2$ %) for various types of strain is also listed

[a] Ref. [79], [b] Ref. [65], [c] Ref. [81], [d] Ref. [80]

biaxial (−*a*−*b*) compression strains, however, shift of bands in biaxial compression strain was more pronounced. For asymmetric biaxial strain (+*a*−*b*), VBM and CBM are found to remain at K point up to 10 % strain [79, 80]. The strain greater than 10 % makes MoS_2 sheet indirect band gap semiconductor and finally metallic at greater than 19 % strain (Table 3.3).

It is very important to mention that very high strains as listed in Table 3.3 are very difficult to achieve in real experimental situations. However, strain engineering has been successfully applied experimentally to the monolayer graphene [85–88]. Experimentally, strain can be introduced to the monolayers in many ways [56–58, 77, 78, 85–89] such as depositing them onto the flexible substrate, by subjecting them to external load, by pseudomorphic growth, and nano indentation of the monolayers. The maximum biaxial tensile strain in the monolayer graphene has been observed experimentally up to 25 % [85] before breaking the C–C bond in the sheet. Many researchers have reported experimental uniaxial tensile strain up to 3.8 % in Ref. [88], 1.3 % in Ref. [87], and 1.3 % in Ref. [86] for monolayer graphene. The experimental compression strain is being reported up to 0.7 % for monolayer graphene [56]. Also, an effective in-plane strain up to 6–11 % has been measured for monolayer MoS_2 [78] by using nano indentation method.

The shift in the bands on the application of the strain is understandable in terms of the change in the atomic distance within the plane as well as out-of-plane [79, 81] as change in the distance between S–Mo–S planes can lead to different superposition of their atomic orbitals with shift in the energies of states. These results clearly suggest that the electronic properties are sensitive to the strain and have implications for how the direct band gap can be maintained if PL feature is desirable.

3.3.2.1 Band Gap Deformation Potentials of Monolayer MoS_2

The change in electronic band structure as a function of strain can be quantified in terms of deformation potential (DP), which is a useful parameter for device modeling [90–92]. The change in the energy of a valence band/conduction band per unit of strain is called valence band deformation potential/conduction band deformation potential [90]. The effect of strain can also be characterized by the change of the transition energy (i.e., energy difference) upon application of strain where the linear coefficient provides the deformation potential [91, 92]. The experimental determination of DP is quite difficult because all strain components cannot be determined accurately without any further approximation [91]. Also experimental DP cannot be isolated from each other. However, in theory, there is a liberty to calculate DP separately for each type of strain. For small strain in the vicinity of equilibrium lattice parameters, the variation of band gap is approximately linear and hence can be expressed as [92]:

$$\Delta E_g = \mathrm{DP} \times e \tag{3.3}$$

where ΔE_g is change in the band gap, "e" is magnitude of strain, and DP is deformation potential. This expression is valid for linear regime where the magnitude of strain in typically of the order of 1–2 %. To calculate the DP at higher value of strain, extracting the slope parameters from strain versus band gap curves may be a useful technique [79]. The calculated values of band gap deformation potential in the linear regime ($e < 2$ %) are given in Table 3.3, which have been calculated by dividing the band gap energy difference (between $e = 0$ and $e = 0.02$) to the difference of strain value (i.e., 0.0–0.02).

It is found that for both uniaxial and biaxial tensile strain in MoS_2, the value of DP is quite high which is attributed due to the change in the nature of band gap from direct to indirect below 2 % of applied tensile strain in monolayer MoS_2 [79]. The positive and negative values of the band gap deformation potential is the indication of the decrease and increase in the band gap, respectively, in linear regime. Large positive/negative magnitude tells about the large change, i.e., decrease/increase in the band gap on the application of applied strain.

3.3.2.2 Carriers Effective-Masses and Exciton Binding Energies of Monolayer MoS_2

As has been discussed above that strain influences the band structure of the studied monolayers, therefore, it is also useful to examine the carrier effective masses of strained monolayers. Furthermore, it is also useful to consider excitonic binding problem [61, 62] in the context of the classical Mott–Wannier model. The exciton binding energy for idealized 2D monolayer is given by the expression [62]:

$$E_b^{ex} = 4 \times \frac{13.6\,\mu_{ex}}{m_e \varepsilon_s^2} \tag{3.4}$$

where $\mu_{ex} = m_e^* m_h^* / (m_e^* + m_h^*)$ is effective exciton mass, m_e is free electron mass, m_e^* is effective electron mass, m_h^* is effective hole mass, and ε_s is the static dielectric constant. The corresponding transition energy (E_T) in Mott–Wannier model is given by:

$$E_T = E_g^{KS} - E_b^{ex} \tag{3.5}$$

where E_g^{KS} is Kohn–Sham band gap energy. The effective mass of electron (m_e^*) and effective mass of hole (m_h^*) can be determined from the curvature of the energy bands at conduction band minimum (CBM) and valence band maximum (VBM), respectively [79]. The effective masses have been calculated by taking average of the curvature ($\partial^2 E/\partial k^2$) in the neighborhood of CBM/VBM and using it in the formula:

$$m^* = \frac{\hbar^2}{\partial^2 E/\partial k^2} \tag{3.6}$$

The DFT/PBE calculated [79] zero strain carrier effective masses, i.e., $m_e^* = 0.177$, $m_h^* = 0.205$; and exciton reduced mass, i.e., $\mu_{ex}=0.094$; for monolayer MoS_2 are found to be consistently lower than the quasiparticle GW calculated values of two different work in Ref. [62] ($m_e^* = 0.60$, $m_h^* = 0.54$, $\mu_{ex}=0.28$) and Ref. [61].($m_e^* = 0.35$, $m_h^* = 0.44$, $\mu_{ex}=0.19$). The DFT/PBE calculated exciton binding energy (E_b^{ex}) within Mott–Wannier model for monolayer MoS_2 comes out to be 0.213 eV which results into lower transition energies (E_T) 1.337 eV as compared to both nonself-consistent G_0W_0 values 1.97 eV [62] and quasiparticle self-consistent GW value 1.86 eV [61]. The underestimation DFT/PBE value is understood in terms of inherent discrepancy in the conventional DFT/PBE approach. However, we can always get a general insight about the carrier effective masses and exciton binding energies on the application of strain, which may be important for the further development of the monolayers MoS_2 for the electronic applications.

The carrier effective masses show change on the application of different types of strain which has been attributed to the change in the curvature of bands at CBM and VBM [79]. The effective masses show abrupt change at the point where CBM/VBM changes from one high symmetry point to another high symmetry point, which is due to the large difference of the carrier masses at different high symmetry point [65, 92]. Exciton reduced masses have been found to show strong dependence on the magnitude of strain as well as on the types of applied strain

[79]. The exciton binding energies are generally high for less than 4–5 % tensile strain, while compression strains as well as asymmetric biaxial strain cause decrease in the exciton binding energies [79]. The small tensile strains are found to increase the exciton binding energies [65, 79] which may be of importance in the applications of optoelectronics.

3.3.3 Mechanical Strain-Mediated Dielectric Properties of Monolayer MoS_2

Since mechanical strain affect the electronic band gap as well as the energy band dispersion in the vicinity of Fermi level [79, 80], therefore it is also expected that mechanical strain may also influence the dielectric properties of monolayer MoS_2.

The mechanical strain-mediated band gap of monolayer MoS_2 can be supported by the change in static dielectric constant (ε_s) on the application of strain [79], since band gap (E_g) of material is related to ε_s by the relation [93]: $\varepsilon_s \approx 1 + (\hbar\omega_p/E_g)^2$. The ε_s is given by the value of real part of dielectric function at zero energy or frequency, i.e., $\varepsilon_1(0)$ [79]. It is worth noting that zero frequency in this context means a frequency low compared to interband transitions but higher than the phonon frequencies. $\varepsilon_1(0)$ has no contributions from lattice vibrations. The calculated value of ε_s for unstrained monolayer MoS_2 is 4.91 which is smaller than the ε_s value 12.8 for corresponding bulk counterpart [59]. The high positive value of ε_s for bulk MoS_2 is consistent with its small band gap and very low positive value ε_s is consistent with the large band gap of monolayer MoS_2. ε_s increases with the increase of tensile strains (both uniaxial and biaxial) and asymmetric biaxial strain, while compression strains show decrease in ε_s for smaller magnitude of strains and increases at higher magnitude of strain [79]. The nature of the response of ε_s (increase/decrease) with applied strains are consistent with the nature of the band gap changes (decrease/increase) when different types of strains are applied to monolayer MoS_2 [79].

The influence of applied strains on the in plane ($E \perp c$) imaginary part of dielectric function (ε_2) is shown in Fig. 3.5. A prominent structure peak in ε_2 of unstrained MoS_2 sheet has been found at 2.70 eV while the weak structure peaks at 3.6 and 4.2 eV have also been observed [79]. It was found that all the structure peaks in the ε_2 show displacement with applied strain, however, for convenience, only the displacement in the prominent structure peak in Fig. 3.5a–e is shown. It is worth identifying the interband transitions in electronic band structure corresponding to the prominent structure peak in ε_2. A prominent structure peak at 2.70 eV in ε_2 of unstrained sheet of MoS_2 has been found to be dominated by the

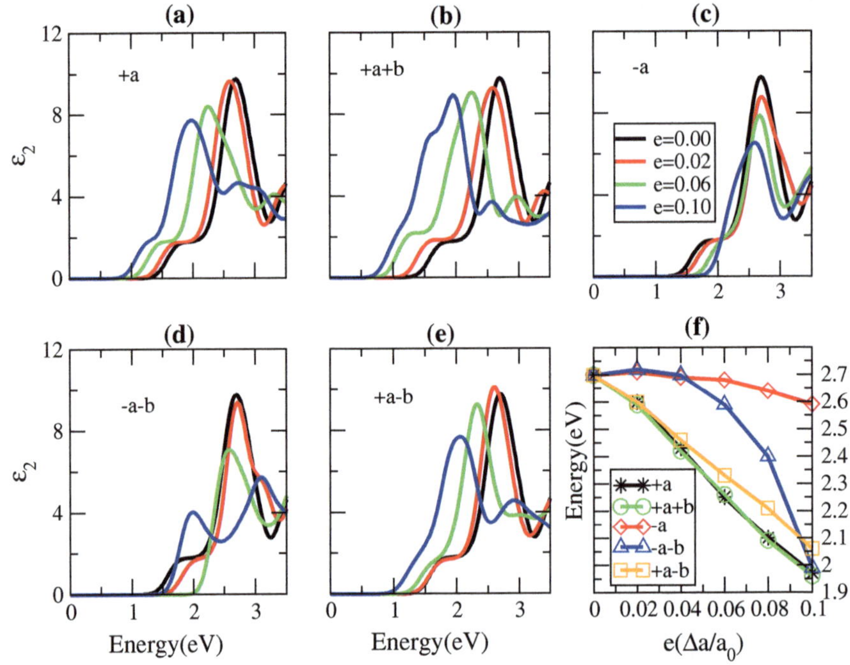

Fig. 3.5 ε_2 corresponding to the applied strain ($e = \Delta a/a_0$) 0.02, 0.06, and 0.1, when different type of strains **a** uniaxial tensile (+*a*) **b** biaxial tensile (+*a*+*b*) **c** uniaxial compression (−*a*) **d** biaxial compression (−*a*−*b*), and **e** asymmetric biaxial strain (+*a*−*b*), are applied to unstrained structure of monolayer MoS_2. The variation in the position of prominent structure peak in ε_2 with applied strains is shown in **f**. Reproduced with permission from Ref. [79]. Copyright 2013 Elsevier Publishing

interband transitions mainly from the valence band 1 (VB1) below the Fermi energy to the conduction band 1 (CB1) above the Fermi energy along Γ-M direction [79]. On the application of different types of strains, the structure peak gets red shifted toward lower energy as shown in Fig. 3.5a–e. It is also evident from the Fig. 3.5f that symmetric tensile strains (+*a* and +*a*+*b*) as well as asymmetric biaxial strain (+*a*−*b*) cause gradual displacement in structure peak energy on increasing the magnitude of strain, while negligible displacement has been caused by compression strains (−*a* and −*a*−*b*) up to 4 % of applied strain.

The displacement in the unstrained structure peak from 2.70 eV can be understood on looking at the dispersion of the first valence band, i.e., VB1 below Fermi energy and first conduction band, i.e., CB1 above the Fermi energy in the vicinity of Γ and M point, along Γ-M direction (see Fig. 3.4 of previous subsection). In the vicinity of Γ and M points along Γ-M direction (in case of both uniaxial and biaxial tensile strain), VB1 shifts toward Fermi level, while CB1 remains almost unshifted. This results into decrease in the interband transition energy (energy difference VB1−CB1 in the vicinity of Γ and M point along Γ-M direction), hence the displacement of the structure peak in ε_2 toward lower energy. In case of compression strains (both uniaxial and biaxial) interband transition in the vicinity of Γ point shifts toward Γ-M direction, VB1 and CB1 in the vicinity of M point (M-Γ direction) remains almost unchanged up to 4 % of applied strain, which results into nearly negligible shift in the structure peak energies in Fig. 3.5f. On applying greater than 4 % compression strain, the resultant band dispersion due to biaxial strain ($-a-b$) changes rapidly as compared to uniaxial ($-a$) strain. In case of asymmetric biaxial strain ($+a-b$), the VB1 and CB1 dispersion shows complex behavior. Both the bands get raised in the vicinity of Γ and M point along Γ-M direction up to 2 % of strain, then get lowered up to 6 % of strain, but the resultant interband transition energies always get lowered on the application of strain.

3.4 Tunable Electronic Properties of Bilayer MoS_2

Similar to the monolayer structure of MoS_2, structural parameters of bilayer MoS_2 remains nearly same as that of bulk MoS_2 and the total density of states around the Fermi level are mainly derived from metal d states along with strong hybridization between metal d and chalcogen p states below the Fermi energy [94]. Bilayer MoS_2 is an indirect band gap semiconductor with valence band maximum (VBM) at Γ point and conduction band minima (CBM) at K point [60, 94]. The GGA-PBE calculated band gap of bilayer MoS_2 is 1.21 eV [94] which is lower than the experimental band gap 1.6 eV [22]. However, bilayer MoS_2 is also tunable with the application of strains.

3.4.1 *Band Gap Engineering of Bilayer MoS_2 by In-plane Strains*

Similar to monolayer MoS_2, the band gap energy of bilayer MoS_2 decreases with the increase of applied in-plane strain, ultimately rendering bilayer MoS_2 into metallic at critical value of strain [94]. The variation of band gap and semiconductor-metal transitions (e_T), however, strongly depends on the type of applied strain as can be seen in Fig. 3.6a. The observation about the band gap engineering with different types of applied strain can be summarized from Fig. 3.6a as follows: (i) the band gap variation with uniaxial tensile strain ($+a$) remains robust against relatively higher value of strain ($e > 0.1$) and gets metallic at applied strain $e = 0.20$ (ii) e_T in case of biaxial tensile strain ($+a+b$) is found to occur at $e = 0.08$, (iii) the compressive strains, irrespective of uniaxial or biaxial, are found to show nearly similar response to band gap variation, (iv) the value of band gap remains nearly constant for small magnitude of asymmetric biaxial strain ($+a-b$), (v) e_T by compression strains is found to occur at $e = 0.16$ and $e = 0.15$,

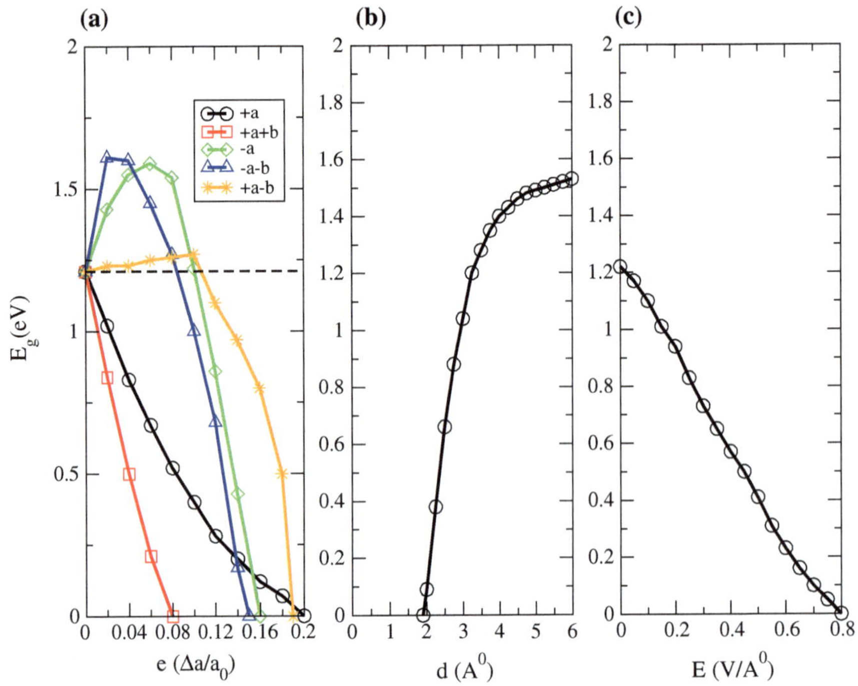

Fig. 3.6 **a** The band gap (E_g) variation at different type of strains, i.e., uniaxial tensile (*+a*), biaxial tensile (*+a+b*), uniaxial compression (*−a*), biaxial compression (*−a−b*), and asymmetric biaxial strain (*+a−b*), applied to bilayer MoS_2. The horizontal *dotted line* indicates the indirect band gap value 1.23 eV for unstrained structure of bilayer MoS_2. **b** The band gap variation of bilayer MoS_2 with changing interlayer distance with respect to the optimized value $d_0 = 3.30$ Å. **c** The band gap variation of bilayer MoS_2 with external electric field. Reproduced with permission from Ref. [94]. Copyright 2013 IOP Publishing

respectively, for uniaxial and biaxial compression, while asymmetric biaxial strain (*+a−b*) shows e_T at 0.19 value of applied strain.

3.4.2 *Band Gap Engineering of Bilayer MoS_2 by Changing Bilayer Separation*

Decrease and increase in the interlayer distance can be regarded as out-of-plane compression and tensile uniaxial strain, respectively [94, 95]. The bilayer separation can be decreased or increased with reference to the optimized value (3.30 Å) [94] of separation between the two layers of unstrained structure. As the bilayer separation of MoS_2 bilayer is decreased, the VB (CB) at Γ (all the high symmetry points) point moves toward the Fermi level while VB at M and K points remain almost unchanged [94]. It was found that CB at K-Γ direction moves faster toward Fermi level as compared to CB at K point at relatively small bilayer separation, that leads to band gap closure between Γ–KΓ direction in the brillouin zone. e_T has been found to occur at critical interlayer distances 1.95 Å as shown in Fig. 3.6b. On the other hand, by increasing the bilayer separation from optimized value, the band gap becomes nearly constant (Fig. 3.6b) due to feeble interactions between the layers that makes the considered structure effectively as monolayers.

3.4.3 Band Gap Engineering of Bilayer MoS_2 by External Electric Field

The influence of external electric field to the electronic band structure can be seen by applying electric field perpendicular to the bilayer sheet [94]. Density functional theory calculations reveal that indirect band gap of bilayer MoS_2 can be continuously driven to zero by external electric field [94, 96, 97]. As the electric field is continuously increased, the band gap closure occurs between VBM and CBM of pristine structures, thereby, rendering the structures into metallic at critical electric fields [94, 96, 97]. It has also been shown that, just before the S–M transition to occur, the band gap decreases slowly with applied field Fig. 3.6c. The critical values of electric field at which semiconductor-metal transition occur is found at 0.8 V/Å.

The calculated value of critical electric field at which e_T occurs in Ref. [94] is found higher than the values (0.2–0.3 V/Å) obtained in Ref. [96]. This mismatch in the critical electric field for e_T is probably due to the constraints imposed in Ref. [96] on the metal–chalcogen bond lengths, interlayer distances, and lattice parameters (a_0 and c). However, in another work [97], the critical value of electric field has been predicted between 1.0 and 1.5 V/Å for various types of stacking in bilayer MoS_2.

Furthermore, applying external electric field to the strained structure can be regarded equivalent to applying external bias voltage to the hetrostructures (lattice mismatch between different types of lattices in hetrostructures may introduce strain) constituting practical devices. It was found that e_T occurs relatively at lower electric field for 4 % tensile-strained bilayer MoS_2 than unstrained bilayer structures [94]. The critical electric field at which e_T occurs is found at 0.4 and 0.3 V/Å for uniaxial (+*a*) and biaxial (+*a*+*b*) strained bilayer MoS_2 as compared to 0.8 V/Å for unstrained structure. Therefore, it can also be seen as an alternate way to tuning the band gap of materials.

Acknowledgments Ashok Kumar is grateful to Council of Scientific and Industrial Research (CSIR), New Delhi, Government of India, for providing financial assistance in the form of Senior Research Fellowship.

References

1. Peierls, R.E.: Ann. Inst. Henri Poincare **5**, 177 (1935)
2. Landau, L.D.: Phys. Z. Sowjetunion **11**, 26 (1937)
3. Lifshitz, E. M., Pitaevskii, L. P.: Statistical physics part I (Chaps. 137 and 138) (Pergamon: Oxford, UK, 1980)
4. Mermin, N.D.: Phys. Rev. **176**, 250 (1968)
5. Novoselov, K., Giem, A., Morozov, S., Jiang, D., Zhang, Y., Dubonos, S., Grigorieva, I., Firsov, A.: Science **306**, 666 (2004)
6. Mohan, B., Kumar, A., Ahluwalia, P.K.: Phys. E **44**, 1670 (2012)
7. Enyashin, A., Gemming, S., Seifert, G.: Eur. Phys. J. **149**, 103 (2007)
8. Xu, M., Liang, T., Shi, M., Chen, H.: Chem. Rev. doi: 10.1021./cr300263a
9. Wang, Q.H., K-Zadeh, K., Kis, A., Coleman, J.N., Strano, M.S.: Natl. Nanotechnol. **7**, 699 (2012)
10. Ataca, C., Sahin, H., Ciraci, S.: J. Phys. Chem. C **116**, 8983 (2012)
11. Huang, X., Zeng, Z., Zhang, H.: Chem. Soc. Rev. **42**, 1934 (2013)
12. Song, X., Hu, J., Zeng, H.: J. Mater. Chem. C **1**, 2952 (2013)
13. Elias, D.C., Nair, R.R., Mohiuddin, T.M.G., Morozov, S.V., Blake, P., Halsall, M.P., Ferrari, A.C., Boukhvalov, D.W., Katsnelson, M.I., Geim, A.K., Novoselov, K.S.: Science **323**, 610 (2009)
14. Nag, A., Raidongia, K., Hembram, K.P.S.S., Dutta, R., Waghmare, U.V., Rao, C.N.R.: ACS Nano **4**, 1539 (2010)
15. Coleman, J.N., Lotya, M., O'Neill, A., Bergin, S.D., King, P.J., Khan, U., Young, K., Gaucher, A., De, S., Smith, R.J., Shvets, I.V., Arora, S.K., Stanton, G., Kim, H.Y., Lee, K., Kim, G.T., Duesberg, G.S., Hallam, T., Boland, J.J., Wang, J.J., Donegan, J.F., Grunlan, J.C., Moriarty, G., Shmeliov, A., Nicholls, R.J., Perkins, J.M., Grieveson, E.M., Theuwissen, K., McComb, D.W., Nellist, P.D., Nicolosi, V.: Science **331**, 568 (2011)
16. Vogt, P., Padova, P.D., Quaresima, C., Avila, J., Frantzeskakis, E., Asensio, M.C., Resta, A., Ealet, B., Lay, G.L.: Phys. Rev. Lett. **108**, 155501 (2012)
17. Bianco, E., Butler, S., Jiang, S., Restrepo, O. D., Windl, W., Goldberger, J. E. : ACS Nano. doi: 10.1021/nn4009406
18. Geim, A.K., Novoselov, K.S.: Natl. Mater. **6**, 183 (2007)
19. Geim, A.K.: Science **324**, 1530 (2009)
20. Rao, C. N. R, Sood, A. K., Subrahmanyam, K. S., Govindraj, A., Angew: Chem. Int. Ed. **48**, 7752 (2009)
21. Ramkrishana Matte, H.S.S., Gomathi, A., Manna, A.K., Late, D.J., Dutta, R., Pati, S.K., Rao, C.N.R.: Angew. Chem. Int. Ed. **49**, 4059 (2010)
22. Mak, K.F., Lee, C., Hone, J., Shan, J., Heinz, T.F.: Phys. Rev. Lett. **105**, 136805 (2010)
23. Kong, B.D., Zeng, C., Gaskill, D.K., Wang, K.L., Kim, K.W.: Appl. Phys. Lett. **101**, 263112 (2012)
24. Buscema, M., Barkelid, M., Zwiller, V., van der Zant, H.S.J., Steele, G.A., Castellanos-Gomez, A.: Nano Lett. **13**, 358 (2013)
25. Parhizgar, F., Rostami, H., Asgari, R.: Phys. Rev. B **87**, 125401 (2013)
26. Cheng, Y.C., Zhu, Z.Y., Mi, W.B., Guo, Z.B., Schwingenschlogl, U.: Phys. Rev. B **87**, 100401 (2013)
27. Sun, Q–.Q., Li, Y.-J., He, J.-L., Yang, W., Zhou, P., Lu, H.-L., Ding, S.-J., Zhang, D.W.: Appl. Phys. Lett. **102**, 093104 (2013)

28. Newaz, A.K.M., Prasai, D., Ziegler, J.I., Caudel, D., Robinson, S., Haglund Jr, R.F., Bolotin, K.I.: Solid State Commun. **155**, 49 (2013)
29. Gao, D., Si, M., Li, J., Zhang, J., Zhang, Z., Yang, Z., Xue, D.: Nano. Res. Lett. **8**, 129 (2013)
30. Radisavljevic, B., Radenovic, A., Brivio, J., Giacometti, V., Kis, A.: Natl. Nanotechnol. **6**, 147 (2011)
31. Kim, S., Konar, A., Hwang, W., Lee, J.H., Lee, J., Yang, J., Jung, C., Kim, H., Yoo, J., Choi, J., Jin, Y.W., Lee, S.Y., Jena, D., Choi, W., Kim, K.: Nature Comm. **3**, 1011 (2012)
32. Liu, H., Gu, J., Ye, P.D.: IEEE Elec. Device. Lett **33**, 1273 (2012)
33. Lee, H.S., Min, S.-W., Chang, Y.-G., Park, M.K., Nam, T., Kim, H., Kim, J.H., Ryu, S., Im, S.: Nano Lett. **12**, 3695 (2012)
34. Yoon, Y., Ganapathi, K., Salahuddin, S.: Nano Lett. **11**, 3768 (2011)
35. Li, H., Yin, Y.Z., He, Q.Y., Li, H., Huang, X., Lu, G., Fam, D.W.H., Tok, A.I.Y., Zhang, Q., Zhang, H.: Small **8**, 63 (2012)
36. Wilson, J., Yoffe, A.D.: Adv. Phys. **18**, 193 (1969)
37. Kam, K., Parkinson, B.: J. Phys. Chem. **86**, 463 (1982)
38. Tenne, R., Wold, A.: Appl. Phys. Lett. **47**, 707 (1985)
39. Coehoorn, R., Hass, C., Degroot, R.A.: Phys. Rev. B **35**, 6203 (1987)
40. Sienicki, W., Hryniewicz, T.: Solar Energy Mater. Solar Cell. **43**, 67 (1996)
41. Wilcoxon, J., Thurston, T., Martin, J.: J. Nano. Mater. **12**, 993 (1999)
42. Matthesis, L.F.: Phys. Rev. Lett. **30**, 784 (1973)
43. A.R Beal, J.C. Knights, W.Y. Liang, J. Phys. C.: Sol. Stat. Phys. **5**, 3540 (1972)
44. Beal, A.R., Hughes, H.P.: J. Phys. C. Sol. Stat. Phys. **12**, 881 (1979)
45. Hughes, H.P., Liang, W.Y.: J. Phys. C. Sol. Stat. Phys. **7**, 1023 (1974)
46. Fortina, E., Raga, F.: Phys. Rev. B **11**, 905 (1975)
47. A. D. Yoffe: http://pubs.rsc.org doi:10.1039/CS9760500051 (1976)
48. Zeppenfeld, K.: Optic. Commun. **1**, 377 (1970)
49. Liang, W.Y., Cundi, S.L.: Phis. Mag. **19**, 1031 (1969)
50. Enyashin, A.N., Yadgarov, L., Houben, L., Popov, I., Weidenbach, M., Tenne, R., bar Sadan, M., Seifert, G.: J. Phys. Chem. C **115**, 24586 (2011)
51. Brivio, J., Alexander, D.T.L., Kis, A.: Nano Lett. **11**, 5148 (2011)
52. Novoselov, K., Jiang, D., Schedin, F., Booth, T., Khotkevich, V., Morozov, S., Geim, A.: Proc. Natl. Acad. Sci. USA **102**, 10451 (2005)
53. Helveg, S., Lauritsen, J., Laegsgaard, E., Stensgaard, I., Norskov, J., Clausen, B., Topsoe, H., Besenbacher, F.: Phys. Rev. Lett. **84**, 951 (2000)
54. Yao, Y., Lin, Z., Li, Z., Song, X., Moon, K.-S., Wong, C.-P.: J. Mater. Chem. **22**, 13494 (2012)
55. Wang, Y., Zhou, C., Wang, W., Zhao, Y.: Ind. Eng. Chem. Res. **52**, 4379 (2013)
56. Lee, Y.H., Zhang, X.Q., Zhang, W., Chang, M.T., Lin, C.T., Chang, K.D., Yu, Y.C., Wang, J.T.W., Chang, C.S., Li, L.J., Lin, T.W.: Adv. Mater. **24**, 2320 (2012)
57. Zhan, Y., Liu, Z., Najmaei, S., Ajayan, P.M., Lou, J.: Small **8**, 966 (2012)
58. Liu, K.K., Zhang, W., Lee, Y.H., Lin, Y.C., Chang, M.T., Su, C.W., Chang, C.S., Li, H., Shi, Y., Zhang, H., Lai, C.S., Li, L.J.: Nano Lett. **12**, 1538 (2012)
59. Kumar, A., Ahluwalia, P.K.: Mater. Chem. Phys. **135**, 755 (2012)
60. Kumar, A., Ahluwalia, P.K.: Eur. Phys. J. B **85**, 186 (2012)
61. Cheiwchanchamnangij, T., Lambrecht, W.R.L.: Phys. Rev. B **85**, 205302 (2012)
62. Ramasubramaniam, A.: Phys. Rev. B **86**, 115409 (2012)
63. Ellis, J.K., Lucero, M.J., Scuseria, G.E.: Appl. Phys. Lett. **99**, 261908 (2011)
64. Kuc, A., Zibouche, N., Heine, T.: Phys. Rev. B **83**, 245213 (2011)
65. Yun, W.S., Han, S.W., Hong, S.C., Kim, I.G., Lee, J.D.: Phys. Rev. B **85**, 033305 (2012)
66. Splendiani, A., Sun, L., Zhang, Y., Li, T., Kim, J., Chim, C.Y., Galli, G., Wang, F.: Nano Lett. **10**, 1271 (2010)
67. Sun, J.B., Hannon, J.B., Tromp, R.M., Johari, P., Bol, A.A., Shenoy, V.B., Pohl, K.: ACS Nano **4**, 7073 (2010)
68. Sedelnikova, O.V., Bulusheva, L.G., Okotrub, A.V.: J Chem. Phys. **134**, 244707 (2011)

69. Klintenberg, M., Lebegue, S., Ortiz, C., Sanyal, B., Fransson, J., Eriksson, O.: J. Phys. Condens. Matter **21**, 335502 (2009)
70. Kumar, A., Ahluwalia, P.K.: Phys. B **407**, 4627 (2012)
71. Johari, P., Shenoy, V.B.: ACS Nano **5**, 5903 (2011)
72. Eda, G., Yamaguchi, H., Voiry, D., Fujita, T., Chen, M., Chhowalla, M.: Nano Lett. **11**, 5111 (2011)
73. Zhou, K., Zhu, Y., Yang, X., Zhou, J., Li, C.: Chem. Phys. Chem. **13**, 699 (2012)
74. Lee, C., Yan, H., Brus, L.E., Heinz, T.F., Hone, J., Ryu, S.: ACS Nano **4**, 2695 (2010)
75. Huang, X., Yin, Z., Wu, S., Qi, X., He, Q., Zhang, Q., Yan, Q., Boey, F., Zhang, H.: Small **7**, 1876 (2011)
76. Yue, Q., Kang, J., Shao, Z., Zhang, X., Chang, S., Wang, G., Qin, S.: Phys. Lett. A **376**, 1166 (2012)
77. Castellanos-Gomez, A., Poot, M., Steele, G.A., van der Zant, H.S.J., Agrait, N., Rubio-Bollinger, G.: Adv. Mater. **24**, 772 (2012)
78. Bertolazzi, S., Brivio, J., Kiss, A.: ACS Nano **5**, 9703 (2011)
79. Kumar, A., Ahluwalia, P.K.: Phys. B **419**, 66 (2013)
80. Johari, P., Shenoy, V.B.: ACS Nano **6**, 5449 (2012)
81. Scalise, E., Houssa, M., Pourtois, G., Afanasev, V., Stesmans, A.: Nano. Res. **5**, 43 (2012)
82. Lu, P., Wu, X., Guo, W., Zeng, X.C.: Phys. Chem. Chem. Phys. **14**, 13035–13040 (2012)
83. Chu, M., Sun, Y., Aghoram, U., Thompson, S.E.: Annu. Rev. Mater. Res. **39**, 203 (2009)
84. Lee, M.L., Fitzgerald, E.A., Bulsara, M.T., Currie, M.T., Lochtefeld, A.: J. Appl. Phys. **97**, 011101 (2005)
85. Lee, C., Wei, X., Kysar, J. W., Hone, J.: Science **321**, 385 (2008) www.sciencemag.org/content/suppl/2008/07/17/321.5887.385.DC1
86. Mohiuddin, T.M.G., Lombardo, A., Nair, R.R., Bonetti, A., Savini, G., Jalil, R., Bonini, N., Basko, D.M., Galiotis, C., Marzari, N., Novoselov, K.S., Geim, A.K., Ferrari, A.C.: Phys. Rev. B **79**, 205433 (2009)
87. Tsoukleri, G., Parthenios, J., Papagelis, K., Jalil, R., Ferrari, A.C., Geim, A.K., Novoselov, K.S., Galiotis, C.: Small **5**, 2397 (2009)
88. Huang, M., Yan, H., Heinz, T.F., Hone, J.: Nano Lett. **10**, 4074 (2010)
89. Fang, J., Qian, X., Huang, C.W., Li, J.: Nature Photon. **6**, 866 (2012)
90. Fischetti, M.V., Laux, S.E.: J. Appl. Phys. **80**, 2234 (1996)
91. Yan, Q., Rinke, P., Scheffler, M., Van de Walle, C.G.: Appl. Phys. Lett. **95**, 121111 (2009)
92. Peelaers, H., Van de Walle, C. G.: Phys. Rev. B **86**, 241401(R) (2012)
93. Penn, D.: Phys. Rev. B **128**, 2093 (1962)
94. Kumar, A., Ahluwalia, P.K.: Model. Simul. Mater. Sci. Eng. **21**, 065015 (2013)
95. Bhattacharyya, S., Singh, A.K.: Phys. Rev. B **86**, 075454 (2012)
96. Ramasubramaniam, A., Naveh, D., Towe, E.: Phys. Rev. B **84**, 205325 (2011)
97. Liu, Q., Li, L., Li, Y., Gao, Z., Chen, Z., Lu, J.: J. Phys. Chem. C **116**, 21556 (2012)

Chapter 4
Ab Initio Study on MoS_2 and Its Family: Chemical Trend, Band Alignment, Alloying, and Gap Modulation

Jun Kang and Jingbo Li

Abstract In this chapter, recent progresses on the theoretical study of low-dimensional semiconducting transition metal dichalcogenides (TMD) MX_2 ($M = Mo, W; X = S, Se, Te$) are reviewed. The chemical trends in basic structural and electronic properties are discussed, and the band offsets between MX_2 monolayers are calculated. A simple model is proposed to interpret the chemical trends of the band offsets. Moreover, the suitable band edge position of MoS_2 monolayer makes it a good candidate for the photo-splitting of water. The cluster expansion method and special quasi-random structure approach are employed to study the properties of MX_2 alloys. It is demonstrated that in (S, Se) alloys, there exist stable-ordered alloy structures even at 0 K, whereas in (Se,Te) and (S,Te) alloys, phase separation into the two constituents will occur at 0 K. Nevertheless, a complete miscibility in these alloys can be achieved by increasing temperature. Finally, we show that the bandgap of MX_2 nanostructures can be efficiently modulated by strain, electronic field, and alloying. By increasing strain or electric field strength, the bandgap of MX_2 can be reduced, and gap closure is achieved when the strain/field strength reaches a critical value. In MX_2 alloys, the bandgap and band edge position varies as the composition changes, and exhibits bowing effect, which is a joint effect of volume deformation, chemical difference, and structure relaxation. More importantly, the direct gap character of MX_2 monolayer is retained in the alloys, making them good candidates for 2D optoelectronics.

J. Kang · J. Li (✉)
State Key Laboratory for Superlattics and Microstructures, Institute of Semiconductors, Chinese Academy of Sciences, P. O. Box 912, Beijing 100083, China
e-mail: jbli@semi.ac.cn

Z. M. Wang (ed.), *MoS_2*, Lecture Notes in Nanoscale Science and Technology 21,
DOI: 10.1007/978-3-319-02850-7_4,

4.1 Introduction

After graphene became experimentally accessible in 2004, two-dimensional (2D) materials have attracted great attention mainly due to their unique physical properties and capability to fulfill the demands of future nanoelectronic industry on flexibility, adaptability, and multifunctionality. Transition metal dichalcogenides MX_2 (where M and X correspond to transition metal and chalcogen, respectively) are members of the layered materials like graphite. Soon after the discovery of graphene, it was demonstrated that monolayer MX_2, such as MoS_2 and $NbSe_2$, could also be achieved using mechanical exfoliation [1]. More recently, various experimental studies have reported the synthesis of MX_2 monolayers, such as MoS_2, WS_2, $MoSe_2$, $MoTe_2$, TiS_2, TaS_2, $TaSe_2$, $NiTe_2$, and ZrS_2 [2–4]. Of particular interest is semiconducting MX_2, which are promising for next generation electronic devices. For example, the existence of the native bandgap provides an excellent current on/off ratio of 10^8 in single-layer MoS_2-based field-effect transistors [5], and the application of monolayer MoS_2 in integrated circuits and logic operations has already been realized [6]. Moreover, bulk MoX_2 and WX_2 are indirect bandgap semiconductors, whereas their monolayers have direct bandgaps and exhibit photoluminescence [7, 8], which are favorable for optoelectronic applications. In addition, the observation of valley polarization [9, 10] in monolayer MoS_2 also demonstrates possible valley-based electronic applications.

Given the already made experimental progresses, theoretical investigations on semiconducting MX_2 monolayers are in urgent need. For instance, the elastic parameters such as in-plane stiffness and Poisson's ratio, as well as band offset between semiconductors, are important parameters in material design. More specifically, the band offset is critical to many properties such as quantum confinement [11], dupability [12], and chemical activity [13]. Therefore, it is necessary to figure out the band offsets between different MX_2, and to understand the corresponding chemical trend [14]. Furthermore, a specific monolayer MX_2 provides a single direct bandgap value. To integrate MX_2 for advanced applications, it is desirable to explore possible methods, such as strain [15, 16] or electric field [17, 18], to extend and tune their bandgaps. Hence, the bandgap modulation of MX_2 nanostructures is a worth discussing topic. Finally, although not been experimentally reported, the idea of MX_2 alloys has been proposed by recent theoretical studies [19, 20]. While alloying in conventional semiconductors is well established, the response of 2D MX_2 to alloying might be quite different due to the low dimensionality and distinct boundary conditions. For example, many 3D semiconductor alloys exhibit spontaneous long-range ordering, but it is unknown whether the 2D MX_2 also display ordered ground-state (GS) configuration. Second, it is intriguing to see how the miscibility in these 2D alloys varies as temperature changes, and at what temperature a complete miscibility can be achieved. Finally, physical properties of alloyed semiconductor often deviate from concentration-averaged values of the constituents, known as the bowing effect.

Fundamentally important for material design of 2D semiconductors, these bowing effects in bandgap and band edge position have yet to be thoroughly analyzed.

In this chapter, we introduce our recent progresses on the theoretical study of semiconducting MX_2 (M = Mo, W; X = S, Se, Te) nanostructures. The calculation methods are briefly described in Sect. 4.2. Section 4.3 presents the basic structural, elastic, and electronic structure of monolayer MX_2. In Sect. 4.4 we discuss the band alignment between different MX_2. In Sect. 4.5 the stability and phase diagram of MX_2 alloys are presented. Section 4.6 discusses the bandgap modulation of MX_2 nanostructures. A summary is given in Sect. 4.7.

4.2 Method of Calculations

In our study, first-principles method within the framework of density functional theory (DFT) was used, as implemented in pseudopotential codes such as VASP [21] and SIESTA [22]. The generalized gradient approximation (GGA) was adopted as the exchange-correlation functional. In the calculation of band structure and alignment of monolayer MX_2, the hybrid HSE method [23] was also employed. In the calculation of band structure and band alignment, spin-orbit interaction (SOI) was taken into consideration, as Mo and W are heavy. However, we also found that when comparing the trend of the band properties, the effect of SOI is quite small. Therefore, SOI was omitted in the study of bandgap modulation. To simulate the 2D and 1D structure, a large vacuum layer was added to avoid interaction between adjacent images, and the vacuum level was taken as a common reference to align the band structures.

In the alloy calculations, we apply the cluster expansion (CE) method [24, 25]. The stability of the $MX_{2(1-x)}X'_{2x}$ (X and X′ are different chalcogens) alloys at 0 K can be evaluated from their formation enthalpies ΔH, which are calculated by:

$$\Delta H = E_{\mathrm{T}}(\mathrm{MX}_2(1-x)X'2x) - xE_{\mathrm{T}}(\mathrm{MX}'_2) - (1-x)E_{\mathrm{T}}(\mathrm{MX}_2) \quad (4.1)$$

where $E_T(MX_2)$, $E_T(MX'_2)$, and $E_T(MX_{2(1-x)}X'_{2x})$ are the total energy of pure MX_2, pure MX'_2 and the mixed alloy, respectively. The value of ΔH depends on the specific atomic arrangement in the $MX_{2(1-x)}X'_{2x}$ alloys. According to the CE formalism, the atomic arrangement can be described by assigning occupation variables σ_i to each anion site i, with σ_i equals −1 or +1 if the site is occupied by X or X′, respectively. Any alloy configuration can be represented by a vector $\vec{\sigma} = (\sigma_1, \sigma_2, \sigma_3, \ldots)$ containing all the σ_i information on the anion lattice. The formation enthalpy of a $MX_{2(1-x)}X'_{2x}$ configuration $\vec{\sigma}$ can be then decomposed into energy contributions of cluster figures (such as pairs, triplets, and quadruplets):

$$\Delta H(\vec{\sigma}) = \sum_{\alpha} m_\alpha J_\alpha \xi_\alpha(\vec{\sigma}), \quad (4.2)$$

where α is a cluster, and the summation is taken over all symmetry-nonequivalent clusters. m_α is the multiplicity which denotes the number of symmetry-equivalent clusters of α, and J_α is the effective cluster interaction (ECI) coefficient. ξ_α is the so-called cluster correlation function [26] for α, which is defined as:

$$\xi_\alpha(\vec{\sigma}) = \left\langle \prod_{i \in \alpha'} \sigma_i \right\rangle. \tag{4.3}$$

The bracket represents the average over all symmetry-equivalent clusters α' of α. For a random alloy, the correlation function is $(2x-1)^k$, with x the concentration and k the number of sites included in the cluster. For each alloy, the formation enthalpies of more than 50 selected alloy configurations up to 24 atoms per cell were evaluated from first-principles calculations, and the cluster expansion was constructed based on the calculated values using the ATAT code [27]. A total of 26 clusters up to quadruplet were included for the fitting of ECIs. The fitted ECIs were then used to predict the formation enthalpies of all symmetry-inequivalent alloy configurations ($\sim$17,000) up to 24 atoms per cell.

To investigate the band structure in $MX_{2(1-x)}X'_{2x}$ random alloy, a 6×6 hexagonal supercell containing 108 atoms was used in the calculation of electronic properties, while the distribution of the X and X′ atoms in the lattice was modeled via the special quasi-random structure (SQS) approach [26]. The most relevant correlation functions of the constructed SQS are close to those of a random alloy, so the physical properties of the random alloy can be well simulated by the SQS.

4.3 Structural, Elastic, and Electronic Properties of MX_2 Monolayers

The monolayers of MX_2 compounds have hexagonal lattices with honeycomb structures, which are similar to graphene. The transition metal atom layer is sandwiched between two chalcogenide atom layers, and the structure has point group symmetry of D_{3h}. The calculated basic properties of MX_2 are listed in Table 4.1. As X goes from S to Se, the lattice constant and bond length of MX_2 increase. This can be understood by the fact that the ionic radius of X increases with increasing atomic number. It can be also seen that the lattice constants of MoX_2 and WX_2 are quite similar, indicating that synthesizing MoX_2–WX_2 heterostructures might allow one to engineer optical and physical properties with minimum structural defects.

Since MX_2 monolayers have hexagonal symmetry, they are homogeneous and isotropic, and their elastic properties can be described by two independent constants, Young's modulus Y and Poisson's ratio v. Due to ambiguous in defining the thickness of monolayer structures, in-plane stiffness C can be used as a measure of the strength, and the Young's modulus can be deduced from C by assuming a

Table 4.1 Calculated properties of MX_2 monolayers: lattice constant a, M–X bond length d, bandgap E_g, spin-orbit splitting Δ_{SO} in valence band, cohesive energy E_C per unit cell, charge transfer $\Delta\rho$ of the M atom, Poisson's ratio v and in-plane stiffness C. a, d, E_g, and Δ_{SO} calculated by HSE06 are also shown

	a (Å)	a^{HSE} (Å)	d (Å)	d^{HSE} (Å)	E_g (eV)	E_g^{HSE} (eV)	Δ_{SO} (eV)	Δ_{SO}^{HSE} (eV)	E_C (eV)	$\Delta\rho$ (e)	v	C (N/m)
MoS_2	3.18	3.16	2.41	2.40	1.59	2.02	0.15	0.20	15.31	1.09	0.25	124.24
$MoSe_{-2}$	3.32	3.29	2.54	2.51	1.33	1.72	0.19	0.27	13.70	0.85	0.23	103.40
$MoTe_2$	3.55	3.52	2.73	2.70	0.94	1.28	0.22	0.35	12.01	0.52	0.24	78.90
WS_2	3.18	3.16	2.42	2.40	1.55	1.98	0.43	0.56	17.28	1.24	0.22	139.54
WSe_{-2}	3.32	3.29	2.55	2.53	1.25	1.63	0.47	0.63	15.45	0.96	0.19	115.52
WTe_2	3.55	3.52	2.74	2.71	0.75	1.03	0.49	0.69	13.51	0.57	0.18	86.93

proper thickness. C is defined as $C = (1/S_0)(\partial^2 E_S/\partial\varepsilon^2)$, where S_0 is the equilibrium area, E_S is the strain energy and ε the strain. The other parameter, Poisson's ratio v, is defined as $v = -\varepsilon_{trans}/\varepsilon_{axial}$, namely, the ratio of the transverse strain to the axial strain. By applying a series of strain ε_a and ε_z along the armchair and zigzag directions of MX_2, we obtained the strain-energy relationship:

$$E_s = a_1\varepsilon_a^2 + a_2\varepsilon_z^2 + a_3\varepsilon_a\varepsilon_z, \quad (4.4)$$

where a_1, a_2, a_3 are fitted parameters. Due to the hexagonal symmetry, a_1 equals a_2, and C and v can be then calculated as $C = (1/S_0)(2a_1 - a_3^2/2a_1)$ and $v = a_3/2a_1$. The calculated in-plane stiffness of MoS_2 is 124.24 N/m. Considering that the distance between the two S layers is 3.13 Å and the effective thickness of MoS_2 monolayer would be larger than this value, the upper limit of the Young's module of MoS_2 monolayer can be estimated to be about 0.4 TPa. A recent experiment shows that the Young's module of few-layer MoS_2 is 0.33 ± 0.07 TPa [28], in line with our calculations. When X goes from S to Te, the in-plane stiffness of MX_2 decreases, whereas when M goes from Mo to W, the in-plane stiffness increases. Such a trend can be attributed to the variations of bonding strength. As shown in Table 4.1, the cohesive energy and charge transfer of MX_2, decreases with increasing atomic number of X, demonstrating that the bonding strength decreases when X goes from S to Te. Thus, the in-plane stiffness also decreases. On the other hand, the cohesive energy and charge transfer in MoX_2 are smaller than those in WX_2, so the bonding strength of the latter is stronger, leading to larger in-plane stiffness. The calculated Poisson's ratios of MX_2 monolayer is around 0.2, and the Poisson's ratio of MoX_2 is slightly larger than that of WX_2.

Bulk MoX_2 and WX_2 are indirect bandgap semiconductors, whereas their monolayers have direct bandgaps, with the conduction band maximum (CBM) and valence band minimum (VBM) located at the K (K′) symmetry points. The band structure of MX_2 monolayers are shown in Fig. 4.1, and the corresponding bandgaps are listed in Table 4.1. The bandgap values are underestimated by GGA, while they are slightly overestimated by HSE06. But regardless of the choice of the method (GGA/HSE06), the observed general trends display similar behavior.

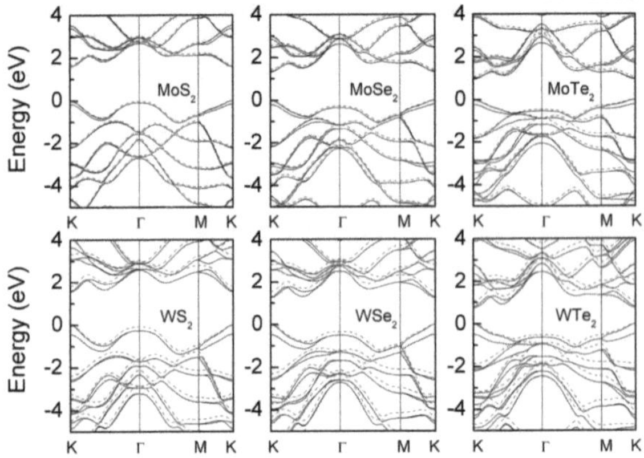

Fig. 4.1 Band structures of MX_2 monolayers calculated by GGA with (*solid lines*) and without (*dashed lines*) SOI

As anion goes from S to Te, the bandgap value decreases. In addition, all MX_2 show significant spin-orbit splitting Δ_{SO} in VBM, which is absent in bulk case. The magnitude of the splitting strongly depends on the type of cation. The GGA calculated Δ_{SO} is around 0.2 eV for MoX_2, and 0.5 eV for WX_2. When the atomic number of X increases, Δ_{SO} also increases, but the variation is small. Such a trend is different from conventional ZB semiconductor, in which the Δ_{SO} of telluride is about twice as large as selenide [29]. Because VBM is dominated by anion *p* states in conventional ZB semiconductor, its Δ_{SO} is mainly determined by the type of anions. Whereas in MX_2, VBM is dominated by cation *d* states, as discussed in the following, and the Δ_{SO} is mostly cation-dependent.

To understand the band structure better, we look at the physical origin of the observed CBM and VBM values. To simplify, the following discussions are made based on the band structure calculated without SOI. Taking MoS_2 as example, the VBM of MoS_2 mainly consists of the $d_{x^2-y^2}$ and d_{xy} orbitals of Mo, with small contribution of the p_x and p_y orbitals of S. In addition, it is found that the 5th band at the K points, which locates about 4 eV lower than the VBM, has similar character as VBM, but the dominate orbital become S p_x and p_y. Therefore, the VBM of MoS_2 originates mainly from the repulsion between the $d_{x^2-y^2}$ and d_{xy} orbitals of Mo and the p_x and p_y orbitals of S. The *d* orbital of Mo is higher than the *p* orbital of S, so it is pushed up by Δ_1, forming the VBM, and the *p* orbital is pushed down by Δ_1', as shown in Fig. 4.2a. The CBM of MoS_2 has the character of the d_{z^2} orbital of Mo and the p_x and p_y orbitals of S, while the 4th band at the K points, located about 4.5 eV lower than the VBM, has the same character. In the CBM state the Mo *d* orbital dominates, where as in the 4th band the S *p* orbital dominates. Thus, the CBM of MoS_2 originates from the repulsion between the d_{z^2} orbital of Mo and the p_x and p_y orbitals of S. The d_{z^2} orbital is pushed up by Δ_2', forming the CBM, and the *p* orbital is pushed down by Δ_2'.

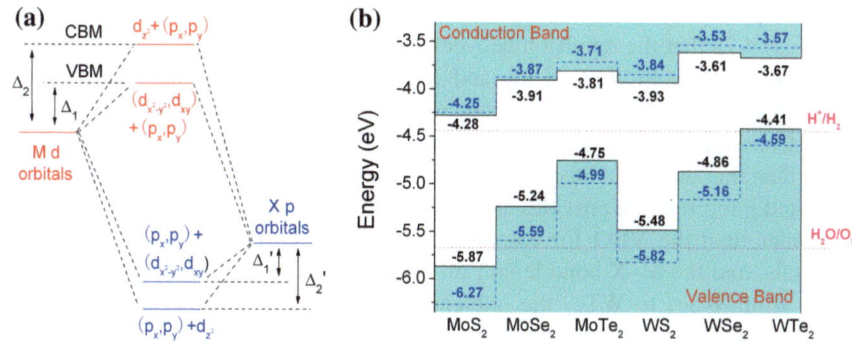

Fig. 4.2 **a** Calculated band alignment for MX_2 monolayers. *Solid lines* are obtained by GGA, and *dashed lines* are obtained by HSE06. The *dotted lines* indicate the water reduction (H^+/H_2) and oxidation (H_2O/O_2) potentials. The vacuum level is taken as zero reference. **b** Schematic of the origin of CBM and VBM in MX_2

4.4 Band Alignments of MX_2 Monolayers

The calculated band alignments for MX_2 monolayers are shown in Fig. 4.2b. In the following, we discuss the results in Fig. 4.2 based on GGA calculations, but similar results can be deduced from the HSE06. Overall, the band offsets show the following trends:

(i) As the atomic number of X increases (from S to Te), the energies of CBM and VBM of MX_2 also increase. Moreover, the conduction band offset (CBO) is smaller than the valence band offset (VBO). More specifically, the VBM of $MoSe_2$ is 0.63 eV higher than that of MoS_2, while its CBM is 0.37 eV higher than that of MoS_2. Here, the only exception is WTe_2 where its CBM is slightly lower (by 0.06 eV) than that of WSe_2.

(ii) For common-X system, the CBM and VBM of WX_2 are higher than those of MoX_2, i.e., MoX_2–WX_2 lateral heterostructures have a type-II band alignment. For example, the VBM of WS_2 is 0.39 eV higher than that of MoS_2, and its CBM is 0.35 eV higher.

The observed band alignment can be understood by the model in Fig. 4.2a. For common-M systems, the VBO and CBO are determined by the repulsion strength Δ_1 and Δ_2 between cation *d* orbitals and anion *p* orbitals. The magnitude of the repulsion solely depends on the overlap integral of *d* and *p* orbitals and their difference in energy. In principle, a larger overlap integral or a smaller energy difference leads to larger Δ value. For increasing X atomic number, its *p* orbitals become shallower. As a result, the anion with shallower *p* orbitals pushes the cation *d* orbitals upwards more than the anion with deeper *p* orbitals. Therefore, Δ_1 and Δ_2 values are larger in systems with larger atomic number of X, and the CBM and VBM would be higher. However, when X goes from S to Te, the M–X bond length also increases. This would lead to a decrease of the overlap integral between the d and p orbitals, and partly counteract the increase of Δ_1 and Δ_2. It can be seen from Fig. 4.2a that the repulsion between M d_{z^2} and X $p_x + p_y$ orbitals is stronger than that between M $d_{x^2-y^2} + d_{xy}$ and X $p_x + p_y$ orbitals, possibly because the former has more σ bonding character, while the latter has more π bonding character. Consequently, the influence of the decrease in overlap integral on Δ_2 is larger than that on Δ_1, and the increase of Δ_2 would be smaller than that of Δ_1. The presented interpretation provides an explanation for the general trend that the CBO is smaller than the VBO for common-cation systems. Particularly, W has high d orbitals, and the M–X bond length of WTe_2 is the largest among all MX_2. As a result, from WSe_2 to WTe_2 the influence of decrease in overlap integral on Δ_2 overrides the influence of increase in the anion p orbital energy, and Δ_2 decreases. So the CBM of WTe_2 is lower than that of WSe_2.

The trends of band offset between common-X systems can be understood by the position of the d orbitals of cation. The energy of the 5d orbital of W is higher than that of the 4d orbital of Mo, therefore, the CBM and VBM of WX_2 are higher than those of MoX_2. Figure 4.2b shows that the band alignment between MoX_2 and WX_2 is type-II. In a type-II heterostructure, free electrons and holes will be spontaneously separated, which is suited for optoelectronics and solar energy conversion. The partial charge densities for the VBM and CBM states of the MoX_2–WX_2 junction are shown in Fig. 4.3. The CBM state, thus the electron, is confined in the MoX_2, while the VBM state, thus the hole, is confined in the WX_2 side. It is therefore expected that spontaneous charge separation occurs when excitons diffuse to the WX_2/MoX_2 junction, a process that is needed for photovoltaics. In addition, superlattices formed by interfacing alternating WX_2 and MoX_2 layers would host mini-bands for quantum devices, and the band offset calculated here would be needed for the design.

It is well established that the bandgap of MX_2 is sensitive to its thickness [7], thus its band offset would also be influenced by the thickness. The VBM and CBM positions of MX_2 for 1, 2, 4 layers and bulk are shown in Fig. 4.4. Generally speaking, as the number of layers is reduced, the VBMs of all MX_2 move downward (lower in energy), whereas the CBMs move upward (higher in energy), both of which can be understood from a simple quantum confinement effect. However, in some cases such as the VBM of WTe_2, this trend is not observed since

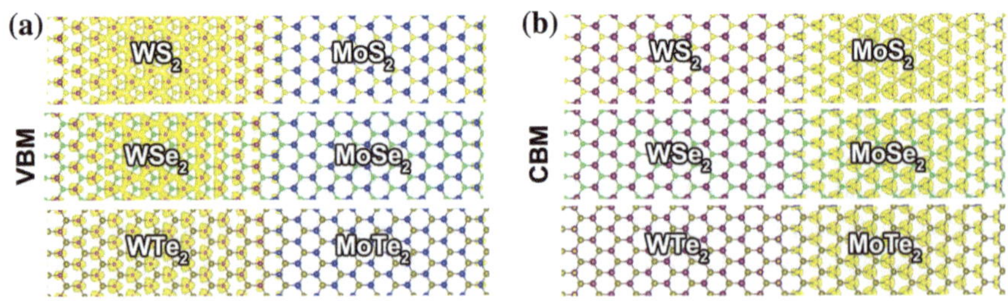

Fig. 4.3 Charge densities of VBM (**a**) and CBM (**b**) states for monolayer WX_2–MoX_2 lateral heterostructures with common-X

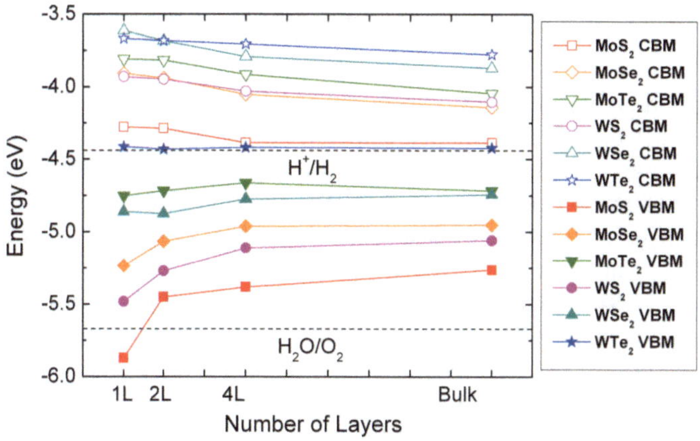

Fig. 4.4 Band edge positions of MX_2 as a function of the number of layers calculated by GGA. The vacuum level is taken as zero reference. The *dashed lines* indicate the water redox potentials

the SOI, which would lift up the VBM, becomes stronger as the thickness decreases. Figure 4.4 clearly shows that the band edge positions of 2D MX_2 are drastically different from the bulk cases, especially for MoS_2 and WS_2. Such differences could lead to possible applications of monolayer MX_2 in fields where their bulk counterparts are not suited. In addition, it is possible to pattern a MX_2 film into structures where different regions have different number of layers, thus achieving functional quantum devices by mere thickness engineering.

Another example is photo-splitting reactions of water. It is well-known that to achieve high efficiency, the photocatalyst should satisfy two main conditions [30]. The first is the presence of an ideal bandgap ($\sim$2.0 eV) that allows absorption of a large portion of the solar spectrum. Among the MX_2 monolayers studied here, MoS_2 and WS_2 meet this condition. The second requirement is that the photocatalyst must have suitable band edges to straddle the redox potentials of water. The standard water redox potentials with respect to the vacuum level are -4.44 eV for the reduction (H^+/H_2), and -5.67 eV for the oxidation (H_2O/O_2) [31]. Bulk MX_2 cannot be applied directly to overall photo-splitting of water because their CBM or VBM positions are not suitable [32]. Our calculations show that indeed the CBMs (VBMs) of bulk MoX_2 and WX_2 are higher than the reduction (oxidation) potential (Fig. 4.4), so they can only be used for water reduction. However, according to Figs. 4.2b, 4.4, monolayer MoS_2 satisfies this band edge requirement. The energy of VBM of MoS_2 monolayer is calculated to be -5.87 eV by GGA and -6.27 eV by HSE06. Since GGA underestimates the bandgap while HSE06 overestimates, it is reasonable to expect that the actual VBM position of MoS_2 monolayer lies between -5.87 and -6.27 eV, which is lower than the water oxidation potential. The CBMs of MoS_2 calculated by GGA and HSE06 are both around -4.3 eV, higher than the water reduction potential. Therefore, MoS_2 monolayer is a good candidate for spontaneous photo-splitting of water.

4.5 Stability and Phase Diagram of MX_2 Alloys

The CE of all six types of $MX_{2(1-x)}X'_{2x}$ alloys (M = Mo, W; X, X′ = S, Se, Te) were constructed, and the formation enthalpies of the alloys were calculated based on the CE. Interestingly, the (S, Se) alloys show a behavior different from the others. The formation enthalpy ΔH of (S, Se) alloys is given in Fig. 4.5. It is seen that for many configurations of these two alloys, ΔH is negative, implying that an ordered alloy can form spontaneously. The solid lines in Fig. 4.5 represent the ground state (GS) configuration. When the calculated ΔH for a particular alloying configuration is on the GS line, that particular configuration is stable even at 0 K. For both $MoS_{2(1-x)}Se_{2x}$ and $WS_{2(1-x)}Se_{2x}$, there are three stable ground states, with concentration x equal to 1/3, 1/2 and 2/3. The structures of the ground states for $MoS_{2(1-x)}Se_{2x}$ are the same as $WS_{2(1-x)}Se_{2x}$, and those of $MoS_{2(1-x)}Se_{2x}$ are shown in Fig. 4.5. In these ground states, clustering of S or Se atoms is not energetically favored, and the S (Se) atoms prefer to occupy the neighbor sites of Se (S) atoms. For instance, in MoSSe, the mirror site of a S (Se) atom with respect to the Mo plane is occupied by a Se (S) atom. The trend of forming dissimilar atom pairs results from the dominance of positive values in the ECIs of short-range pairs. Positive pair interaction leads to repulsion between anions with the same species, thus clustering is prevented.

Among the ground states of $MoS_{2(1-x)}Se_{2x}$, MoSSe has the lowest ΔH, and the stability follows MoSSe > $MoS_{4/3}Se_{2/3}$ > $MoS_{2/3}Se_{4/3}$, whereas in $WS_{2(1-x)}Se_{2x}$ alloys, $WS_{4/3}Se_{2/3}$ is the most stable, and the stability follows $WS_{4/3}Se_{2/3}$ > WSSe > $WS_{2/3}Se_{4/3}$. The magnitude of the ΔH of these ground states is rather small, about −3 meV per anion. In addition, the ΔH for random $MoS_{2(1-x)}Se_{2x}$ and $WS_{2(1-x)}Se_{2x}$ alloys are very close to zero. Therefore, complete miscibility in these alloys can be achieved at very low temperatures.

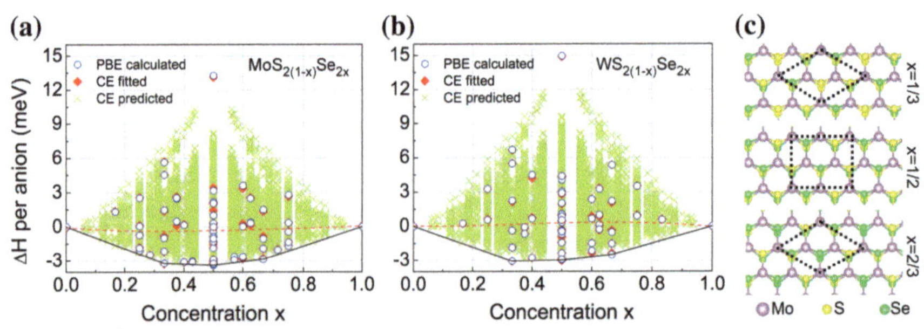

Fig. 4.5 **a**, **b** First-principles calculated formation enthalpies for $MoS_{2(1-x)}Se_{2x}$ and $WS_{2(1-x)}Se_{2x}$, respectively, along with the CE fitted results. The formation enthalpies predicted by CE for all symmetry-inequivalent alloy configurations (~17,000) up to 24 atoms per cell are also given. **c** The structure of ground states of $MoS_{2(1-x)}Se_{2x}$ at 0 K. The *dashed lines* indicate the corresponding supercells

The presence of negative ΔH in (S, Se) alloys is somewhat unexpected, as in most semiconductor alloys ΔH is always positive. Here, we take the MoSSe as an example to make further discussion. The formation of the alloy can be decomposed into three steps [25]: (i) volume deformation (VD): compress or expand the equilibrium lattice constant and the bond length of the constituents to the corresponding concentration-averaged values; (ii) charge exchange (CEX): construct unrelaxed alloy supercell using the deformed cell in (i) and mix the two types of anions into the lattice; and (iii) structural relaxation (SR): fully relax the alloy supercell using quantum mechanical forces. The formation enthalpy ΔH can be then decomposed as: $\Delta H = \Delta H_{VD} + \Delta H_{CEX} + \Delta H_{SR}$. The calculated ΔH_{VD}, ΔH_{CEX} and ΔH_{SR} for MoSSe are 70.0, −13.9, and −59.5 meV, respectively. The result reveals that the residual strain energy in MoSSe is small, as the positive ΔH_{VD} is largely compensated for by ΔH_{SR}, which can be easily understood by the small lattice mismatch (~4 %) between MS_2 and MSe_2. However, $\Delta H_{VD} + \Delta H_{SR}$ is still positive, indicating the SR process alone cannot stabilize the ordered structure. Hence, the negative ΔH_{CEX} is another key role in the formation of ordered alloy. In contrast, many conventional semiconductor alloys have positive ΔH_{CEX}, which leads to destabilization. [25] According to a excess Madelung energy model, the sign of ΔH_{CEX} for a ternary alloy A(B,C) depends on the additional charge exchange ($\Delta Q/\Delta q$) [25], where $\Delta q = q_B(AB) - q_C(AC)$ is the charge difference between B and C in the two constituents, and $\Delta Q = q_B[A(B,C)] - q_C[A(B,C)]$ is the corresponding difference in the alloy. ΔH_{CEX} become negative when $\Delta Q/\Delta q$ is large and exceeds a critical value (typically smaller than 2) [25]. The calculated $\Delta Q/\Delta q$ for MoSSe is 1.92, which is likely to result in the negative ΔH_{CEX}. Overall, the formation of the ordered (S, Se) alloys can be attributed to the small lattice mismatch between the constituents and the large additional charge exchange in the alloys.

The formation enthalpies of (Se, Te) and (S, Te) alloys are shown in Fig. 4.6. The ΔH for all alloy configurations (~17,000) are positive, which differs from the cases of (S, Se) alloy. The positive ΔH can be attributed to the relatively larger lattice mismatch between the two constituent compounds of the alloy. Although the ΔH_{CEX} is also negative for some specific configurations as in the case of (S, Se) alloy, it is overcompensated by the large strain energy. Moreover, the size difference between S and Te is larger than that between Se and Te, thus the ΔH in (S, Te) alloy is larger than that in (Se, Te) alloy. It is also observed that the Mo alloys have lower ΔH than the corresponding W alloys, which is consistent with the smaller in-plane stiffness (thus smaller strain energy) of MoX_2 compared to WX_2.

The positive ΔH indicates that there is no stable configuration for (Se, Te) and (S, Te) alloys at 0 K, and phase separation will occur. However, the solubility in these alloys can be improved by increasing the temperature. In addition, as seen from Fig. 4.6, the magnitude of ΔH for these alloys in random configuration is very small, on the order of 10 meV per anion; hence a complete miscibility is possible at experimentally achievable temperatures. Using the constructed CE, the critical temperature for complete miscibility (T_{CM}) of (Se, Te) and (S, Te) alloys

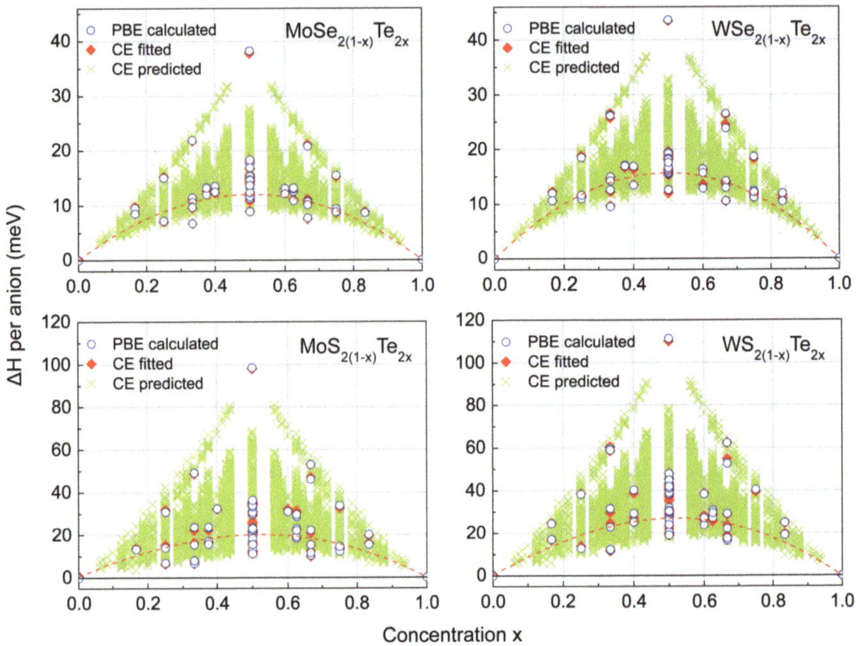

Fig. 4.6 First-principles calculated, CE fitted, and CE predicted formation enthalpies for $MoSe_{2(1-x)}Te_{2x}$, $WSe_{2(1-x)}Te_{2x}$, $MoS_{2(1-x)}Te_{2x}$, and $WS_{2(1-x)}Te_{2x}$

can be estimated analytically. The free energy *F*(*x*, *T*) of a random alloy at temperature *T* with concentration *x* is defined as:

$$F(x, T) = \Delta H(x) - TS(x), \tag{4.5}$$

where Δ*H*(*x*) is the formation enthalpy, which can be calculated by the fitted CE and correlation functions of the random alloy. *S(x)* is the entropy, and can be estimated within a mean-field approximation [33, 34],

$$S(x) = -k_B[x \ \ln x + (1-x)\ln(1-x)] \tag{4.6}$$

with k_B being the Boltzman constant. The binodal line on the phase diagram can then be derived by applying the common tangent construction to the *F*–*x* curves at different temperatures. [35] The calculated phase diagrams are shown in Fig. 4.7. The binodal lines of (Se, Te) alloys are nearly symmetric, which can be attributed to their negligible point and triplet ECIs. However, an asymmetric solubility is predicted for (S, Te) alloys. Therefore, at a given temperature, mixing Te atoms into MS_2 is easier than mixing S atoms into MTe_2, and this asymmetry is enhanced with increasing temperature. This occurs because the point and triplet ECIs cannot be neglected in (S, Te) alloys. The point ECIs in (S, Te) alloys are found to be positive, which make the dissolution of Te easier. The estimated T_{CM} for $MoSe_{2(1-x)}Te_{2x}$, $WSe_{2(1-x)}Te_{2x}$, $MoS_{2(1-x)}Te_{2,x}$ and $WS_{2(1-x)}Te_{2x}$ are about 279,

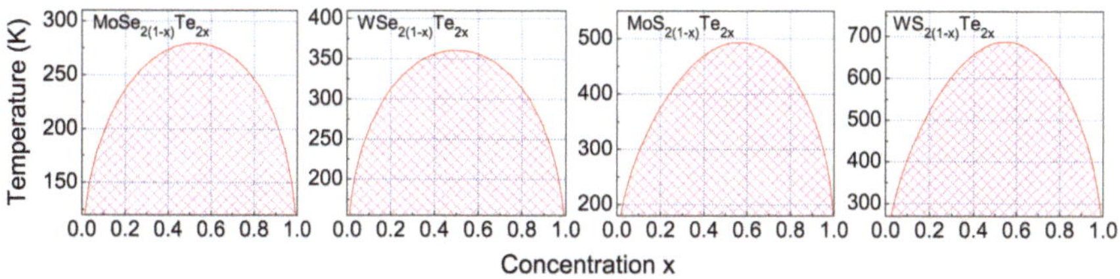

Fig. 4.7 Phase diagrams for $MoSe_{2(1-x)}Te_{2x}$, $WSe_{2(1-x)}Te_{2x}$, $MoS_{2(1-x)}Te_{2x}$ and $WS_{2(1-x)}Te_{2x}$. The areas of phase separation are *shaded*

360, 493, and 687 K, respectively. The values of T_{CM} follow the trend of formation enthalpies: T_{CM} of (S, Te) alloys is higher than that of (Se, Te) alloys, and T_{CM} of W alloys is higher than that of Mo alloys. $WS_{2(1-x)}Te_{2x}$ has the highest T_{CM} of ~687 K, which is experimentally easily accessible. It should also be noted that lattice vibration is neglected in the mean-field approximation for entropy, therefore the solubility is underestimated, and the actual T_{CM} would be even lower than the estimated values. Therefore, a class of complete miscible $MX_{2(1-x)}X'_{2x}$ alloys is experimentally achievable in principle.

4.6 Bandgap Modulation of MX_2

4.6.1 Strain Engineering

To integrate MX_2 for advanced applications, it is desirable to extend and tune their physical properties especially the bandgaps. One of the possible ways is strain engineering. Compared with bulk materials, much larger strain can be applied to low-dimensional materials due to their flexibility. Hence, their properties can be tuned by strain to a large extent. Figure 4.8a shows the response of bandgap of MoS_2 to biaxial tensile strain. While the strain ε reaches 0.10, the bandgap vanishes and the edges of valence and conduction bands overlap the Fermi level. Furthermore, this variant course is accompanied by two characteristic transitions. Upon $\varepsilon = 0.0$, the monolayer MoS_2 shows a behavior of semiconductor with direct gap at the K point, while it transits into indirect semiconductor within region $0.01 \leq \varepsilon \leq 0.09$. In the indirect gap region, the CBM remains at the K point, but the VBM transfers from the K point to the Γ point. The highest valence band (HVB) at the Γ point originates from the Mo d_{z^2} and S p_z orbitals, which differs from the HVB at the K point. Because of the different orbital characters, these two states response differently to strain, leading to the change of VBM position. With increasing strain value, the bandgap decreases, and finally vanishes. The monolayer then behaves as metal in region $\varepsilon \geq 0.1$. Two typical band structures corresponding to (a) $\varepsilon = 0.05$, (b) $\varepsilon = 0.10$ cases are plotted in Fig. 4.8b, and they

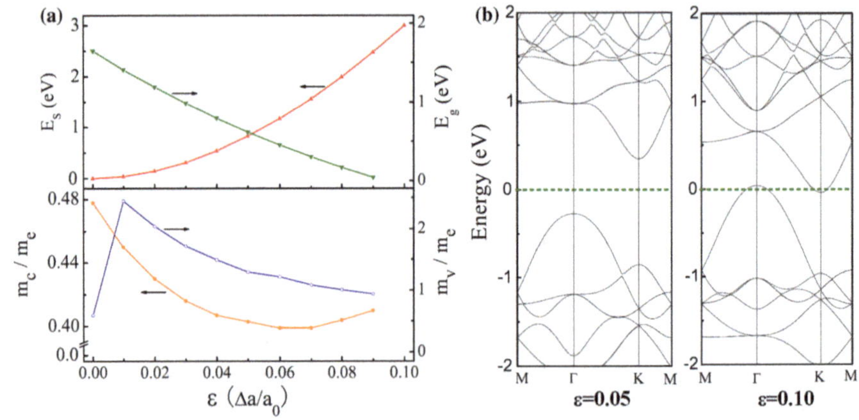

Fig. 4.8 **a** (*Upper*) The variations of strain energy and bandgap of MoS_2 with strain, represented by the *red up-triangles line* and *green down-triangles line*. (*Lower*) The variations of effective mass of carriers in MoS_2 with strain. m_c (m_v) is the effective mass of electrons (*holes*) at the CBM (VBM), denoted by *solid* (*open*) *circles*. m_e is the electron mass. **b** Band structures of MoS_2 for $\varepsilon = 0.05$ and 0.10. The Fermi level is set at 0

clearly illustrates these transitions. In other MX_2 monolayers, similar strain effect is also reported [15]. Therefore, not only can the bandgap of monolayer MoX_2 be tuned by strain, but also the direct-to-indirect and semiconductor-to-metal transitions are controlled.

With the applied strain influencing the band structure of monolayer MoS_2, the effective mass of carriers is also changed. The calculated effective mass of electrons at CBM and holes at VBM in strained monolayer MoS_2 is shown in Fig. 4.8a. It is apparent that the applied strain engenders an asymmetric influence on electrons and holes. That is, the effective mass of holes experiences an oscillation as the strain increases, whereas that of electrons displays a stable trend. The oscillation is caused by the VBM jumping from K to Γ point during strain induced deformation of band structure. Since the effective mass of holes and electrons is determined by the dispersion of energy at the VBM and CBM respectively, the jumping of VBM leads to a dramatic increase of associated effective mass value. Besides, the plots of effective mass of carries also provide meaningful insight into the transport performance of strained monolayer MoS_2 structure. When strain is zero and the bandgap is direct, effective mass of electrons is slightly larger than that of holes. After strain is applied, the bandgap becomes indirect, and the effective mass of electrons becomes smaller. It is well known that smaller effective mass can lead to larger carrier mobility. Therefore, *n*-typed devices based on strained monolayer MoS_2 should be advantageous than *p*-typed devices in view of effective mass.

4.6.2 Electric Field Control

Another method to tune the bandgap of MX_2 is applying external electric field. In this part, we illustrate this by showing the effect of electric field to the band structure of mono- and multilayer armchair MoS_2 nanoribbons ($AMoS_2NR$).

Monolayer $AMoS_2NR$ exhibits semiconducting characteristics with a direct bandgap. The bandgaps of 10-, 11-, and 12-$AMoS_2NRs$ without electric field are 0.74, 0.72, and 0.76 eV. When external transverse electric fields are applied, the gaps of monolayer $AMoS_2NRs$ with different widths vary as a function of field strength, as shown in Fig. 4.9a. The gap decreases almost linearly with increasing electric field. While the electric field is continuously enhanced, the bandgap reduces smoothly and eventually closes when the field strength reaches a critical value. Then the ribbon renders metallic properties. For instance, the semiconductor-to-metal transition in 11-$AMoS_2NR$ occurs when the transverse critical field strength reaches 3 V/nm. Moreover, the critical field strength for gap closure decreases with increasing ribbon width. In the case of 20-$AMoS_2NR$, the bandgap decreases rapidly and closes at 1.5 V/nm, which is half of the critical field strength for 11-$AMoS_2NR$. Thus, one can expect that the bandgap of a wider ribbon is more sensitive to the electric field, and the semiconductor-metal transition in a sufficient wide ribbon can be observed at experimentally achievable field strength.

The decreasing bandgap with increasing field strength originates from the redistribution of the wavefunction of the nanoribbon under an external electric field, namely the so-called giant Stark effect (GSE) [36, 37]. Figure 4.10 illustrates the band structures and charge densities of VBM and CBM when the field strengths are 0, 1, and 2 V/nm, respectively. All the bandgaps are direct at the Γ point. In the 0 V/nm case, the CBM and VBM states distributes symmetrically on both sides of the ribbon (Fig. 4.10d). Upon application of external electric field, the electrostatic potential symmetry over the ribbon is broken and the charges are redistributed. As a result, the charge density of VBM is localized at the right side

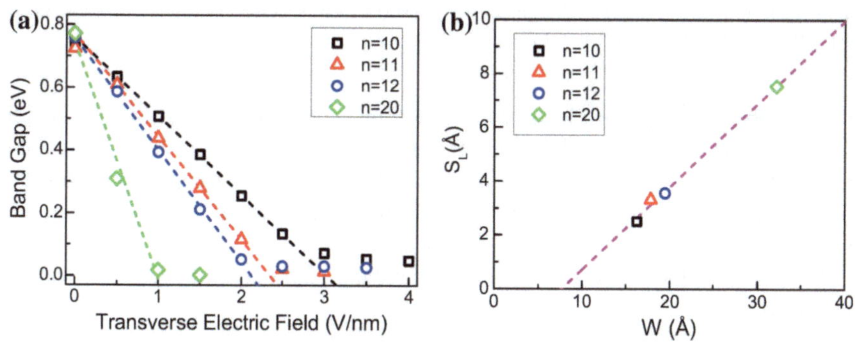

Fig. 4.9 **a** Variation of bandgap of monolayer $AMoS_2NR$ versus transverse electric field, represented by open symbols. The *dashed lines* are fitting results. **b** The linear coefficient of the giant Stark effect S_L versus the ribbon width W. The *fitted line* (*dashed*) is also given

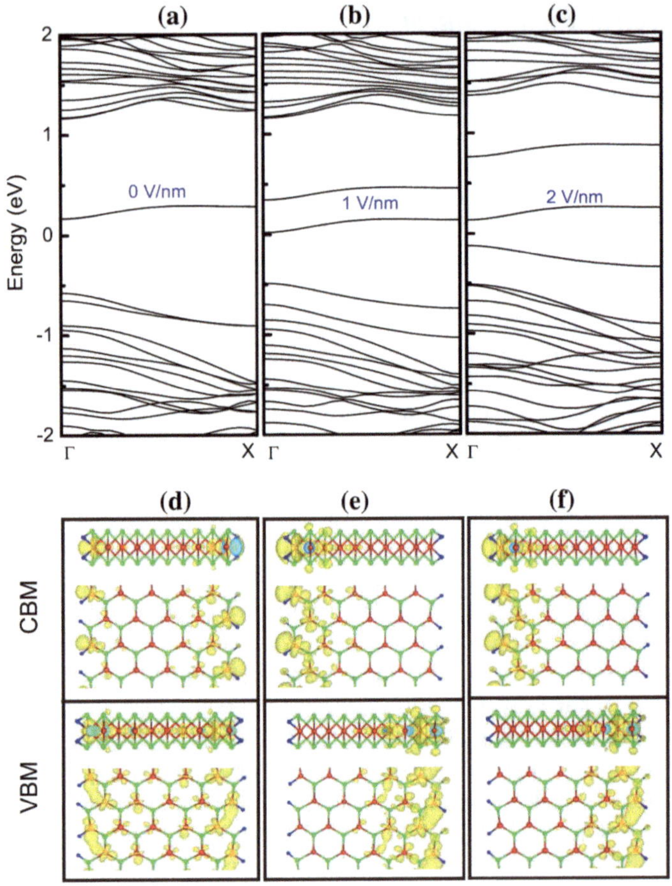

Fig. 4.10 Band structures and corresponding charge densities for CBM and VBM of the monolayer 10-AMoS$_2$NR under varied transverse electric fields. **a**, **d** Correspond to 0 V/nm, **b**, **e** correspond to 1 V/nm, while **c**, **f** correspond to 2 V/nm. Each rectangle contains the side view (*upper*) and top view (*lower*) of charge density iso surfaces of specified states at CBM and VBM, respectively. The Fermi level is set at zero

of the ribbon, where the electrostatic potential is raised. In contrast, the charge density of CBM is localized at the left side of the ribbon, where the electrostatic potential is lowered, as shown in Fig. 4.10e, f. As the field strength increases, the magnitude of the corresponding potential increase or decrease becomes larger, and the bandgap decreases and finally vanishes.

The slope of the linear portion of the bandgap (E_g) curve in Fig. 4.10a can be described as:

$$\frac{\mathrm{d}E_g}{\mathrm{d}E} = -eS_L, \tag{4.7}$$

where E is the external electric field, e is the electron charge, and S_L is the linear GSE coefficient which indicates the ability of electric field to tune the bandgap. Since the potential of the transverse field is eE, the bandgap change $\Delta E_g = eEr_c - eEr_v$, where r_c and r_v are the centers of the CBM and VBM along the direction of the electric field. The distance between r_c and r_v would be proportional to the ribbon width W, then $\Delta E_g \approx eE(\alpha W + C)$. Hence, the GSE coefficient S_L would linearly depend on the ribbon width W as

$$S_L = \alpha W + C_1 \tag{4.8}$$

The calculated S_L for 10, 11, 12, and 20-$AMoS_2NRs$ are 2.51, 3.33, 3.57, and 7.53 Å, respectively. These values are well fitted to the formula, $S_L = 0.31\ W - 2.38$, as plotted in Fig. 4.9b. The fitted line with a slope of 0.31 proves S_L obeys a linear scaling law on W. Moreover, it indicates that the transverse electric field exhibits a stronger ability to modify the bandgap in a wider monolayer $AMoS_2NR$.

The influence of electric field on $AMoS_2NR$ also depends on the thickness of the ribbon. Here, we take 10-$AMoS_2NR$ as an example. Similar to the monolayer ribbon, the ground state of multilayer ribbon remains semiconducting with a direct bandgap at the Γ point. Nevertheless, the multilayer ribbon has a smaller bandgap (0.55, 0.43, and 0.38 eV for bi/tri/tetralayer ribbons, respectively) than monolayer ribbon (0.74 eV), and the bandgap decreases while the number of layers increases. When transverse electric field is applied, the bandgap of multilayer $AMoS_2NR$ is reduced and eventually driven to zero, as shown in Fig. 4.11a. However, the field-induced modulation effect on multilayer ribbon is weaker than the monolayer case in terms of the descending rate of the gap curve. Moreover, when the electric field strength becomes larger, the gap curves of multilayer ribbons gradually approach each other. Besides, the influence of perpendicular electric field on the gap in multilayer 10-$AMoS_2NR$ is also examined, as shown in Fig. 4.11b. The gap decreases as the perpendicular electric field increases. In addition, the critical field strength for gap closure decreases as the number of layers increases. For example, the critical strength in the tetralayer case reduces to 1.5 V/nm, whereas it is about 6 V/nm in bilayer case. Comparing with the results presented in Fig. 4.11a, it is

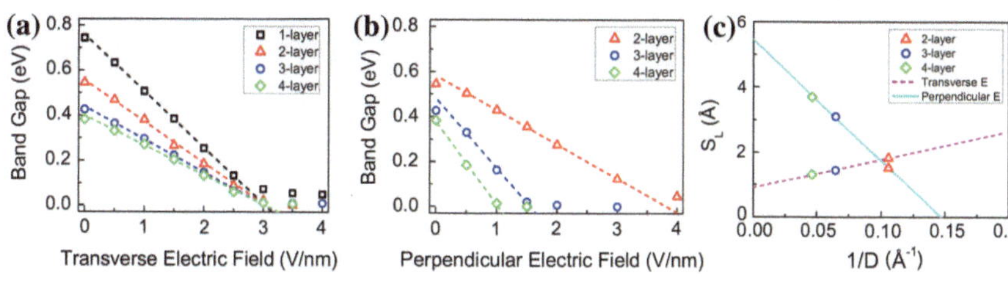

Fig. 4.11 Variation of the bandgap of multilayer $AMoS_2NR$ versus **a** transverse electric field, **b** perpendicular electric field, represented by open symbols. *Dashed lines* are linear fitting. **c** The linear coefficient of the giant Stark effect S_L as a function of the reciprocal of ribbon thickness 1/D. The *dashed* and *solid lines* correspond to transverse and perpendicular field cases, respectively

clear that the two kinds of electric fields present different modulation effects on the gaps of multilayer ribbons. For ribbons with the same width, the critical strength of perpendicular field for gap closure decreases rapidly with increasing ribbon thickness, whereas the critical strength of transverse field is almost independent of it.

Figure 4.12 displays the band structures and charge distributions for the CBM and VBM states of bilayer 10-AMoS_2NR with and without external electric field. When transverse electric field is applied, charge density redistributes, as shown in Fig. 4.12a. The density of state at the VBM (CBM) localizes on the right (left) edge of the bilayer ribbon, owing to the raised (lowered) electrostatic potential at the right (left) edge. Figure 4.12f shows the charge density distribution with perpendicular field applied. It is found that the density of state at the VBM (CBM) localizes on the lower (upper) layer of the bilayer ribbon, where the electrostatic potential rises (lowers). The two electric fields induce different density

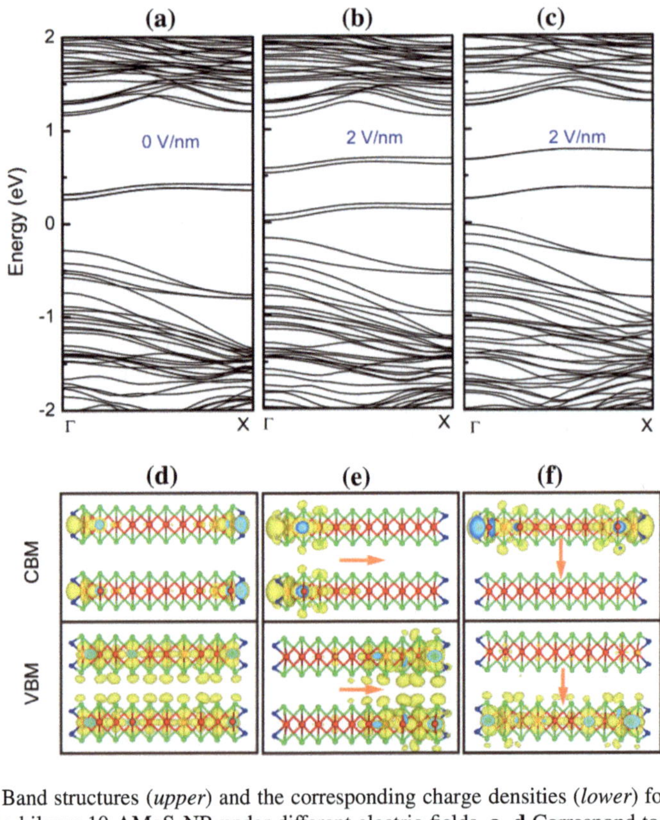

Fig. 4.12 Band structures (*upper*) and the corresponding charge densities (*lower*) for CBM and VBM of the bilayer 10-AMoS_2NR under different electric fields. **a**, **d** Correspond to 0 V/nm. **b**, **e** corresponds to the transverse electric field (2 V/nm), while **c**, **f** correspond to the perpendicular electric field (2 V/nm). *Arrows* in (**e**) and (**f**) indicate the directions of applied electric fields. The Fermi level is set at zero

distributions at the VBM and CBM in multilayer structure. Nevertheless, the modulation mechanisms behind these results are the same, namely, breaking the electrostatic potential symmetry of the structure which leads to the GSE. Still, we can calculate the linear GSE coefficient S_L for multilayer 10-AMoS$_2$NR. Here, the S_L obeys a linear scaling law on the reciprocal of ribbon thickness (D):

$$S_L = \beta \frac{1}{D} + C_2. \tag{4.9}$$

Figure 4.11c plots S_L versus $1/D$ together with their fits corresponding to both the transverse and perpendicular electric fields. For the transverse field case, S_L decreases with D increasing, and the fitted line is $S_L = 8.69(1/D) + 0.91$. The result indicates that the gap modulation capacity of transverse electric field gets weaker in multilayer AMoS$_2$NR when its thickness increases. As thicker multilayer ribbons have smaller bandgap, the gap curves would approach each other within a larger electric field region (Fig. 4.11a). In contrast, for the perpendicular field case, S_L presents an inverse variation versus D, namely, the gap modulation capacity of the field gets stronger with increasing ribbon thickness. Similarly, considering the linear variation region, a fitted $S_L = -37.12(1/D) + 5.46$ is obtained, which has a larger slope than the former. More interestingly, there exists an intersection point between the two fits, indicating a competition between the two electric fields. While the transverse field demonstrates slightly better capacity to modulate the gap of bilayer ribbon, the perpendicular field exhibits a stronger ability to tune the gap of other multilayer ribbons.

4.6.3 Alloying

Historically, semiconductor science and technology have been greatly advanced by alloying different semiconductors to achieve continuously tunable properties. In most cases, a quadratic rule can be used to describe the relationship between the physical property P and concentration x of the alloy $A_{(1-x)}B_x$, namely,

$$P(x) = (1-x)E_g(A) + xE_g(B) - bx(1-x) \tag{4.10}$$

where b is the bowing parameter. Figure 4.13a shows the lattice constants of $MX_{2(1-x)}X'_{2x}$ random alloys. They vary almost linearly with the concentration x. Similarly, the bandgap of MX_2 can also be continuously tuned by alloying. Figure 4.13b shows the bandgaps of the alloys. In pure MX_2 monolayer, the bandgap is direct, with the CBM and VBM located at the K point. The calculated bandgaps for $MX_{2(1-x)}X'_{2x}$ alloys are also direct. By projecting the band edge states into the atomic orbitals, it is found that the CBM and VBM of the alloys have the same orbital characters as those of pure MX_2, i.e., the CBM state consists mainly of the d_{z^2} orbital of cations and the p_x and p_y orbitals of anions, while the VBM state consists of the $d_{x^2-y^2}$ and d_{xy} orbitals of cations and the p_x and p_y orbitals of anions.

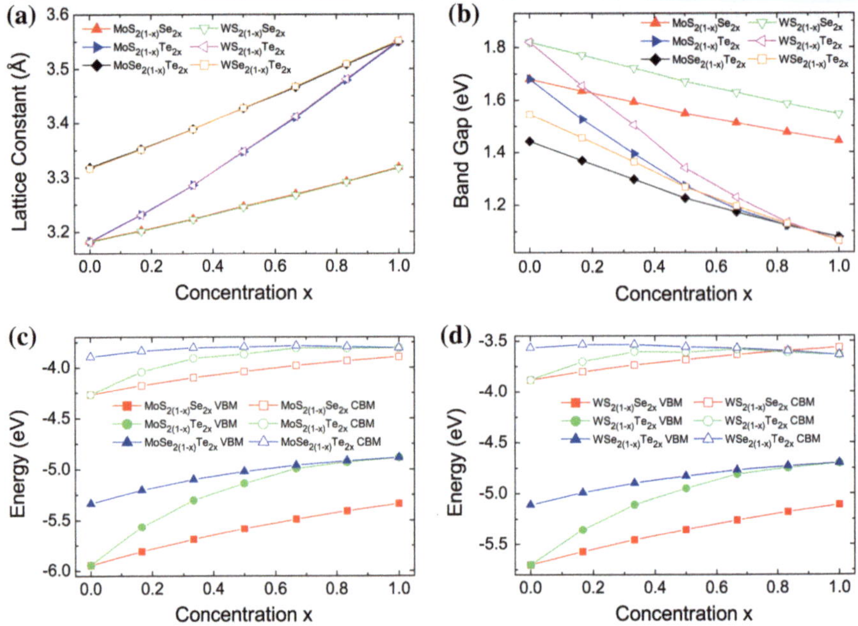

Fig. 4.13 **a** Lattice constant of $MX_{2(1-x)}X'_{2x}$ alloys as a function of concentration x. **b** The bandgap of $MX_{2(1-x)}X'_{2x}$ alloys as a function of concentration. **c–d** The VBM and CBM positions of Mo and W alloys as a function of concentration. The vacuum level is taken as zero energy reference

Therefore, the bandgap value of MX_2 monolayers can be tuned over a wide range by alloying, whereas the direct gap character is maintained, which is favorable for optoelectronic applications. In contrast, although the bandgap of MX_2 can be also tuned by strain, the range of strain that retains the direct gap is quite limited as shown above. The bowing parameters of bandgaps (b_{Gap}) for the alloys are listed in Table 4.2. All b_{Gap} are positive, and their magnitude is typically less than 0.5 eV. As the size and chemical differences between the two constituent compounds increase, the gap bowing effect in the alloy becomes more significant. For example, $MoS_{2(1-x)}Se_{2x}$ has a quite small b_{Gap} of 0.05 eV, whereas b_{Gap} of $MoS_{2(1-x)}Te_{2x}$ increases to 0.41 eV.

The CBM and VBM positions of the alloys are also shown in Fig. 4.13, and the bowing parameters are listed in Table 4.2. The trends of b_{VB} and b_{CB} are similar to b_{Gap}, i.e., larger size and chemical differences lead to larger bowing. Both b_{VB} and b_{CB} are negative, namely, VBM and CBM show upward bowing, and the bowing in the band edge positions is much more significant than in the bandgap. The weak bowing in bandgap is a consequence of partial mutual cancelation of the strong CBM and VBM bowing. In contrast, in most traditional semiconductor alloys such as ZnSTe [38], GaNAs [39], and SnGe [40], VBM shows upward bowing while CBM shows downward bowing. This stark difference possibly originates from the

Table 4.2 Bowing parameters (in eV) for the bandgap (b_{Gap}), VBM position (b_{VB}), and CBM position (b_{VB}) of $MX_{2(1-x)}X'_{2x}$ random alloys, as well as the decomposed bowing parameters in the VD, CEX, and SR processes

	Total			VD			CEX			SR		
	b_{Gap}	b_{VB}	b_{CB}	b_{Gap}	b_{VB}	b_{CB}	b_{Gap}	b_{VB}	b_{CB}	b_{Gap}	b_{VB}	b_{CB}
$MoS_{2(1-x)}Se_{2x}$	0.05	−0.22	−0.18	0.02	0.08	0.10	0.05	−0.10	−0.05	−0.03	−0.20	−0.23
$MoSe_{2(1-x)}Te_{2x}$	0.12	−0.36	−0.24	0.07	0.20	0.27	0.12	−0.20	−0.08	−0.07	−0.36	−0.43
$MoS_{2(1-x)}Te_{2x}$	0.41	−1.17	−0.76	0.19	0.52	0.70	0.33	−0.61	−0.28	−0.11	−1.08	−1.19
$WS_{2(1-x)}Se_{2x}$	0.04	−0.21	−0.16	0.02	0.09	0.11	0.04	−0.10	−0.06	−0.02	−0.20	−0.22
$WSe_{2(1-x)}Te_{2x}$	0.12	−0.32	−0.20	0.10	0.23	0.33	0.09	−0.20	−0.11	−0.06	−0.35	−0.41
$WS_{2(1-x)}Te_{2x}$	0.36	−1.06	−0.70	0.24	0.55	0.79	0.24	−0.56	−0.32	−0.12	−1.05	−1.17

unique character of band edge states in $MX_{2(1-x)}X'_{2x}$ distinctly different from traditional semiconductors. The band bowing in semiconductor alloys is typically caused by different deformation potentials of the constituents, as well as coupling of states through the potential difference ΔV between the alloy potential and the average potential of the constituents. [29, 41] ΔV induces intraband coupling within the conduction band (CB) and valence band (VB), and interband coupling between them. The interband coupling raises the CBM and lowers the VBM, where as the intraband coupling lowers the CBM and raises the VBM. [29] In traditional semiconductors, VBM is mostly p-like and CBM is mostly s-like. Because of the very different orbital characters between VBM and CBM, the intraband coupling is much stronger than the interband coupling, leading to the upward (downward) bowing of VBM (CBM). However, in $MX_{2(1-x)}X'_{2x}$ alloys both VBM and CBM are d-like, and the interband coupling is much enhanced, leading to the upward bowing of CBM.

Similar to other semiconductor alloys, the bowing of bandgap and band edge position in $MX_{2(1-x)}X'_{2x}$ can also be decomposed to the above-mentioned three parts VD, CEX, and SR [42]. The bowing parameter for each process is calculated, and listed in Table 4.2, and the total bowing parameter is given by $b = b_{VD} + b_{CEX} + b_{SR}$. The following results are observed:

(i) The sign of the VBM bowing parameters is the same as the CBM bowing in all three processes. The VD-induced VBM and CBM bowing is positive; the positive b_{VD} is partially compensated by the CEX and SR processes, where b_{VB} and b_{CB} are negative, and the total b_{VB} and b_{CB} become negative. The bandgap bowing in each process is much smaller, resulting from the partial cancelation of VBM and CBM bowing.
(ii) In the VD and SR processes, the CBM bowing is more obvious than the VBM bowing. These two processes displace relative positions of anions and cations, and the CBM state is more sensitive to such displacements, so its bowing is more distinct. This contributes to positive and negative gap bowing in VD and SR, respectively. In the CEX process, the VBM bowing is more significant, and this contributes to a positive bandgap bowing. This occurs because: first of all, the VBM state of the MX_2 has more anion p character than the CBM state, resulting in larger influence on VBM than on CBM when different anions are mixed. Second, the original valence band offset between different MX_2 is larger than the conduction band offset; therefore, the chemical difference between VBM is larger than between CBM.
(iii) The values of the bowing parameters for the VD, CEX, and SR processes are on the same order, indicating that both lattice mismatch and chemical potential difference are important factors responsible for the bandgap and band edge position bowing in $MX_{2(1-x)}X'_{2x}$ alloys. The contribution of SR to the VBM and CBM bowing is the largest, indicating a unusually large structure relaxation in $MX_{2(1-x)}X'_{2x}$ alloys, which is enhanced by their 2D nature as discussed above.

(iv) All bowing parameters follow the trend $|b(\text{S, Te})| > |b(\text{Se, Te})| > |b(\text{S, Se})|$. Such an order is not surprising given that the size mismatch and chemical potential difference increase from S–Se, Se–Te, to S–Te.

4.7 Conclusions

In summary, the elastic properties, band structure, band alignment, alloying, and gap modulation of low-dimensional MX_2 were investigated from first principles calculations. The in-plane stiffness of the single-layer MX_2 decreases as the atomic number of X increases and M decreases as a result of changes in the bonding strength. Calculated band offsets between MX_2 reveal that as the atomic number of M or X increases, the CBM and VBM energy also increases, and the VBO is larger than the CBO for common-cation pairs. A simple model is proposed to interpret the chemical trends of the band offsets. Moreover, the suitable band edge position of MoS_2 monolayer makes it a good candidate for the photo-splitting of water. In alloying calculations, we show that there exist stable ordered (S, Se) alloy structures even at 0 K, whereas in (Se, Te) and (S, Te) alloys, phase separation into the two constituents will occur at 0 K. Nevertheless, a complete miscibility in these alloys can be achieved by increasing temperature. Finally, we show that the bandgap of MX_2 nanostructures can be efficiently modulated by strain, electronic field and alloying. The bandgap decreases with increasing strain or electric field strength, and gap closure can be achieved when the strain/field strength is sufficient large. In $MX_{2(1-x)}X'_{2x}$ alloys, The bandgap and band edge position varies as a function of the concentration x, and exhibits bowing effect, which is a joint effect of volume deformation, chemical difference, and structure relaxation. More importantly, the direct gap character of MX_2 monolayer is retained in the alloys. Therefore, the $MX_{2(1-x)}X'_{2x}$ alloys are good candidates for 2D optoelectronics.

Acknowledgments J. Li gratefully acknowledges financial support from the National Science Fund for Distinguished Young Scholar (Grant No. 60925016). This work is partially supported by the National Basic Research Program of China (Grant No. 2011CB921901).

References

1. Novoselov, K.S., Jiang, D., Schedin, F., Booth, T.J., Khotkevich, V.V., Morozov, S.V., Geim, A.K.: Two-dimensional atomic crystals. Proc. Natl. Acad. Sci. USA. **102**, 10451 (2005)
2. Coleman, J.N., Lotya, M., O'Neill, A., Bergin, S.D., King, P.J., Khan, U., Young, K., Gaucher, A., De, S., Smith, R.J., Shvets, I.V., Arora, S.K., Stanton, G., Kim, H.-Y., Lee, K., Kim, G.T., Duesberg, G.S., Hallam, T., Boland, J.J., Wang, J.J., Donegan, J.F., Grunlan, J.C., Moriarty, G., Shmeliov, A., Nicholls, R.J., Perkins, J.M., Grieveson, E.M., Theuwissen, K., McComb, D.W., Nellist, P.D., Nicolosi, V.: Two-dimensional nanosheets produced by liquid exfoliation of layered materials. Science **331**, 568 (2011)

3. Tongay, S., Zhou, J., Ataca, C., Lo, K., Matthews, T.S., Li, J., Grossman, J.C., Wu, J.: Thermally driven crossover from indirect toward direct bandgap in 2D semiconductors: $MoSe_2$ versus MoS_2. Nano Lett. **12**, 5576 (2012)
4. Zeng, Z., Yin, Z., Huang, X., Li, H., He, Q., Lu, G., Boey, F., Zhang, H.: Single-layer semiconducting nanosheets: high-yield preparation and device fabrication. Angew. Chem. Int. Ed. **50**, 11093 (2011)
5. Radisavljevic, B., Radenovic, A., Brivio, J., Giacometti, V., Kis, A.: Single-layer MoS_2 transistors. Nature Nanotech. **6**, 147 (2012)
6. Radisavljevic, B., Whitwicj, M.B., Kis, A.: Integrated circuits and logic operations based on single-layer MoS_2. ACS. Nano. **5**, 9934 (2011)
7. Mak, K.F., Lee, C., Hone, J., Shan, J., Heinz, T.F.: Atomically thin MoS_2: a new direct-gap semiconductor. Phys. Rev. Lett. **105**, 136805 (2010)
8. Splendiani, A., Sun, L., Zhang, Y., Li, T., Kim, J., Chim, C.-Y., Galli, G., Wang, F.: Emerging photoluminescence in monolayer MoS_2. Nano Lett. **10**, 1271 (2010)
9. Cao, T., Wang, G., Han, W., Ye, H., Zhu, C., Shi, J., Niu, Q., Tan, P., Wang, E., Liu, B., Feng, J.: Valley-selective circular dichroism of monolayer molybdenum disulphide. Nat. Commun. **3**, 887 (2012)
10. Zeng, H., Dai, J., Yao, W., Xiao, D., Cui, X.: Valley polarization in MoS_2 monolayers by optical pumping. Nat. Nanotechnol. **7**, 490 (2012)
11. Li, J., Wang, L.-W.: First principle study of core/shell structure quantum dots. Appl. Phys. Lett. **85**, 2929 (2004)
12. Zhang, S.B., Wei, S.H., Zunger, A.: Microscopic origin of the phenomenological equilibrium doping limit rule in *n*-type III-V semiconductors. Phys. Rev. Lett. **84**, 1232 (2000)
13. Kavan, L., Grätzel, M., Gilbert, S.E., Klemenz, C., Scheel, H.J.: Electrochemical and photoelectrochemical investigation of single-crystal anatase. J. Am. Chem. Soc. **118**, 6716 (1996)
14. Kang, J., Tongay, S., Zhou, J., Li, J., Wu, J.: Band offsets and heterostructures of two-dimensional semiconductors. Appl. Phys. Lett. **102**, 012111 (2013)
15. Johari, P., Shenoy, V.B.: Tuning the electronic properties of semiconducting transition metal dichalcogenides by applying mechanical strains. ACS. Nano. **6**, 5449 (2012)
16. Yue, Q., Kang, J., Shao, Z., Zhang, X., Chang, S., Wang, G., Qin, S., Li, J.: Mechanical and electronic properties of monolayer MoS_2 under elastic strain. Phys. Lett. A **376**, 1166 (2012)
17. Ramasubramaniam, A., Naveh, D., Towe, E.: Tunable band gaps in bilayer transition-metal dichalcogenides. Phys. Rev. B **84**, 205325 (2011)
18. Yue, Q., Chang, S., Kang, J., Zhang, X., Shao, Z., Qin, S., Li, J.: Bandgap tuning in armchair MoS_2 nanoribbon. J. Phys. Condens. Matter **24**, 335501 (2012)
19. Komsa, H.-P., Krasheninnikov, A.V.: Two-dimensional transition metal dichalcogenide alloys: stability and electronic properties. J. Phys. Chem. Lett. **3**, 3652 (2012)
20. Kang, J., Tongay, J., Li, J., Wu, J.: Monolayer semiconducting transition metal dichalcogenide alloys: stability and band bowing. J. Appl. Phys. **113**, 143703 (2013)
21. Kresse, G., Furthmuller, J.: Efficient iterative schemes for ab initio total-energy calculations using a plane-wave basis set. Phys. Rev. B **54**, 11169 (1996)
22. Soler, J.M., Artacho, E., Gale, J.D., García, A., Junquera, J., Ordejón, P., Sánchez-Portal, D.: The SIESTA method for ab initio order-*N* materials simulation. J. Phys. Condens. Matter **14**, 2745 (2002)
23. Heyd, J., Scuseria, G.E., Ernzerhof, M.: Hybrid functionals based on a screened Coulomb potential. J. Chem. Phys. **118**, 8207 (2003)
24. Sanchez, J., Ducastelle, F., Gratias, D.: Generalized cluster description of multicomponent systems. Phys. A **128**, 334 (1984)
25. Ferreira, L.G., Wei, S.-H., Zunger, A.: First-principles calculation of alloy phase diagrams: The renormalized-interaction approach. Phys. Rev. B **40**, 3197 (1989)
26. Zunger, A., Wei, S.-H., Ferreira, L.G., Bernard, J.E.: Special quasirandom structures. Phys. Rev. Lett. **65**, 353 (1990)

27. Van de Walle, A., Asta, M., Ceder, G.: The alloy theoretic automated toolkit: a user guide. Calphad **26**, 539 (2002)
28. Castellanos-Gomez, A., Poot, M., Steele, G.A., Van der Zant, H.S.J., Agraït, N., Rubio-Bollinger, G.: Elastic properties of freely suspended MoS_2 nanosheets. Adv. Mater. **24**, 772 (2012)
29. Wei, S.-H., Zunger, A.: Band offsets and optical bowings of chalcopyrites and Zn-based II-VI alloys. J. Appl. Phys. **78**, 3846 (1995)
30. Gai, Y., Li, J., Li, S–.S., Xia, J.-B., Wei, S.-H.: Design of narrow-gap TiO_2: a passivated codoping approach for enhanced photoelectrochemical activity. Phys. Rev. Lett. **102**, 036402 (2009)
31. Chakrapani, V., Angus, J.C., Anderson, A.B., Wolter, S.D., Stoner, B.R., Sumanasekera, G.U.: Charge transfer equilibria between diamond and an aqueous oxygen electrochemical redox couple. Science **318**, 1424 (2007)
32. Jiang, H.: Electronic band structures of molybdenum and tungsten dichalcogenides by the GW approach. J. Phys. Chem. C **116**, 7664 (2012)
33. Lambrecht, W.R.L., Segall, B.: Electronic structure and equilibrium properties of $Ga_xAl_{1-x}N$ alloys. Phys. Rev. B **47**, 9289 (1993)
34. Alling, B., Ruban, A.V., Karimi, A., Peil, O.E., Simak, S.I., Hultman, L., Abrikosov, I.A.: Mixing and decomposition thermodynamics of c-$Ti_{1-x}Al_xN$ from first-principles calculations. Phys. Rev. B **75**, 045123 (2007)
35. Teles, L.K., Furthmüller, J., Scolfaro, L.M.R., Leite, J.R., Bechstedt, F.: First-principles calculations of the thermodynamic and structural properties of strained InxGa1-xN and $Al_xGa_{1-x}N$ alloys. Phys. Rev. B **62**, 2475 (2000)
36. Zhang, Z., Guo, W.: Energy-gap modulation of BN ribbons by transverse electric fields: first-principles calculations. Phys. Rev. B **77**, 075403 (2008)
37. Zheng, F., Liu, Z., Wu, J., Duan, W., Gu, B.-L.: Scaling law of the giant Stark effect in boron nitride nanoribbons and nanotubes. Phys. Rev. B **78**, 085423 (2008)
38. Wu, J., Walukiewicz, W., Yu, K.M., Ager, J.W., Haller, E.E., Miotkowski, I., Ramdas, A.K., Su, C.-H., Sou, I.K., Perera, R.C.C., Denlinger, J.D.: Origin of the large band-gap bowing in highly mismatched semiconductor alloys. Phys. Rev. B **67**, 035207 (2003)
39. Wu, J., Walukiewicz, W., Yu, K.M., Denlinger, J.D., Shan, W., Ager, J.W., Kimura, A., Tang, H.F., Kuech, T.F.: Valence band hybridization in N-rich $GaN_{1-x}As_x$ alloys. Phys. Rev. B **70**, 115214 (2004)
40. Yin, W.-J., Gong, X.-G., Wei, S.-H.: Origin of the unusually large band-gap bowing and the breakdown of the band-edge distribution rule in the Sn_xGe_{1-x} alloys. Phys. Rev. B **78**, 161203 (2008)
41. Wei, S.-H., Zunger, A.: Disorder effects on the density of states of the II-VI semiconductor alloys $Hg_{0.5}Cd_{0.5}Te$, $Cd_{0.5}Zn_{0.5}Te$, and $Hg_{0.5}Zn_{0.5}Te$. Phys. Rev. B **43**, 1662 (1991)
42. Bernard, J.E., Zunger, A.: Optical bowing in zinc chalcogenide semiconductor alloys. Phys. Rev. B **34**, 5992 (1986)

Chapter 5
MoS_2: A First-Principles Perspective

Yingchun Cheng and Udo Schwingenschlögl

Abstract The structural, electronic, and vibrational properties of MoS_2 together with other semiconducting transition metal dichalcogenides (TMDCs) are investigated based on first principles calculations. Due to its layered structure, single-layer MoS_2 can be fabricated by the mechanical exfoliation method. The band structure of MoS_2 shows an indirect-to-direct semiconductor transition from bulk to single layer because of a lack of interlayer interaction. Giant spin splitting at the K point of the Brillouin zone is predicted for single-layer MoS_2 and other TMDCs, due to the intrinsic strong spin–orbit coupling and the absence of inversion symmetry. Moreover, enhancement of the Rashba splitting is predicted for polar single-layer TMDCs. Two-dimensional dilute magnetic semiconductors are proposed for substitution of Mo by other transition metal atoms, such as Mn, Fe, and Co. Experimentally observed anomalous vibrational properties can be attributed to reduction of the interlayer interaction and strengthening of the intralayer interaction from bulk to single layer. It is demonstrated that strain plays an important role for the electronic and vibrational properties of single-layer MoS_2. The electronic states of one-dimensional structures (nanoribbon, nanotube, etc.) are sensitive to the edge structure, charity, and strain. Under sulfur-rich conditions, zero-dimensional MoS_2 shows a Mo-edge triangular structure with sulfur saturation.

5.1 Introduction

Simulations can be seen as a bridge between experiment and theory. In particular, simulations based on first principles methods play an important role, because they not only help to explain experimental results but also can predict properties of new materials or new structures with high accuracy. Transition metal dichalcogenides

Y. Cheng · U. Schwingenschlögl (✉)
Physical Sciences and Engineering Division, KAUST, Thuwal 23955-6900, Saudi Arabia
e-mail: udo.schwingenschlogl@kaust.edu.sa

Z. M. Wang (ed.), *MoS_2*, Lecture Notes in Nanoscale Science and Technology 21,
DOI: 10.1007/978-3-319-02850-7_5,

(TMDCs) are a class of layered materials with strong in-plane and weak out-of-plane bonding, enabling exfoliation into two-dimensional layers (thickness of a single unit cell). Some TMDCs exhibit superconductivity [1], such as NbS_2 and $NbSe_2$, some show a charge density wave, such as $TiSe_2$ [2], and some are semiconducting with typical bandgaps between 1 and 2 eV [3, 4], such as MoS_2, $MoSe_2$, WS_2, and WSe_2. In addition, the bandgaps of semiconducting TMDCs can change from indirect to direct in single layers [5, 6], allowing applications such as transistors [7], phototransistors [8, 9], and valleytronic devices [10]. One member of the TMDCs, MoS_2 is a traditional industrial catalysis. Recently, single-layer MoS_2 has been fabricated and a direct bandgap of 1.9 eV has been observed. It has been demonstrated that the mobility at room temperature is at least 2000 $cm^2V^{-1}s^{-1}$, the current on/off ratios is about 1×10^8, and the standby power dissipation is ultralow in field effect transistors based on single-layer MoS_2 [7]. Due to the intrinsic strong spin–orbit coupling and the absence of inversion symmetry, a giant spin splitting at the K point of the Brillouin zone is predicted for single-layer MoS_2 and other TMDCs, which is very promising for the field of valleytronics [10–13]. In this chapter, we first focus on the electronic and vibrational properties of single-layer MoS_2 together with other semiconducting TMDCs, using first principles calculations. One-dimensional (such as nanoribbons and nanotubes) and zero-dimensional structures based on MoS_2 are also investigated.

5.2 Theoretical Background

Solids consist of atomic nuclei and electrons whose behavior is governed by the laws of quantum mechanics. Beginning with Schrödinger's equation (without making any assumption, such as the introduction of fitting parameters) the first principles method gives access to the complete many electron system. However, the number of particles in a solid (both nuclei and electrons) makes a full quantum mechanical treatment impossible. A first approximation, the Born–Oppenheimer approximation assumes that the motion of the electrons can be separated from that of the massive nuclei. Therefore, the energy of the many-body system simplifies to the sum of the kinetic energies of the electrons and the columbic energies of the repulsive electron–electron and attractive electron–nucleus interactions. The Hamiltonian for a given system becomes:

$$\mathrm{H} = \sum_{i=1}^{N}\left(-\frac{\hbar^2}{2m}\nabla_i^2 - Ze^2\sum_I \frac{1}{|\boldsymbol{r}_i - \boldsymbol{R}_I|}\right) + \frac{1}{2}\sum_{i \neq j}\frac{e^2}{|\boldsymbol{r}_i - \boldsymbol{r}_j|} \tag{5.1}$$

Here, $\boldsymbol{r}_i$ denotes the position of an electron and $\boldsymbol{R}_I$ the position of a nucleus. An analytical solution to this many-body Schrödinger equation is impossible to be obtained, because there are many electrons ($N \sim 10^{28}$ in one mole of solid) and each electron contains $3N$ degrees of freedom. In the Hartree–Fock approach, the full electron wave function of the system is constructed from the product of individual electron wave functions and then optimized. In addition, for a crystalline solid, the Bloch theorem can be used to impose boundary conditions for the ion potentials and electron wave function, resulting in a set of coupled second-order differential equations, each describing an electron moving in an average potential due to the other electrons. This set of equations can be solved by an iterative procedure.

Another method for addressing the problem is the application of density functional theory [14, 15]. In the basic Heisenberg–Kohn theorem [14] all ground state observables are functions of the electron density n. For the total energy we have

$$E[n] = \int n(\boldsymbol{r})V_{\text{ext}}(\boldsymbol{r})d\boldsymbol{r} + F[n] \tag{5.2}$$

where F is a universal functional of the electron density, and independent of $V_{\text{ext}}(\boldsymbol{r})$. It was shown by Kohn and Sham [15] that the many-body electron problem can be transformed into multiple single particle electron problems by treating the electron density as basic variable:

$$E[n] = \int n(\boldsymbol{r})V_{\text{ext}}(\boldsymbol{r})\mathrm{dr} + \frac{1}{2}\iint \frac{n(\boldsymbol{r})n(\boldsymbol{r}')}{|\boldsymbol{r} - \boldsymbol{r}'|}\mathrm{drdr'} + T[n(\boldsymbol{r})] + E_{\text{XC}}[n(\boldsymbol{r})] \tag{5.3}$$

The first term is the Coulomb interaction between electrons and ions, the second term is the Coulomb interaction among the electrons, the third term is the kinetic energy of a system of noninteracting electrons, and the fourth term is the exchange–correlation potential affecting the electrons. The many-body system is then mapped onto a system of noninteracting single particles and the solutions to the Kohn–Sham equations of all electrons in the system are used to build the electron density. The effective potential depends on the electron density that is determined by the Kohn–Sham orbitals, which in turn depend on the effective potential. Thus, the problem of solving the Kohn–Sham equations is iterative with the requirement that the effective potential and the resulting charge density are self-consistent. To solve the Kohn–Sham equations, an explicit form is required to approximate the exchange–correlation functional. A first approximation is the local density approximation where the exchange–correlation energy of the interacting electrons at every point in space is taken to be the same as for the uniform electron gas, regardless of the inhomogeneity of the real charge density. Better approximations, such as the generalized gradient approximation, which take into account the gradient and higher spatial derivatives of the total charge density have proven to be superior in various respects and therefore nowadays are almost universally adopted in electronic structure calculations.

It is possible to solve the Kohn–Sham equations for solid state systems employing Bloch's theorem, by a plane wave expansion of the wave function and k-point sampling. A plane wave basis set is usually poorly suited for expanding electron wave functions, because it is difficult to accurately describe the rapid oscillations in the core regions. Usually, the valence electrons determine the physical properties of a solid. For this reason, the pseudopotential approximation can be introduced, which removes the core electrons and replaces the strong nuclear potential with a weaker pseudopotential. Ultrasoft pseudopotentials [16] are among the most widely used pseudopotentials.

We study the electronic and vibrational properties of single-layer TMDCs by the Quantum-ESPRESSO package [17] in the framework of density functional theory. We use ultrasoft pseudopotentials [16] as well as the generalized gradient approximation in the Perdew-Burke-Ernzerhof parameterization [18]. Phonons for single-layer MoS_2 are calculated by density functional perturbation theory [19]. A 10 Å thick vacuum is adopted in the slab calculations to avoid self-interaction of the MoS_2 single layer. Analogous supercells are used for the one-dimensional and zero-dimensional structures based on MoS_2.

5.3 Application

5.3.1 *Structure*

TMDCs are a class of layered materials with strong in-plane and weak out-of-plane bonding, which enables exfoliation into two-dimensional layers of single unit cell thickness. The chemical formula of such a layer in general is MX_2, where M is a transition metal element from group IV (Ti, Zr, Hf, etc.), group V (V, Nb, Ta), or group VI (Mo, W), and X is a chalcogen (S, Se, Te). Most layered TMDCs crystallize in the 2*H* ($P6_3$/mmc) or 1*T* ($P\bar{3}$m1) prototype structure, where most semiconducting TMDCs prefer the 2*H* prototype, such as MoS_2, $MoSe_2$, $MoTe_2$, WS_2, and WSe_2. The absence of 2*H* WTe_2 in nature is due to an intrinsic instability of this structure. On the other hand, single-layer WTe_2 has been predicted to be stable and can be fabricated by chemical vapor deposition or molecular epitaxial growth [20]. For MoS_2 there is another layered structure reported, the 3*R* (*R*3m) phase [21]. The unit cells of 2*H* and 3*R* MoS_2 are shown in Fig. 5.1a. Because the total energy of

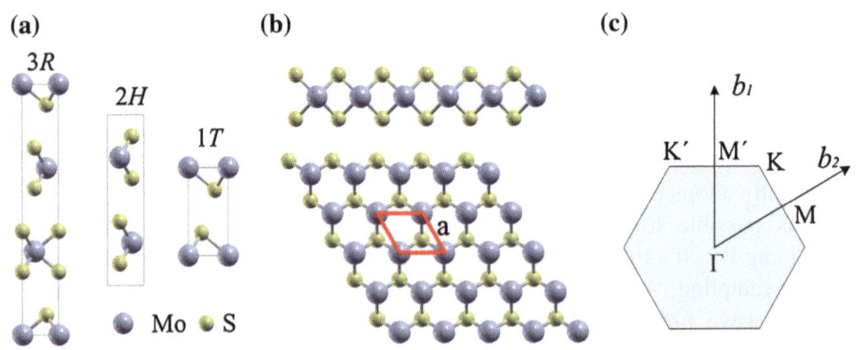

Fig. 5.1 **a** Side views of the 3*R*, 2*H*, and 1*T* phases of MoS_2 (unit cells). **b** Side and top views of single-layer MoS_2. **c** Brillouin zone of single-layer MoS_2

bulk 2*H* (3*R*) MoS_2 is 0.193 (0.187) eV per formula unit lower than that of single-layer MoS_2, formation of the latter is more difficult. In addition, the ground state energy of 2*H* MoS_2 is smaller than that of 3*R* MoS_2 by only 0.06 eV per formula unit, which explains the combined growth of the two phases, where the 3*R* portion is smaller than the 2*H* portion [22]. Metallic 1*T* MoS_2 has been observed in transmission electron microscopy measurements [23, 24]. Its unit cell is also shown in Fig. 5.1a. Single-layer MoS_2 is semiconducting and can be mechanically exfoliated from 3*R* and 2*H* MoS_2. Side and top views of single-layer MoS_2 (1*H*) are given in Fig. 5.1b and the corresponding Brillouin zone is shown in Fig. 5.1c. It should be noted that the K and K′ high symmetry points shown in Fig. 5.1c are not equivalent due to the threefold symmetry of single-layer MoS_2.

5.3.2 Band Structures of Bulk to Few-Layer MoS_2 Without Spin–Orbit Coupling

Figure 5.2a displays orbital-projected electronic band structures of bulk MoS_2, which is an indirect bandgap semiconductor with the valence band maximum (VBM) at the Γ-point and the conduction band minimum (CBM) along K-Γ. The lowest conduction states at the K point and the VBM are primarily composed of the Mo d_{z^2} and S p_z states, while the CBM and the highest valence states at the K point are dominated by the Mo $d_{x^2-y^2,\,xy}$ and S $p_{x,\,y}$ states. The Mo $d_{xz,\,yz}$ states are

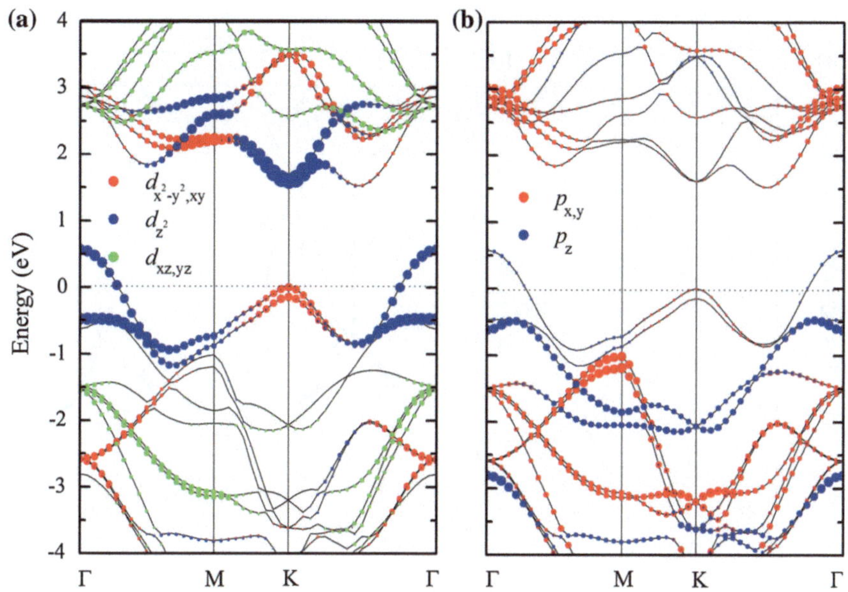

Fig. 5.2 **a** Mo d and **b** S *p* -projected band structures of bulk MoS_2

found far away from the Fermi energy. Few-layer MoS_2 has been characterized by absorption, photoluminescence, and photoconductivity experiments [5, 6]. It has been found that with decreasing thickness the indirect bandgap shifts upward in energy by more than 0.6 eV, which leads to a crossover to a direct bandgap in the single-layer limit. Figure 5.3a shows the direct bandgap for single-layer MoS_2 (without taking into account the spin–orbit coupling). It is not intuitively clear why only single-layer MoS_2 is a direct bandgap semiconductor, while thicker layers of MoS_2 still exhibit an indirect bandgap.

Figure 5.4a shows the variations of the highest valence and lowest conduction states as functions of the number of S-Mo-S layers. The direct excitonic transition energy at the K point barely changes with the thickness, whereas the indirect bandgap increases for thinner layers. Moreover, there appears a sudden decrease of the VBM at the Γ point from bilayer to single-layer MoS_2. In the single-layer system there exists, of course, no interlayer coupling, whereas interlayer coupling still is present in few-layer MoS_2. We can conclude that the absence of interlayer coupling is the origin of the direct bandgap of single-layer MoS_2.

To illustrate the influence of the interlayer coupling on the electronic states of few-layer MoS_2, the variation of the band structure as a function of the distance d between the two layers of bilayer MoS_2 is examined in Fig. 5.4b. Significant changes take place at the top of the valence band, where the VBM at the Γ point drops substantially when d increases. For decreasing interlayer interaction, i.e., for increasing d, the Mo d_{z^2} and S $p_z$states shift to lower energy, which not only results in an increase of the bandgap but also changes its nature from indirect to direct.

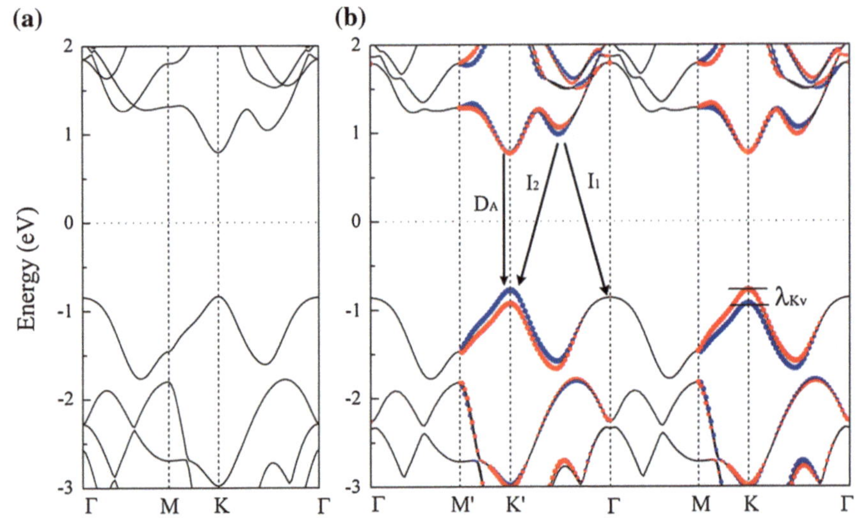

Fig. 5.3 Band structure of single-layer MoS_2 **a** without and **b** with spin–orbit coupling. *Red* and *blue color* indicates the spin *up* and *down* states, respectively

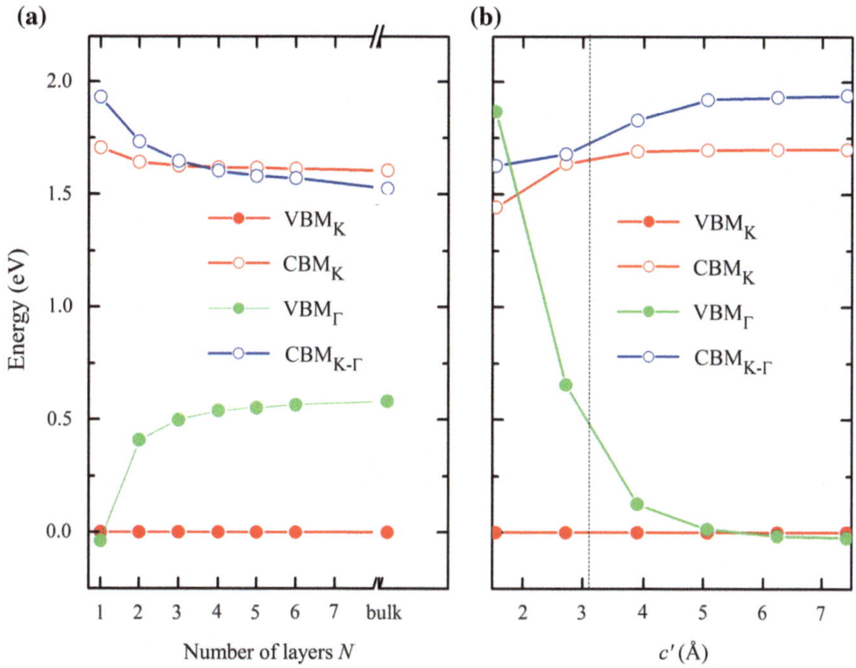

Fig. 5.4 Variation of the VBM and CBM as a function of **a** the number of layers N and **b** the distance between the two layers of bilayer MoS_2. VBM_K and CBM_K refer to the K point, VBM_Γ to the Γ point, and $CBM_{K\text{-}\Gamma}$ to the path K-Γ

Moreover, the highest valence band of bulk MoS_2 at the Γ point, see Fig. 5.2, shows a strong splitting, which indicates that the interlayer interaction is mainly due to a strong coupling of the out-of-plane Mo d_{z^2} orbitals. Recall that for graphene the highest valence band belongs to the out-of-plane C p_z orbitals. Due to weak interaction between the Mo d_{z^2} and C p_z orbitals the direct bandgap nature of MoS_2 is retained for MoS_2/graphene hybrid systems [25]. However, the bandgap of the MoS_2/WS_2 hybrid system shows a strong interaction of the Mo and W d_{z^2} orbitals [26], which induces a complex band structure.

5.3.3 Band Structure of Single-Layer MoS_2 with Spin–Orbital Coupling

Combination of time-reversal $[E^\uparrow(\boldsymbol{k}) = E^\downarrow(-\boldsymbol{k})]$ and inversion symmetry $[E^\uparrow(\boldsymbol{k}) = E^\uparrow(-\boldsymbol{k})]$ establishes Kramer's degeneracy $[E^\uparrow(\boldsymbol{k}) = E^\downarrow(\boldsymbol{k})]$ in bulk MoS_2 when the spin–orbit coupling is considered. Due to the absence of inversion symmetry in single-layer MoS_2 the bands exhibit spin splitting [13], except for points with special symmetry in the Brillouin zone. As is shown in Fig. 5.3b, the

Table 5.1 Structural parameters a (Å) and d (Å), bandgap Δ (eV) with spin–orbit coupling, spin splitting for the highest valence band at the K point λ_{Kv} (meV), spin splitting for the lowest conduction band at the K point λ_{Kc} (meV), Rashba spin splitting for highest valence band at the Γ point α_R (meVÅ), bandgap type (direct or indirect) with spin–orbit coupling, and single-layer structure stability

	a	d	Δ	λ_{Kv}	λ_{Kc}	α_R	Type	Stable
MoS_2	3.21	3.16	1.65	150	3	0	Direct	Yes
$MoSe_2$	3.34	3.34	1.38	188	20	0	Direct	Yes
$MoTe_2$	3.57	3.62	1.00	219	34	0	Direct	Yes
WS_2	3.19	3.14	1.68	431	29	0	Direct	Yes
WSe_2	3.34	3.35	1.32	473	34	0	Direct	Yes
WTe_2	3.57	3.63	0.85	493	51	0	Direct	No
MoSSe	3.21	3.28	1.55	170	13	2	Direct	Yes
MoSTe	3.37	3.39	1.04	139	68	4	Indirect	No
MoSeTe	3.45	3.49	1.19	204	30	12	Direct	No
WSSe	3.24	3.26	1.53	440	34	5	Indirect	Yes
WSTe	3.37	3.40	1.14	426	29	10	Indirect	Yes
WSeTe	3.44	3.50	1.15	463	43	14	Indirect	Yes

spin–orbit coupling breaks the spin degeneracy of the valence and conduction bands along the line Γ-K. Zero spin splitting is seen along the high-symmetry line Γ-M. The maximum splitting for the valence band appears at the K and K′ points (150 meV). In addition, the splitting at these points is not spin degenerate, i.e., the highest valence band at K belongs to the up spin while at K′ it belongs to the down spin. The spin splitting for the lowest conduction band at K and K′ is tiny (3 meV). The maximal spin splitting of the VBM at the K point amounts to 148, 188, 219, 431, 473, and 493 meV for single-layer MoS_2, $MoSe_2$, $MoTe_2$, WS_2, WSe_2, and WTe_2, respectively. The maximal spin splitting of the CBM at the K point amounts to 3, 20, 34, 29, 34, and 51 meV, respectively. Because the spin–orbit coupling is stronger, the spin splitting at the K point is enhanced for TMDCs with heavier atoms. Values for different TMDCs are listed in Table 5.1.

From an application point of view, the valence band is more promising than the conduction band due to the spin splitting. The breaking of the inversion symmetry in single-layer MoS_2 together with the strong spin–orbit coupling leads to coupled spin and valley physics and enables a spin and valley control [10]. The polarization selection rules of optical interband transitions depend on the frequency, which paves the way to a selective photoexcitation of carriers with various combinations of valley and spin indices [11, 12, 27]. Photoinduced spin Hall and valley Hall effects can generate long lived spin and valley accumulations on sample boundaries [10], which provides a route to the integration of valleytronics and spintronics in single-layer MoS_2.

5.3.4 Rashba Splitting for Single-Layer MoS_2

In a two-dimensional electron gas, due to the relativistic effects, an electric field **E** is seen by moving electrons as a magnetic field that couples to the spin in the *xy*-plane. This fact is described by the Bychkov-Rashba Hamiltonian [28]

$$\mathrm{H}_R = \alpha_R(|E|)\boldsymbol{\sigma} \cdot (\boldsymbol{k}_\parallel \times \boldsymbol{e}_z), \tag{5.4}$$

where $\boldsymbol{k}_\parallel$ is the momentum of an electron, $\boldsymbol{\sigma}$ the vector of Pauli matrices, and $\boldsymbol{e}_z = (0, 0, 1)$. Thus, the spin degeneracy is lifted and the energy dispersion has the form

$$E_{\mathrm{RSO}} = \frac{\hbar^2}{2m^*}(k_\parallel \pm \Delta k) - E_{\mathrm{SO}}, \tag{5.5}$$

which describes two parabolas that are shifted by $\pm\Delta k$ from the origin. For a surface, the origin of the electric field can be the potential gradient normal to the surface. This effect has been observed in two-dimensional geometries, such as semiconductor heterostructures metal surfaces, and metallic surface alloys. In general, research on two-dimensional materials has been boosted by the discovery of graphene in 2004. Due to the weak spin–orbit coupling in C and Si, there is still no experimental observation of Rashba splitting in pristine graphene or silicene. A lack of semiconductors with an enhanced Rashba effect in fact hinders realization of spintronics devices.

In MX_2 compounds, there is no inversion symmetry $(x, y, z) \rightarrow (-x, -y, -z)$, which leads to giant spin splitting [13]. However, there is no Rasha splitting of the VBM at the Γ point due to the mirror symmetry $(x, y, z) \rightarrow (x, y, -z)$. In general, Rashba splitting can be introduced by applying an asymmetric potential in the direction perpendicular to the basal plane. To follow this route, we investigate MXY compounds. Figure 5.5a exhibits an apparent Rashba splitting of the VBM for WSSe. The Rashba effect leads to a shifting of oppositely spin-polarized bands by the momentum k_R in opposite directions. The key ingredient here is the magnitude of the spin splitting due to spin–orbit coupling, characterized by the Rashba energy of the split states, E_R, and the Rashba coupling parameter, $\sigma_R = 2E_R/k_R$, which measures the strength of the spin splitting. We obtain values of 2, 12, 4, 5, 14, and 10 meVÅ for MoSSe, MoSTe, MoSeTe, WSSe, WSTe, and WSeTe, respectively, compare Table 5.1.

For single-layer MXY (C_{3v}) the mirror symmetry is broken, which leads to a local potential gradient normal to the basal plane. According to the two-dimensional electron gas model [28], this gives rise to an in-plane spin orientation. Taking into account, the point group 3 m of single-layer MXY, the spin polarization at a fixed energy is given by

$$\vec{P}(\theta) = (\alpha \sin\theta, -\alpha\cos\theta, \beta\cos 3\theta), \tag{5.6}$$

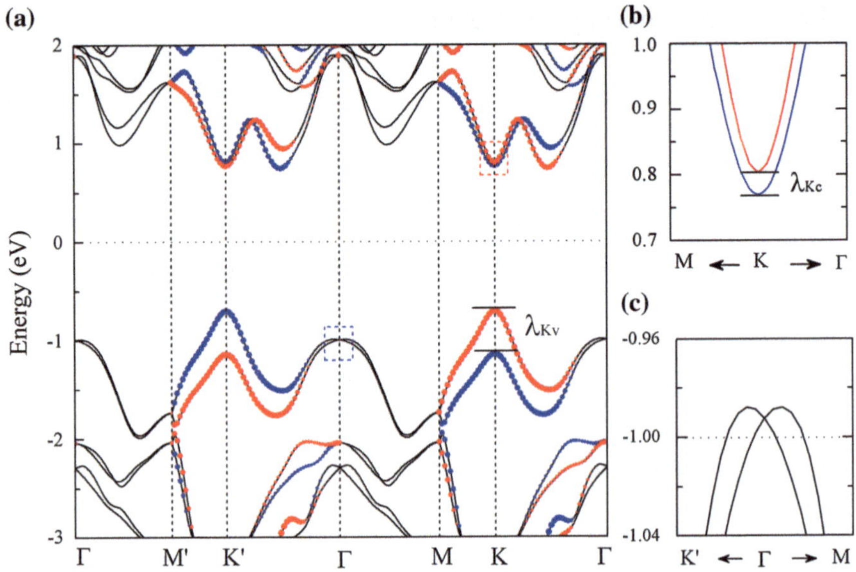

Fig. 5.5 **a** Band structure of WSSe with spin–orbit coupling. *Red* and *blue colors* indicate spin up and down states, respectively. **b** Magnified view of the conduction band around the K point, corresponding to the *dashed red box* in (**a**). **c** Magnified view of the valence band around the Γ point, corresponding to the *blue-dashed box* in (**a**)

where θ is the azimuthal angle [29]. For the Γ-M direction we have, for example, $\theta = \pi/4$ and $\theta = 0$ for Γ-K. The in-plane spin polarization rotates clockwise (anticlockwise) for $\alpha > 0$ ($\alpha < 0$) around the z-axis. The nonzero P_z reflects directly the threefold symmetry of single-layer MXY.

The Rashba splitting can be enhanced by a stronger spin–orbit coupling of the involved atoms (M from Mo to W, X/Y from S to Te), which has been demonstrated above. However, an enhanced local potential gradient also can lead to a stronger Rashba splitting. It therefore can be expected that there is a detectable Rashba splitting in MX_2 on a substrate that introduces an asymmetric potential in the out-of-plane direction. In addition, for bulk MoSSe in ABAB stacking, in which we have such an asymmetric potential, the electronic structure shows the bulk Rashba splitting at the Γ point for the highest valence band.

5.3.5 Two-Dimensional Diluted Magnetic Semiconductors

Chemical doping is a promising approach to tailor the electronic properties of semiconductors. It has been demonstrated that H, B, C, N, and O can be chemically absorbed on single-layer MoS_2, leading to total magnetic moments of 1, 1, 2, 1, and 1 μ_B, respectively. Also, *n*-type and *p*-type MoS_2 semiconductors can be realized by absorbing H and N atoms, respectively [30]. It has been reported that a S vacancy is easier to form in MoS_2 than a Mo vacancy under electron irradiation [31], which indicates that substitution can be achieved in specific gas atmosphere under electron irradiation. F, Cl, Br, and I substitution will lead to *n*-type doping, while N, P, As, and Sb substitution will lead to *p*-type doping [31]. However, the impurity atoms for both the chemical absorption and substitution are located outside of the MoS_2 layer, enabling further pollution in atmosphere. This situation is similar to chemical doping for graphene and single-layer hexagonal boron nitride [32]. Due to the sandwich structure of MoS_2, it is interesting to study the substitution of Mo by other transition metal species. A schematic view of such a substitution is shown in Fig. 5.6.

We assume here that the Mo site is already substituted by the impurity transition metal atom. Binding energies are calculated to check the stability of the different substitutions. The binding energy is defined as $E_b = E_v + \mu - E_d$, where E_v is the energy of a relaxed MoS_2 single-layer with one Mo vacancy, μ is the energy of the doped transition metal atom, and E_d is the energy of a relaxed MoS_2 single-layer with one Mo atom replaced by the transition metal atom. The value of E_b for different dopants varies strongly from 0.2 to 16 eV. Mo self-substitution yields the binding energy of a Mo atom, which is 13.5 eV. The binding energy increases with the period for the IVB, VB, VIB, VIII8, and VIII9 groups, while it decreases for the VIII10, IB, and IIB groups. Substitutional doping of transition metal atoms from the IIIB to VIII10 groups is stable, whereas doping of transition metal atoms from the IB and IIB groups is not stable, because of the low binding energies (especially for Hg).

Single-layer MoS_2 doped by Mn, Fe, Co, Zn, Cd, and Hg is ferromagnetic. The magnetic moments are 1, 2, 3, 2, 2, and 2 μ_B., respectively. The binding energies and magnetic moments indicate that Mn, Fe, Co, and Zn doping of single-layer MoS_2 are very promising approaches to achieve two-dimensional diluted magnetic semiconductors. The density of states in Fig. 5.7a shows a bandgap of 1.7 eV for the pristine single-layer MoS_2. This is only an underestimation of 0.2 eV as compared to the values obtained experimentally and by GW calculations, indicating the validity of the generalized gradient approximation for MoS_2. Neither

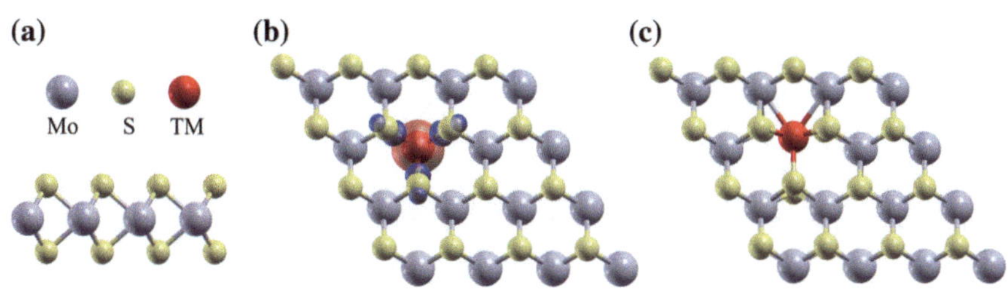

Fig. 5.6 **a** Sandwich structure of single-layer MoS_2. Spin density for **b** Mn doping with C_{3v} symmetry and **c** Ni doping with C_1 symmetry. Spin densities of ± 0.002 e/Bohr^{-3} are visualized by *red* and *blue* isosurfaces, respectively

Mo vacancies nor S vacancies can introduce magnetism in MoS_2, which is different from the intrinsic magnetism in defective graphene and hexagonal boron nitride [32].

With respect to the structure and calculated magnetism of (Mo, X)S_2, the transition metals can be distinguished into three classes: the IIIB to VIB groups, the VII8 to IB groups, and the IIB group. We take Sc as a prototypical element from the first class. The symmetry of (Mo, Sc)S_2 is C_{3v} and the density of states shows spin degeneracy. We note that an isolated Sc atom has a $3d^1 4s^2$ configuration and three valence electrons less than Mo. The fact that the impurity states are close to the valence bands indicate p-type doping. (Mo, Mn)S_2 is a prototypical compound from the second class, for which Fig. 5.7b shows spin polarization. An isolated Mn atom has a $3d^5 4s^2$ configuration, i.e., one additional valence electron as compared to Mo. The spin density isosurface in Fig. 5.7b shows that the magnetic moment is mainly due to the Mn atom. Because of the C_{3v} symmetry, the Mn $3d$ orbitals are split into a single a (d_{z^2}) state and twofold degenerate e_1 ($d_{xy,\, x^2-y^2}$) and e_2 ($d_{xz,\, yz}$) states, with a total of 6 valence electrons.

Structure relaxation shows that the symmetry of (Mo, Ni)S_2 is reduced to C_1 due to a Jahn–Teller distortion, see Fig. 5.6c. The energy difference of the C_{3v} and C_1 structures of (Mo, Ni)S_2 is 383 meV/cell. Due to the Jahn–Teller distortion, the

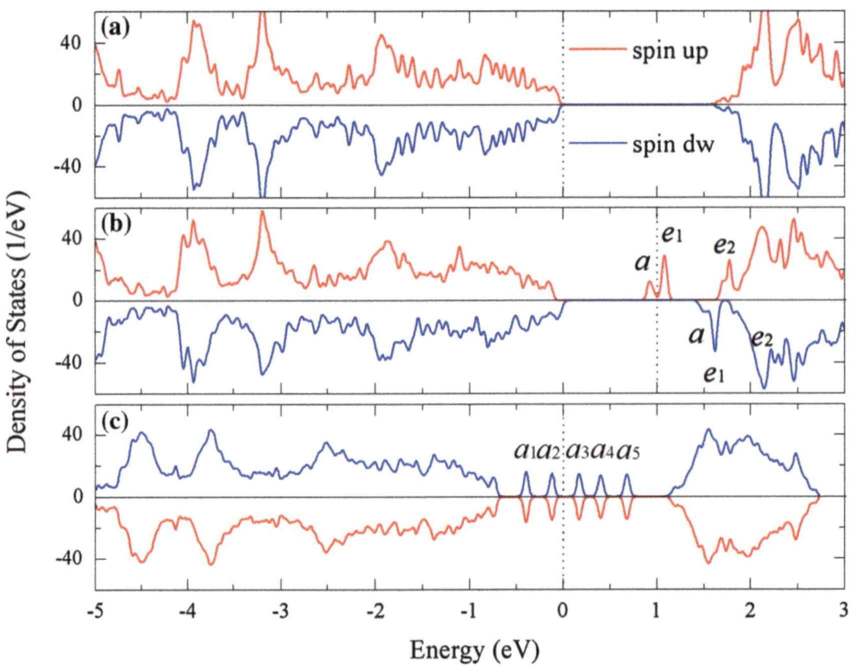

Fig. 5.7 Density of states of single-layer MoS_2: **a** pristine, **b** Mn doped, and **c** Ni doped. The energy axis is calibrated with respect to the core states. The *dashed vertical lines* indicate the Fermi energy

degenerate e_1 and e_2 states split into four a states. Therefore, four electrons occupy the a_1 and a_2 states, see Fig. 5.7c. All transition metals from the second class, except for Mn, Fe, and Co, show a Jahn–Teller distortion and consequently no magnetism. Though the binding energy for elements from the third class is lower than for the other two classes, (Mo, X)S_2 with X = Zn, Cd, Hg develops magnetism with a moment of 2 μ_B.

To find the magnetic ground state, both ferromagnetic and antiferromagnetic coupling between the dopants (concentration 6.25 %) are considered for the transition metals Mn, Fe, Co, Zn, Cd, and Hg. For Fe and Co doping the antiferromagnetic state is energetically favorable, while Mn, Zn, Cd, and Hg doping result in ferromagnetic ground states, which are 0.02, 3.30, 2.88, and 1.85 meV per cell lower in energy than the respective antiferromagnetic states. An important parameter of a diluted magnetic semiconductor is the Curie temperature T_C below which the system develops long range ferromagnetic ordering. We estimate T_C based on the mean-field theory and Heisenberg model using the relation $k_B T_C = (2/3)\Delta E$ [33]. For Mn we find $T_C = 0.2$ K, which is much less than room temperature. For a smaller dopant concentration of 4 % a significantly enhanced Curie temperature of $T_C = 92.7$ K is obtained.

5.3.6 Vibrational Properties

Figure 5.8a shows the phonon dispersion of single-layer MoS_2 calculated by density functional perturbation theory [19] using the generalized gradient approximation for the exchange–correlation potential. For the calculation of the dynamical matrix, the atomic displacements are taken as a perturbation potential and the resulting changes in the electron density and energy are calculated self-consistently by a system of Kohn–Sham-like equations. Within this approach, the phonon frequency can be obtained for arbitrary q considering only a single unit cell. Since bulk MoS_2 is a slightly polar material, certain infrared active phonon modes at the Γ point give rise to a macroscopic electric field, which breaks the degeneracy of the longitudinal optical and transversal optical modes in the limit $q \rightarrow 0$ [34]. The splitting for the E_{1u} mode amounts to 2.8 cm^{-1} for the bulk [34], while in the case of a single-layer the effect is even smaller.

The phonon dispersion of single-layer MoS_2 shows three acoustic modes. The in-plane modes (longitudinal acoustic and transversal acoustic) have a linear dispersion and higher energy than the out-of-plane acoustic mode. The latter displays a q^2 dependence, which is a feature of two-dimensional materials [35]. For few-layer and bulk MoS_2 there are low-frequency optical modes around 40 cm^{-1} that correspond to rigid-layer shear and vertical motions, respectively. The low-frequency optical modes are also features of two-dimensional materials. The high-frequency optical modes are separated from the low-frequency modes by a gap of about 50 cm^{-1}, see Fig. 5.8b. The single-layer, few-layer, mode of bulk dispersions have common features. Due to interlayer interaction, all single-layer

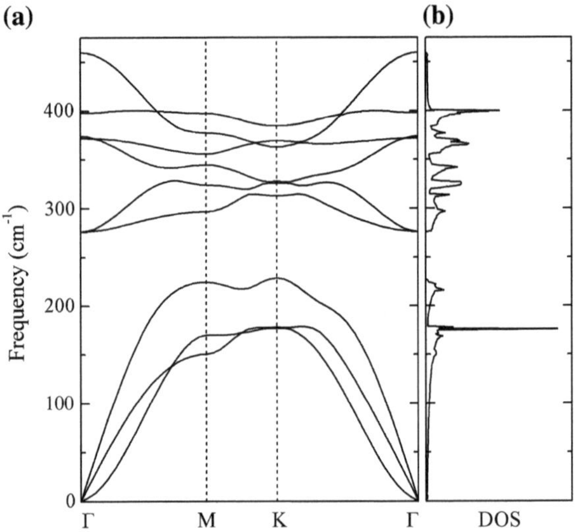

Fig. 5.8 **a** Phonon dispersion and **b** phonon density of states for single-layer MoS_2

modes are split into a few branches, but since the interlayer interaction is weak, the splitting is small. It should be noted that in the single-layer the low-frequency optical modes are absent.

For bulk MoS_2 there are 18 normal modes of vibration at the Γ point with the irreducible representation $\Gamma \equiv A_{1g} \times 2A_{2u} \times 2B_{2g} \times B_{1u} \times E_{1g} \times 2E_{2g} \times 2E_{1u} \times E_{2u}$. Here, one A_{2u} and one E_{1u} modes are translational acoustic modes, the other A_{2u} and E_{1u} modes are infrared active modes, A_{1g}, E_{1g}, and E_{2g} are Raman active modes, and finally $2B_{2g}$ and B_{1u} are silent modes. The atomic displacements for the Raman and infrared active modes are shown in Fig. 5.9. The calculated frequencies of the bulk MoS_2 Raman modes based on density functional perturbation theory and the generalized gradient approximation without considering layer–layer van der Waals interaction are 11.8, 276.4, 371.7, and 398.5 cm^{-1} for E^2_{2g}, E_{1g}, E^1_{2g}, and A_{1g}, respectively. There are nine normal modes of vibration at the Γ point with the irreducible representation $\Gamma \equiv 2A''_2 \times A'_1 \times 2E' \times E''$. Here, one A''_2 and one E' modes are translational acoustic modes, the other E' mode is both Raman and infrared active modes, the other A''_2 mode is infrared active mode, and the A'_1 and E'' modes are Raman active modes. According to a symmetry analysis, the A'_1 (399.3 cm^{-1}), E' (376.0 cm^{-1}), and E'' (277.4 cm^{-1}) modes correspond to the A_{1g}, E_{2g}, and E_{1g} modes, respectively. The shear E^2_{2g} mode of bulk and few-layered MoS_2 is absent in a single-layer [36]. Moreover, frequencies calculated in the generalized gradient approximation are smaller than those calculated in the local density approximation, which is due to the well-known fact that these approximations describe the exchange correlation in more soft and rigid manners, respectively. Comparison of the theoretical values with experimental data, as listed in Table 5.2, shows that the local density approximation describes the interlayer interaction better than the generalized gradient approximation.

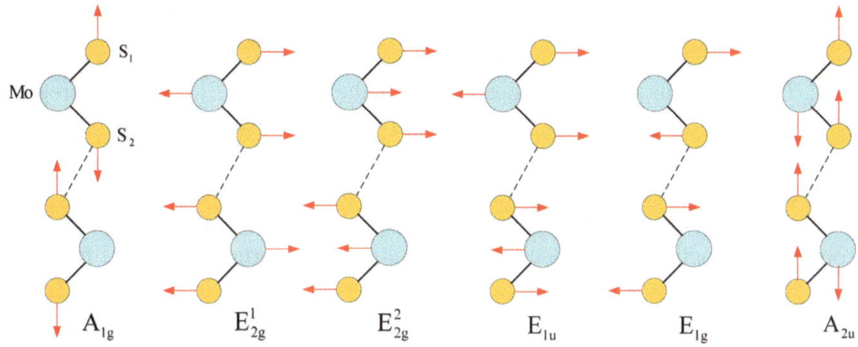

Fig. 5.9 Atomic displacements of the four Raman active modes $\left(A_{1g},\ E_{2g}^{1},\ E_{2g}^{2},\ \text{and}\ E_{1g}\right)$ and the two infrared active modes (E_{1u} and A_{2u}) of bulk MoS_2, viewed along the [1000] direction

Table 5.2 Phonon symmetry representations of single-layer (point group D_{3h}) and bulk (point group D_{6h}) MoS_2 as well as theoretical and experimental phonon frequencies

D_{3h}	D_{6h}	Character	Generalized gradient approximation [37]		Local density approximation [34]		Experiment [36, 45, 62]	
–	E_{2g}^2	Raman		11.0		35.2		32
	B_{2g}	Silent		18.7		55.7		
E''	E_{1g}	Raman	277.4	277.8	289.2	288.7		287
	E_{2u}	Silent		277.8		287.1		287
E'	E_{2g}^1	Raman	375.3	372.2	391.7	387.8	384.3	383
	E_{1u}	Infrared		372.2		391.2		384
A'_1	B_{1u}	Silent	399.3	398.1	410.3	407.8	403.0	409
	A_{1g}	Raman		398.7		412.0		409
A''_2	B_{2g}	Silent	460.0	459.7	476.0	473.2		470
	A_{2u}	Infrared		459.8		469.4		470

The Raman spectra of few-layer MoS_2 exfoliated on a SiO_2/Si substrate indicate that the two Raman modes E_{2g}^1 and A_{1g} depend remarkably on the thickness N. The frequency of the E_{2g}^1 mode increases and that of A_{1g} mode decreases with decreasing N. The atom–atom interaction force model with two intralayer and one interlayer force constants highlights the dependence of the E_{2g}^1 and A_{1g} frequencies on the lattice parameters and force constants. In contrast to the experimental observations, a classical model for coupled harmonic oscillators predicts that both modes become stiffer when additional layers are added. This discrepancy is explained by an enhancement of the dielectric screening of the long-range Coulomb interaction between the effective charges with a growing number of layers [34]. It can also be explained by the reduced interlayer coupling for decreasing slab thickness [37]. In the low-frequency range, there appear two Raman modes with opposite thickness dependence. With increasing N the shear mode E_{2g}^2 significantly increases in frequency, while the other compression mode follows a 1/N trend [36]. Therefore, the thickness dependence of the E_{2g}^1, A_{1g}, E_{2g}^2, and compressive modes provides a convenient and reliable way to determine the layer thickness with atomic precision. By the theoretical concern that both the local density approximation and generalized gradient approximation miss the interlayer van der Waals interaction, it is desirable to develop efficient van der Waals exchange correlation functionals to calculate the vibrational properties of layered materials with enhanced accuracy.

5.3.7 Mechanical Properties

The application of strain in low-dimensional materials can be used to tailor many materials properties, which is important in the context of nanoelectronic devices, particularly in the search for flexible electronics [38–40]. Moreover, the most promising method to grow single-layer MoS_2, chemical vapor deposition [41], will induce residual strain in the MoS_2 film due to the different lattice expansion coefficients [22]. Therefore, it is essential to understand the effects of strain on the vibrational and electronic properties as well as the elastic limit of single-layer MoS_2.

Experimentally, it has been reported that single-layer MoS_2 breaks at a tensile stress of 22 ± 4 GPa and the critical tensile strain was estimated to be 6–11 % by assuming a linear stress–strain relationship [42]. Theoretically, the stress–strain relation was predicted for the zigzag direction, the armchair direction, and biaxial tension in Ref. [38]. The corresponding elastic moduli are $E_x = 197.9 \pm 4.3$ GPa, $E_y = 200.3 \pm 3.7$ GPa, and $E_{2D} = 250.2 \pm 5.8$ GPa. The calculated ideal tensile stress–strain relations indicate different behaviors for different strain direction. Both the maximum stress and strain in the armchair direction are higher than those in the zigzag direction, which suggests that single-layer MoS_2 is stronger in the former case. Phonon dispersions for single-layer MoS_2 under strain have been calculated to check the dynamical stability when the strain approaches the maximum [38]. One phonon branch shows negative frequencies near the M point at a uniaxial tensile strain of $\varepsilon_{yy} = 28$ %. Similarly, at a biaxial tensile strain of $\varepsilon_{2D} = 19.5$ %, one phonon branch becomes unstable near the K point. Examination of the eigenvectors of the unstable phonon modes indicates that both branches represent out-of-plane acoustic modes. In contrast, these phonon modes remain stable under uniaxial tension along the x-direction when the maximum tensile stress is reached. There is a soft phonon near the Γ point at a critical tensile strain of $\varepsilon_{xx} = 36$ %, which reflects that the single-layer becomes unstable with respect to uniform tensile strain along the x-direction. In other words, single-layer MoS_2 reaches its ideal tensile strength under uniaxial tension along the x-direction, where the failure mechanism is the loss of elastic stability [38], similar to graphene. Under a generic tension the strength of graphene is also limited by a soft mode instability [43], with an in-plane soft phonon. In single-layer MoS_2 the out-of-plane phonon mode plays the key role in the failure mechanism upon tension along the armchair direction [38]. The difference in the response to uniaxial tensile strain is attributed to the different atomic structures, i.e., graphene is a truly two-dimensional system, while single-layer MoS_2 is composed of three atomic layers.

5.3.8 Strain Effects on the Vibrational Properties

With increasing tensile strain, the frequencies of the A_{1g} and E^1_{2g} modes decrease due to the softening of the atom–atom interaction. It is reported that the A_{1g} mode shifts slightly at a rate of -0.4 cm^{-1}/% and the E^1_{2g} mode at -2.1 cm^{-1}/% for single-layer MoS_2 under uniaxial tensile strain. The experimental shifts are consistent with density functional theory (-0.55 and -2.22 cm^{-1}/%, respectively [44]). As is shown in Fig. 5.9, the A_{1g} mode corresponds to out-of-plane displacements of the S atoms, while the E^1_{2g} mode corresponds to in-plane displacements. Under tensile strain the in-plane interaction between the atoms decreases more than the out-of-plane interaction, which explains that the shift of the A_{1g} mode is much smaller than that of the E^1_{2g} mode.

Figure 5.10a demonstrates shifts of the A_{2u}, A_{1g}, E^1_{2g}, and E_{1g} modes at rates of 4.5, 1.6, 4.8, and 2.8 cm^{-1}/% under biaxial tensile strain, respectively. Previously, the frequency difference between the A_{1g} and E^1_{2g} modes has been reported as 18.9, 21.7, and 25.2 cm^{-1} for single-layer, bilayer, and bulk MoS_2, respectively [45].

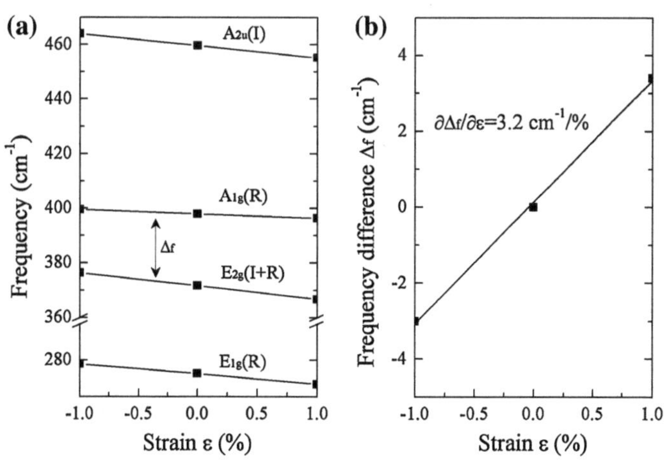

Fig. 5.10 **a** Frequency shifts of optical modes (E_{1g}, E^1_{2g}, A_{1g}, and A_{2u}) as a function of biaxial strain. **b** Frequency difference of the A_{1g} and E^1_{2g} modes as a function of biaxial strain

The frequency difference between the A_{1g} and E_{2g}^1 modes for single-layer MoS_2 under biaxial strain is addressed in Fig. 5.10b. We observe a slope of 3.2 cm^{-1}/%, which indicates that under tensile strain up to about 1 % the frequency difference is similar to that of bilayer MoS_2. Hence, it is not sufficient to identify the layer thickness of MoS_2 under strain.

5.3.9 Strain Effects on the Electronic Properties

We denote by D_A the direct gap transition at the K point, by I_1 the indirect gap transition from the Γ-K line to the Γ point, and by I_2 the indirect gap transition from the Γ-K line to the K point, see Fig. 5.3b. The D_A, I_1, I_2 excitation energies as function of the biaxial strain are shown in Fig. 5.11. We find that the direct bandgap nature of single-layer MoS_2 turns indirect under a biaxial tensile strain of 0.8 % or a biaxial compressive strain of −1.2 %. Both the D_A and I_1 energies show a red shift for increasing tensile strain, while the I_2 energy is subject to a blue shift. The shift rates for the three excitation energies are, respectively, −114, −242, and 85 meV/%. The transition from direct to indirect bandgap is related to variations in the orbital composition of the states at the VBM [39]. In direct bandgap case, these states originate from bonding Mo $d_{x^2-y^2,xy}$ and S $p_{x,y}$ orbitals at the K point, whereas in the indirect bandgap case they originate from bonding Mo d_{z^2} and S p_z orbitals at the Γ point. On the other hand, the orbital composition at the CBM does not change: antibonding Mo d_{z^2} and S $p_{x,y}$ orbitals at the K point. When tensile strain is applied to single-layer MoS_2, the distance between the Mo and S atoms increases. Consequently, the overlap of the Mo d_{z^2} and S p_z orbitals as well as that of the Mo d_{z^2} and S

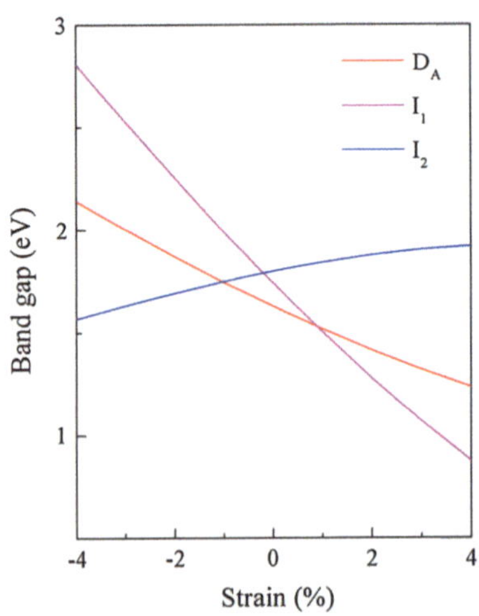

Fig. 5.11 Excitation energies as a function of biaxial strain

$p_{x,y}$ decreases. As a consequence, the bonding-antibonding splitting of the states is reduced and the VBM shifts up, the CBM shifts down, and the bandgap decreases. Under larger biaxial strain ($\sim$30 %) single-layer MoS_2 becomes a semimetal [39].

The bandgap of bilayer MoS_2 and other semiconducting TMDCs has been studied by first principles in Refs. [46, 47], finding that it can be reduced smoothly by vertical compressive pressure. The materials undergo a universal reversible semiconductor-to-metal transition at some critical pressure, which depends on the stacking sequence (AA or AB) [46]. The transition can be attributed to a lifted degeneracy of the bands at the Fermi energy, caused by interlayer interaction [28]. Using bandgap corrections by hybrid functionals or the GW method, the semiconductor-to-metal transition can be reproduced theoretically. The ability to tune the bandgap of TMDCs in a controlled fashion over a wide energy range opens the possibility for usage in a range of applications. For example, it has been proposed that a photovoltaic device made from strain-engineered single-layer MoS_2 will capture a broad range of the solar spectrum [48].

5.4 Electronic Properties of MoS_2 Nanostructures

5.4.1 Nanoribbon Under Strain

MoS_2 nanoribbons can be directly obtained by cutting single-layer MoS_2. With respect to the termination, there exist two kinds of nanoribbons: armchair and zigzag. And with respect to the S or Mo layer, the armchair nanoribbons can be symmetric or asymmetric. On the other hand, there are various types of zigzag nanoribbon, considering the atoms at the edges and the symmetry. Figure 5.12a and b shows an asymmetric armchair nanoribbon and one type of zigzag nanoribbon, respectively. Geometry optimization leads to strong distortions of the atoms at the edges, while the lattice parameters within the ribbons remain almost unchanged. The armchair nanoribbon is nonmagnetic and has a direct bandgap at the Γ point, see Fig. 5.12c, whereas the zigzag nanoribbon is ferromagnetic, see Fig. 5.12d. It has been found that the electronic and magnetic properties of zigzag nanoribbons depend on the chirality [49]. The ferromagnetic nature of zigzag nanoribbons originates from unpaired edge electrons.

The electronic and magnetic properties of MoS_2 nanoribbons under strain have been studied by first principles calculations in Refs. [49, 50]. The bandgap of armchair nanoribbons decreases under tensile strain, which is called strain-induced bandgap narrowing. Localized CBM states and decoupled VBM states are induced by extended bonds at large strain, which reduces the overlap of the electron wave functions. Zigzag nanoribbons maintain magnetic under strain. Interestingly, it was found that the energy differences between the nonmagnetic and magnetic states of zigzag nanoribbons are enhanced by strain, indicating that the magnetic states become more stable. Moreover, the magnetic moments are found to be very

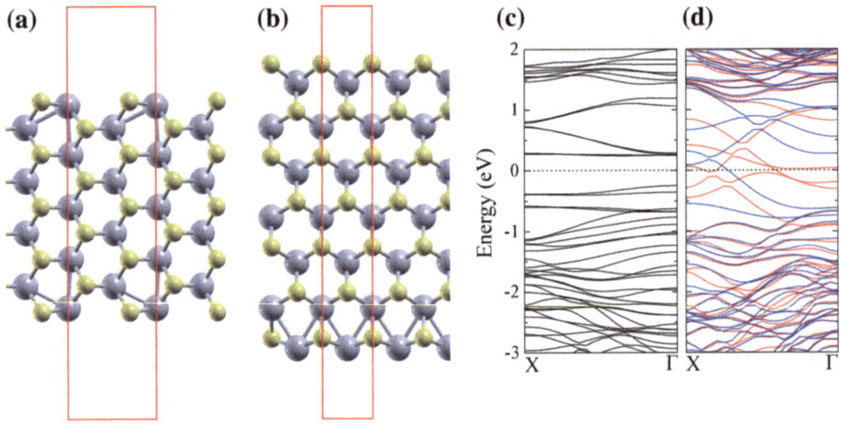

Fig. 5.12 **a** Armchair and **b** zigzag MoS_2 nanoribbons. The corresponding band structures calculated without spin–orbit coupling are shown in (**c**) and (**d**)

sensitive to both the edge structure and the magnitude of the strain. The appearance of magnetization can be related to changes in the coordination numbers of unsaturated S and Mo atoms on the zigzag edges [50]. H-saturated zigzag nanoribbons are ferromagnetic with enhanced magnetic moments. Interestingly, the metallic state of zigzag nanoribbons can be transformed into a semiconducting (*n*- or *p*-type) or half-metallic state by controlling the dangling edge bonds with H. MoS_2 nanoribbons with controlled edge structure, chirality, and saturation may find applications in nanodevices, spintronics, and photovoltaic cells [51].

It has been reported that application of an external transversal electric field can significantly reduce the bandgap of armchair MoS_2 nanoribbons and even lead to a metal-to-insulator transition beyond a critical value [52]. The critical field for gap closure decreases when the nanoribbon width increases and for bilayer nanoribbons the bandgap can be changed from indirect to direct [53]. However, there appears no bandgap modulation in a perpendicular field. In multilayer armchair nanoribbons, in contrast, the bandgap can be effectively reduced by both transversal and perpendicular fields. The critical perpendicular field for gap closure decreases with the thickness, while the critical transversal field hardly changes.

Zigzag MoS_2 nanoribbons have been proposed for cathode materials for rechargeable Li/Mg batteries [54, 55]. The Mo top site at the edge of the nanoribbon is favorable for Mg ions. In zigzag nanoribbons, a maximal theoretical capacity of 223.2 $mAhg^{-1}$ could be achieved by double-side Mg adsorption. The Mg diffusion pathway on the zigzag nanoribbon passes two adjacent top sites mediated by the hollow site between them. The activation barrier of this process is only 0.48 eV and thus much less than the 2.61 eV of the bulk interlayer migration. Although the Li mobility can be significantly enhanced in MoS_2 nanosheets, decreased Li binding energies limit cathode applications. Because of the special edge states, zigzag MoS_2 nanoribbons show a remarkably enhanced binding interaction with Li without sacrificing the Li mobility. They thus are promising for cathode materials of Li-ion batteries with a high power density and fast charge/discharge rates [54].

5.4.2 Nanotube

The discovery of fullerenes and carbon nanotubes has opened a challenging new field in solid state physics, chemistry, and materials science. Moreover, tubular structures for more complex layered compounds such as MoS_2 and WS_2 have been reported [56]. In general, the rolling up direction of the tube can be described in terms of primitive two-dimensional lattice vectors $\boldsymbol{a}$ and $\boldsymbol{b}$ together with two integer indices: $\boldsymbol{B} = \mathrm{n}\boldsymbol{a} + \mathrm{m}\boldsymbol{b}$. In this way three classes of nanotubes can be distinguished: the "armchair" nanotubes with $n = m$, the "zigzag" nanotubes with $n \neq 0$, $\mathrm{m} = 0$, and the "chiral" nanotubes with $\mathrm{n} \neq m$. Figure 5.13a and b shows two typical nanotubes: the (8, 0) nanotube, which possesses a small direct bandgap, and the (8, 8) nanotube, which exhibits a small indirect as well as a moderate direct bandgap. It has been reported that all MoS_2 nanotubes are semiconducting and stable with strain energies similar to carbon nanotubes, which decrease as the diameter grows [57]. This suggests that zigzag nanotubes can be used for optoelectronics, in contrast to carbon nanotubes. Because of the structural similarity between MoS_2 and other semiconducting TMDCs similar electronic and mechanical properties can be expected.

For armchair MoS_2 nanotubes tensile/compressive axial strain reduces/enlarges the bandgap linearly. It can be fully closed for nanotubes with relatively small

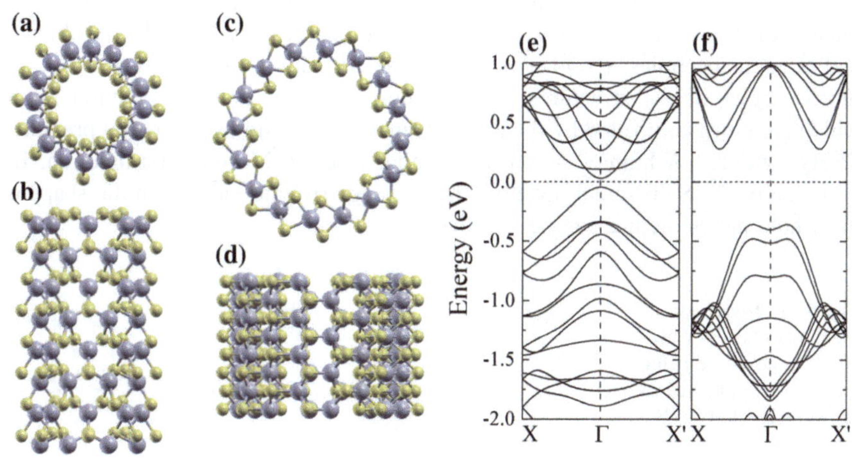

Fig. 5.13 **a** *Top* and **b** *side views* of (**a**) (8, 0) MoS_2 nanotube. **c** *Top* and **d** *side views* of (**a**) (8, 8) MoS_2 nanotube. Band structures of the (8, 0) and (8, 8) nanotubes are shown in (**e**) and (**f**), respectively

diameter and under large tensile strain. For zigzag MoS_2 nanotubes the strain effects are nonlinear. Tensile strain reduces the bandgap, whereas compressive strain initially enlarges and then also reduces it. Strain induced changes in the Mo orbital energy a modified coupling between the Mo *d* and S *p* states can explain the strong strain effects on the bandgap and magnetic properties [58]. Again, similar behaviors can be expected for other TMDCs.

5.4.3 Nanoplate

By sulfurization of Mo clusters (using H_2S in ultrahigh vacuum), single-layer triangular MoS_2 nanostructures have been fabricated on Au(111) substrate [59]. MoS_2 is traditionally used as solid state lubricant and industrial catalyst for hydrodesulphurization and hydrogen evolution [60]. Determination of the morphology and electronic properties of single-layer nanosize MoS_2 particles is of considerable interest for a better understanding of the active phase of hydrotreating catalysts. There are two types of triangular MoS_2 nanostructures distinguished by Mo and S edge termination, see Fig. 5.14a and b, respectively. The structure and electronic properties of such nanostructures have been extensively studied by first principles calculations in Ref. [61]: The Mo edge triangle with 100 % S coverage exhibits S_2 dimers after relaxation and the outer Mo atoms are sixfold coordinated, while for 50 % S coverage S atoms are found in a bridge position and for the clean surface with 0 % S coverage the Mo atoms are fourfold coordinated. For the S edge triangle with 100 % S coverage, the coordination of the Mo atoms in the outermost Mo-S chain is six and the S atoms occupy a bridge position close to the bulk terminated case with pairing of the S_2 dimers at the three corners. Reduction of the S coverage to 50 % leads to a zigzag configuration in which the S atoms lie outside of the Mo plane. Moreover, the Mo atoms are only fourfold coordinated and S again occupies a bridge position. The formation energy reveals that the morphologies of the structures are determined by the chemical potential of S, which is in turn controlled by the temperature and the ratio of the partial pressures of H_2S and H_2. A high S chemical potential leads to Mo edge termination. By variation of the S chemical potential the Mo edge structure and even the shape of the whole nanostructure can be modified.

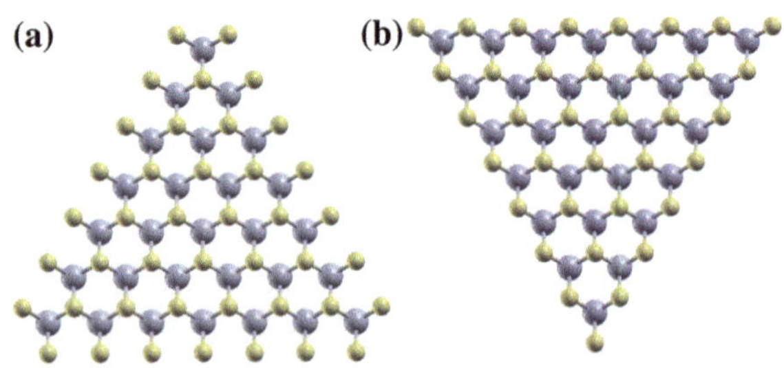

Fig. 5.14 *Top views* of triangular clusters with 100 % S coverage before relaxation, terminated by the (**a**) Mo edge and (**b**) S edge

5.5 Conclusions

Employing first-principles calculations, we have investigated the structural, electronic, and vibrational properties of single-layer MoS_2 and of one- and zero-dimensional nanostructures based on MoS_2. Symmetry breaking together with spin–orbit coupling plays an important role for the electronic states. We have proposed a two-dimensional dilute magnetic semiconductor to be obtained by substitution of Mo by other transition metal atoms, such as Mn, Fe, and Co. The experimental observation of an anomalous vibrational behavior can be attributed to the reduction of interlayer interaction and strengthening of intralayer interaction from bulk to the single layer. Strain is demonstrated to be an effective method to tune the electronic structures of MoS_2 based systems. The edges of zero-dimensional MoS_2 structures are found to be sensitive to the S chemical potential. In our discussions we have focused on various aspects of MoS_2, but due to the structural similarity most of the derived conclusions are also valid for other semiconducting TMDCs.

References

1. Castro Neto, A.H.: Charge density wave, superconductivity, and anomalous metallic behavior in 2D transition metal dichalcogenides. Phys. Rev. Lett. **86**, 4382–4385 (2001)
2. Zhu, Z.Y., Cheng, Y.C., Schwingenschlögl, U.: Origin of the charge density wave in 1T-$TiSe_2$. Phys. Rev. B **85**, 245133 (2012)
3. Mattheiss, L.: Energy bands for 2H-$NbSe_2$ and 2H-MoS_2. Phys. Rev. Lett. **30**, 784–787 (1973)
4. Coehoorn, R., Haas, C., de Groot, R.: Electronic structure of $MoSe_2$, MoS_2, and WSe_2. II. The nature of the optical band gaps. Phys. Rev. B **35**, 6203–6206 (1987)
5. Splendiani, A., Sun, L., Zhang, Y., Li, T., Kim, J., Chim, C.Y., Galli, G., Wang, F.: Emerging photoluminescence in monolayer MoS_2. Nano Lett. **10**, 1271–1275 (2010)
6. Mak, K., Lee, C., Hone, J., Shan, J., Heinz, T.: Atomically thin MoS_2: a new direct-gap semiconductor. Phys. Rev. Lett. **105**, 136805 (2010)
7. Radisavljevic, B., Radenovic, A., Brivio, J., Giacometti, V., Kis, A.: Single-layer MoS_2 transistors. Nat. Nanotechnol. **6**, 147–150 (2011)
8. Yin, Z.Y., Li, H., Li, H., Jiang, L., Shi, Y.M., Sun, Y.H., Lu, G., Zhang, Q., Chen, X.D., Zhang, H.: Single-layer MoS_2 phototransistors. ACS Nano **6**, 74–80 (2012)
9. Choi, W., Cho, M.Y., Konar, A., Lee, J.H., Cha, G.B., Hong, S.C., Kim, S., Kim, J., Jena, D., Joo, J., Kim, S.: High-detectivity multilayer MoS_2 phototransistors with spectral response from ultraviolet to infrared. Adv. Mater. **24**, 5832–5836 (2012)
10. Xiao, D., Liu, G.-B., Feng, W., Xu, X., Yao, W.: Coupled spin and valley physics in monolayers of MoS_2 and other group-VI dichalcogenides. Phys. Rev. Lett. **108,** 196802 (2012)
11. Zeng, H.L., Dai, J.F., Yao, W., Xiao, D., Cui, X.D.: Valley polarization in MoS_2 monolayers by optical pumping. Nat. Nanotechnol. **7**, 490–493 (2012)
12. Cao, T., Wang, G., Han, W.P., Ye, H.Q., Zhu, C.R., Shi, J.R., Niu, Q., Tan, P.H., Wang, E., Liu, B.L., Feng, J.: Valley-selective circular dichroism of monolayer molybdenum disulphide. Nat. Comm. **3**, 887 (2012)

13. Zhu, Z.Y., Cheng, Y.C., Schwingenschlögl, U.: Giant spin-orbit-induced spin splitting in two-dimensional transition-metal dichalcogenide semiconductors. Phys. Rev. B **84**, 153402 (2011)
14. Hohenberg, P., Kohn, W.: Inhomogeneous Electron Gas. Phys. Rev. B. **136**, B864–B871 (1964)
15. Kohn, W., Sham, L.J.: Self-consistent equations including exchange and correlation effects. Phys. Rev. **140**, 1133–1138 (1965)
16. Vanderbilt, D.: Soft self-consistent pseudopotentials in a generalized eigenvalue formalism. Phys. Rev. B. **41**, 7892–7895 (1990)
17. Giannozzi, P., Baroni, S., Bonini, N., Calandra, M., Car, R., Cavazzoni, C., Ceresoli, D., Chiarotti, G.L., Cococcioni, M., Dabo, I., Dal Corso, A., de Gironcoli, S., Fabris, S., Fratesi, G., Gebauer, R., Gerstmann, U., Gougoussis, C., Kokalj, A., Lazzeri, M., Martin-Samos, L., Marzari, N., Mauri, F., Mazzarello, R., Paolini, S., Pasquarello, A., Paulatto, L., Sbraccia, C., Scandolo, S., Sclauzero, G., Seitsonen, A.P., Smogunov, A., Umari, P., Wentzcovitch, R.M.: Quantum-ESPRESSO: a modular and open-source software project for quantum simulations of materials. J. Phys. Condens. Mat. **21**, 395502 (2009)
18. Perdew, J.P., Burke, K., Ernzerhof, M.: Generalized gradient approximation made simple. Phys. Rev. Lett. **77**, 3865–3868 (1996)
19. Baroni, S., de Gironcoli, S., Dal Corso, A., Giannozzi, P.: Phonons and related crystal properties from density-functional perturbation theory, Rev. Mod. Phys. **73**, 515–562 (2001)
20. Cheng, Y.C., Zhu, Z.Y., Tahir, M., Schwingenschlögl, U.: Spin-orbit induced spin splitting in polar transition metal dichalcogenide monolayers. Europhys. Lett. **102**, 57001 (2013)
21. Bell, R.E., Herfert, R.E.: Preparation and characterization of a new crystalline form of molybdenum disulfide. J. Am. Chem. Soc. **79**, 3351–3354 (1957)
22. Cheng, Y.C., Yao, K.X., Yang, Y., Li, L., Yao, Y.B., Zhang, X.X., Han, Y., Schwingenschlögl, U.: Van der Waals epitaxial growth of MoS_2 on SiO_2/Si by Chemical Vapor Deposition. RSC Adv **3**, 17287–17293 (2013)
23. Eda, G., Fujita, T., Yamaguchi, H., Voiry, D., Chen, M.W., Chhowalla, M.: Coherent Atomic and Electronic Heterostructures of Single-Layer MoS_2. ACS Nano **6**, 7311–7317 (2012)
24. Enyashin, A.N., Yadgarov, L., Houben, L., Popov, I., Weidenbach, M., Tenne, R., Bar-Sadan, M., Seifert, G.: New route for stabilization of 1T-WS_2 and MoS_2 phases. J. Phys. Chem. C **115**, 24586–24591 (2011)
25. Ma, Y.D., Dai, Y., Guo, M., Niu, C.W., Huang, B.B.: Graphene adhesion on MoS_2 monolayer: An ab initio study. Nanoscale **3**, 3883–3887 (2011)
26. Kosmider, K., Fernandez-Rossier, J.: Electronic properties of the MoS_2-WS_2 heterojunction. Phys. Rev. B **87**, 075451 (2013)
27. Mak, K.F., He, K.L., Shan, J., Heinz, T.F.: Control of valley polarization in monolayer MoS_2 by optical helicity. Nat. Nanotechnol. **7**, 494–498 (2012)
28. Bychkov, Y.A., Rashba, E.I.: Properties of a 2d Electron-Gas with lifted spectral degeneracy. JETP Lett. **39**, 78–81 (1984)
29. Henk, J., Ernst, A., Bruno, P.: Spin polarization of the L-gap surface states on Au (111). Phys. Rev. B **68**, 165416 (2003)
30. He, J.G., Wu, K.C., Sa, R.J., Li, Q.H., Wei, Y.Q.: Magnetic properties of nonmetal atoms absorbed MoS_2 monolayers. Appl. Phys. Lett. **96**, 082504 (2010)
31. Komsa, H.-P., Kotakoski, J., Kurasch, S., Lehtinen, O., Kaiser, U., Krasheninnikov, A.: Two-dimensional transition metal dichalcogenides under electron irradiation: defect production and doping. Phys. Rev. Lett. **109**, 035503 (2012)
32. Huang, B., Lee, H.: Defect and impurity properties of hexagonal boron nitride: a first-principles calculation. Phys. Rev. B **86**, 245406 (2012)
33. Kudrnovský, J., Turek, I., Drchal, V., Máca, F., Weinberger, P., Bruno, P.: Exchange interactions in III-V and group-IV diluted magnetic semiconductors. Phys. Rev. B **69**, 115208 (2004)
34. Molina-Sanchez, A., Wirtz, L.: Phonons in single-layer and few-layer MoS_2 and WS_2. Phys. Rev. B **84**, 155413 (2011)

35. Cheng, Y.C., Zhu, Z.Y., Schwingenschlögl, U.: Doped silicene: evidence of a wide stability range. EPL **95,** 17005 (2011)
36. Zeng, H., Zhu, B., Liu, K., Fan, J., Cui, X., Zhang, Q.M.: Low-frequency Raman modes and electronic excitations in atomically thin MoS_2 films. Phys. Rev. B **86**, 241301 (2012)
37. Cheng, Y.C., Zhu, Z.Y., Schwingenschlögl, U.: Role of interlayer coupling in ultra thin MoS_2. RSC Adv. **2**, 7798–7802 (2012)
38. Li, T.S.: Ideal strength and phonon instability in single-layer MoS_2. Phys. Rev. B **85**, 235407 (2012)
39. Yue, Q., Kang, J., Shao, Z.Z., Zhang, X.A., Chang, S.L., Wang, G., Qin, S.Q., Li, J.B.: Mechanical and electronic properties of monolayer MoS_2 under elastic strain. Phys. Lett. A **376**, 1166–1170 (2012)
40. Cooper, R.C., Lee, C., Marianetti, C.A., Wei, X., Hone, J., Kysar, J.W.: Nonlinear elastic behavior of two-dimensional molybdenum disulfide. Phys. Rev. B **87**, 035423 (2013)
41. Lee, Y.H., Zhang, X.Q., Zhang, W.J., Chang, M.T., Lin, C.T., Chang, K.D., Yu, Y.C., Wang, J.T.W., Chang, C.S., Li, L.J., Lin, T.W.: Synthesis of large-Area MoS_2 atomic layers with chemical vapor deposition. Adv. Mater. **24**, 2320–2325 (2012)
42. Bertolazzi, S., Brivio, J., Kis, A.: Stretching and breaking of ultrathin MoS_2. ACS Nano **5**, 9703–9709 (2011)
43. Marianetti, C.A., Yevick, H.G.: Failure mechanisms of graphene under tension. Phys. Rev. Lett. **105**, 245502 (2010)
44. Rice, C., Young, R.J., Zan, R., Bangert, U., Wolverson, D., Georgiou, T., Jalil, R., Novoselov, K.S.: Raman-scattering measurements and first-principles calculations of strain-induced phonon shifts in monolayer MoS_2. Phys. Rev. B **87**, 081307 (2013)
45. Lee, C., Yan, H., Brus, L.E., Heinz, T.F., Hone, J., Ryu, S.: Anomalous lattice vibrations of single- and 5-layer MoS_2. ACS Nano **4**, 2695–2700 (2010)
46. Bhattacharyya, S., Singh, A.K.: Semiconductor-metal transition in semiconducting bi-layer sheets of transition-metal dichalcogenides. Phys. Rev. B **86**, 075454 (2012)
47. Scalise, E., Houssa, M., Pourtois, G., Afanas'ev, V.V., Stesmans, A.: Strain-induced semiconductor to metal transition in the two-dimensional honeycomb structure of MoS_2. Nano Res. **5**, 43–48 (2012)
48. Feng, J., Qian, X., Huang, C.-W., Li, J.: Strain-engineered artificial atom as a broad-spectrum solar energy funnel. Nat. Photonics **6**, 866–872 (2012)
49. Pan, H., Zhang, Y.W.: Tuning the electronic and magnetic properties of MoS_2 nanoribbons by Strain Engineering. J. Phys. Chem. C **116**, 11752–11757 (2012)
50. Shidpour, R., Manteghian, M.: A density functional study of strong local magnetism creation on MoS_2 nanoribbon by sulfur vacancy. Nanoscale **2**, 1429–1435 (2010)
51. Pan, H., Zhang, Y.W.: Edge-dependent structural, electronic and magnetic properties of MoS_2 nanoribbons. J. Mater. Chem. **22**, 7280–7290 (2012)
52. Dolui, K., Das Pemmaraju, C., Sanvito, S.: electric field effects on armchair MoS_2 nanoribbons. Acs Nano **6**, 4823–4834 (2012)
53. Yue, Q., Chang, S.L., Kang, J., Zhang, X.A., Shao, Z.Z., Qin, S.Q., Li, J.B.: Bandgap tuning in armchair MoS_2 nanoribbon. J. Phys. Condens. Mat. **24**, 335501 (2012)
54. Li, Y.F., Wu, D.H., Zhou, Z., Cabrera, C.R., Chen, Z.F.: Enhanced li adsorption and diffusion on MoS_2 zigzag nanoribbons by edge effects: a computational study. J. Phys. Chem. Lett. **3**, 2221–2227 (2012)
55. Yang, S.Q., Li, D.X., Zhang, T.R., Tao, Z.L., Chen, J.: First-principles study of zigzag MoS_2 nanoribbon as a promising cathode material for rechargeable Mg batteries. J. Phys. Chem. C **116**, 1307–1312 (2012)
56. Tenne, R., Margulis, L., Genut, M., Hodes, G.: Polyhedral and cylindrical structures of tungsten disulfide. Nature **360**, 444–446 (1992)
57. Seifert, G., Terrones, H., Terrones, M., Jungnickel, G., Frauenheim, T.: Structure and electronic properties of MoS_2 nanotubes. Phys. Rev. Lett. **85**, 146–149 (2000)

58. Lu, P., Wu, X.J., Guo, W.L., Zeng, X.C.: Strain-dependent electronic and magnetic properties of MoS_2 monolayer, bi-layer, nanoribbons and nanotubes. Phys. Chem. Chem. Phys. **14**, 13035–13040 (2012)
59. Helveg, S., Lauritsen, J.V., Laegsgaard, E., Stensgaard, I., Norskov, J.K., Clausen, B.S., Topsoe, H., Besenbacher, F.: Atomic-scale structure of single-layer MoS_2 nanoclusters. Phys. Rev. Lett. **84**, 951–954 (2000)
60. Karunadasa, H.I., Montalvo, E., Sun, Y., Majda, M., Long, J.R., Chang, C.J.: A molecular MoS_2 edge site mimic for catalytic hydrogen generation. Science **335**, 698–702 (2012)
61. Schweiger, H., Raybaud, P., Kresse, G., Toulhoat, H.: Shape and edge sites modifications of MoS_2 catalytic nanoparticles induced by working conditions: a theoretical study. J. Catal. **207**, 76–87 (2002)
62. Wieting, T.J., Verble, J.L.: Infrared and Raman studies of long-wavelength optical phonons in hexagonal MoS_2. Phys. Rev. B **3**, 4286–4292 (1971)

Chapter 6
Mechanical Properties and Electric Field Screening of Atomically Thin MoS_2 Crystals

Jorge Quereda, Gabino Rubio-Bollinger, Nicolás Agraït and Andres Castellanos-Gomez

Two-dimensional atomically thin crystals have recently emerged as a broad and an interesting field of research due to their technological applicability, due to their exceptional electrical and mechanical properties, and to the challenge they represent for theoretical condensed matter physics. Although graphene has been the most widely studied prototype of 2D material [1, 2], in the last few years other two-dimensional crystals (MoS_2 among them [3–13]), have been investigated in search of different properties, which could lead to a broadening of their technological applicability. For instance, single layer MoS_2 crystals have been used to elaborate field-effect transistors, with current on/off ratios exceeding 10^8 [4], as well as other devices and sensors [14–18], due to their large intrinsic band gap (1.8 eV) [19, 20], carrier mobility (up to 500 $cm^2V^{-1}s^{-1}$) [21, 22] and Seebeck coefficient ($-4{\cdot}10^2$ and $-1{\cdot}10^5$ $\mu V\,K^{-1}$) [17]. Additionally, MoS_2 thin crystals have been also proposed as a promising material for the development of valleytronic devices [5, 7, 13].

In this chapter, we will show how the micromechanical cleavage technique, originally developed for graphene, can be used to easily obtain thin MoS_2 thin flakes (even down to a single layer) deposited on SiO_2 substrates. Also, in Sects. 6.2, 6.3 and 6.4, some of the most technologically relevant properties of such crystals will be explored (i.e., optical properties [23, 24], elasticity [11], and electric field screening [25]).

J. Quereda · G. Rubio-Bollinger · N. Agraït
Departamento de Física de la Materia Condensada, Universidad Autónoma de Madrid, Campus de Cantoblanco, E-28049 Madrid, Spain
e-mail: gabino.rubio@uam.es

N. Agraït
Instituto Madrileño de Estudios Avanzados en Nanociencia IMDEA-Nanociencia, E-28049 Madrid, Spain

A. Castellanos-Gomez (✉)
Kavli Institute of Nanoscience, Delft University of Technology, Lorentzweg 1, 2628 CJ Delft, The Netherlands
e-mail: a.castellanosgomez@tudelft.nl

Z. M. Wang (ed.), *MoS_2*, Lecture Notes in Nanoscale Science and Technology 21,
DOI: 10.1007/978-3-319-02850-7_6,

6.1 Micromechanical Exfoliation

MoS_2 crystals, as well as other transition metal dichalcogenide crystals, are composed of stacks of X–M–X layers, being X sulfur or selenium and M a transition metal (molybdenum in this case). While atoms within a layer are strongly bound, the interaction between different layers is due to van der Waals forces, making these materials prone to mechanical cleavage.

The micromechanical cleavage technique is an easy and a widely used method for obtaining graphene monolayers [2, 26] and can be applied without important modifications to other easy-to-peel materials such as mica [27–29] and some transition metal dichalcogenide crystals [2, 3, 30–32]. For instance, it can be used to easily obtain atomically thin MoS_2 crystals. The technique consists in repeatedly exfoliating the layered bulk crystal using adhesive tape, and subsequently transferring the extracted material on top of a substrate (usually a silicon wafer with a SiO_2 capping layer of controlled thickness).

In the original method, Scotch tape was used to extract layers from the bulk crystal by exfoliation. As a result, highly crystalline transferred flakes were not only obtained, but also residues of the adhesive substance were deposited on the substrate. In order to avoid this contamination, viscoelastic poly-dimethilsyloxane (PDMS) stamps can be used to perform the exfoliation [33] (see Fig. 6.1). Viscoelastic materials behave like a solid substance for fast deformation rates but tend to flow like a very viscous liquid for slow deformation rates. Therefore, placing a viscoelastic stamp on a bulk MoS_2 crystal and waiting enough time (minutes), one allows the stamp material to microscopically flow, accommodating to the shape of

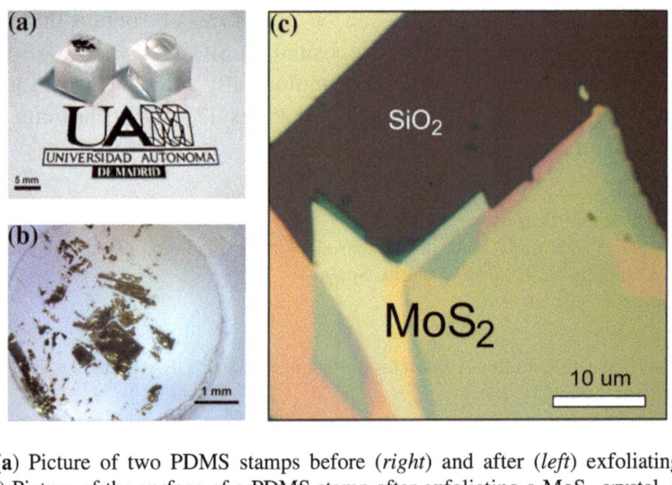

Fig. 6.1 (**a**) Picture of two PDMS stamps before (*right*) and after (*left*) exfoliating a $NbSe_2$ crystal. (**b**) Picture of the surface of a PDMS stamp after exfoliating a MoS_2 crystal. (**c**) Optical micrograph of a MoS_2 flake of different thicknesses deposited on a Si–SiO_2 substrate by micromechanical cleavage using a PDMS stamp. Part of this figure has been adapted from Ref. [23] with permission of the American Institute of Physics

the crystal. If now one separates the stamp quickly (seconds), it will grab and take away some crystals from the bulk. Lately, the stamp can be placed on a substrate and then be slowly separated, allowing the viscoelastic stamp material to flow and cleanly drop the crystals on the substrate. Resulting samples, as the one shown in Fig. 6.1 contain atomically thin flakes similar in shape and number to those obtained using adhesive tape, but without traces of remaining adhesives.

Figure 6.2 shows optical micrographs and contact mode AFM topography images of a $NbSe_2$ and a MoS_2 flake deposited on a 90 nm SiO_2–Si substrate [23]. Contact mode AFM, instead of dynamic mode, has been used to avoid possible artifacts in the topography image [34]. As shown in the histograms in Fig. 6.2, both the $NbSe_2$ and the MoS_2 flakes contain single-layer and bilayer regions. The thickness of each region can be obtained by comparison of the relative position of the peaks in the histogram. In the case of MoS_2, we obtain thicknesses of $d = 0.54 \pm 0.18$ nm for the monolayer and $d = 1.1 \pm 0.2$ nm for the bilayer, which is consistent with the bulk interlayer distance (0.6 nm).

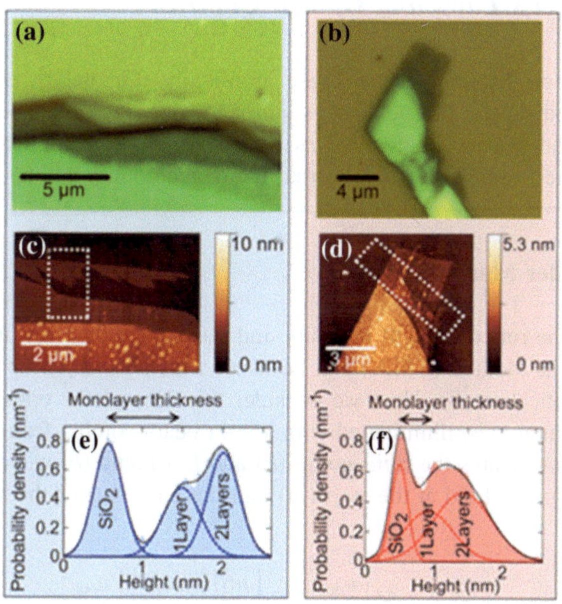

Fig. 6.2 (*Color online*) Optical micrographs under *white* light illumination of $NbSe_2$ (**a**) and MoS_2 (**b**) ultrathin crystals on a 90 nm SiO_2/Si surface. (**c**) and (**d**) are contact mode AFM topography images of the same flakes shown in (**a**) and (**b**), respectively. Regions containing a monolayer and a bilayer are marked by rectangles in (**c**) and (**d**). A height histogram of these zones is shown in (**e**) and (**f**). A multigaussian fit (*solid line*) has been performed to determine the thicknesses of the different regions. Reprinted from Ref. [23] with permission of the American Institute of Physics

6.2 Optical Identification of Atomically Thin MoS_2 Crystals

As shown in Fig. 6.1, the method of micromechanical cleavage explained in the previous section yields the deposition of flakes of a wide variety of thicknesses, making difficult to identify the thinner ones. Although scanning probe microscopies enable very accurate thickness measurements, their low imaging speed and the low density of thin flakes on the substrate make unviable the application of those techniques to discriminate thin flakes from thicker ones. Scanning electron microscopy is characterized by large scanning area and a high operation speed, but nonetheless it is also an inadequate technique for this purpose, as the exposition to the electron beam tends to increase the contamination of the sample [35].

In this section, we illustrate how optical microscopy can be used in combination with atomic force microscopy (AFM) to determine in a fast, nondestructive way the number of layers of MoS_2 atomically thin flakes, as reported in literature for graphene [36], mica [27], and other transition metal dichalcogenides [23, 24].

6.2.1 Fresnel's Laws and Optical Contrast

To study the optical contrast of these atomically thin flakes, one can use an approach based on the Fresnel's laws similar to the one developed by P. Blake et al. [37]. The system is modeled as a stack of different optical media under monochromatic illumination in a normal incidence configuration.

6.2.1.1 Transfer Matrix Method

To calculate the reflected light intensity and the optical contrast of the flakes the Transfer Matrix Method (TMM) can be used, which yields compact expressions for stacked optical media. Here we consider the slabs of the optical media stack along a horizontal axis, illuminated with a light beam coming from the left. There is matrix M that relates the right going (a) and left going (b) wave amplitudes of the electromagnetic field at the start (in) and the end (out) of the stack of optical media:

$$\begin{bmatrix} a_{\text{in}} \\ b_{\text{in}} \end{bmatrix} = M \begin{bmatrix} a_{\text{out}} \\ b_{\text{out}} \end{bmatrix} = \begin{bmatrix} m_{11} & m_{12} \\ m_{21} & m_{22} \end{bmatrix} \begin{bmatrix} a_{\text{out}} \\ b_{\text{out}} \end{bmatrix} \tag{6.1}$$

6.2.1.2 Scattering Matrix

The relation between the amplitudes of the right going (a) and the left going (b) wave amplitudes at the two sides of an interface between media i and j can be written in matrix notation as

$$\begin{bmatrix} a_i \\ b_i \end{bmatrix} = S_{ij} \times \begin{bmatrix} a_i \\ b_i \end{bmatrix} = \begin{bmatrix} s_{11} & s_{12} \\ s_{21} & s_{22} \end{bmatrix} \begin{bmatrix} a_i \\ b_i \end{bmatrix}, \tag{6.2}$$

where S_{ij} is the scattering matrix, whose elements can be obtained from the Fresnel's relations for the refractive indexes of the two media, n_i and n_j. According to Fresenel's laws, the transmission (t_{ij}) and reflection (r_{ij}) coefficients under normal incidence are:

$$t_{ij} = \frac{2n_i}{n_i + n_j}, \tag{6.3}$$

$$r_{ij} = \frac{n_i - n_j}{n_i + n_j} = -r_{ji}. \tag{6.4}$$

The outgoing wave amplitudes a_j and b_i can be obtained in terms of the incoming wave amplitudes a_i and b_j:

$$a_j = t_{ij}a_i + r_{ji}b_j, \tag{6.5}$$

$$b_i = r_{ij}a_i + t_{ji}b_j. \tag{6.6}$$

Expressing both equations with a_i and b_i as independent variables and using the property $r_{ij} = -r_{ji}$ we obtain

$$a_i = \frac{1}{t_{ij}}a_j + \frac{r_{ij}}{t_{ij}}b_j, \tag{6.7}$$

$$b_i = \frac{r_{ij}}{t_{ij}}a_j + \left(\frac{r_{ij}^2}{t_{ij}} + t_{ji}\right)b_j. \tag{6.8}$$

The coefficient of b_j in Eq. 6.8 can be simplified as follows:

$$\begin{aligned} \frac{r_{ij}^2}{t_{ij}} + t_{ji} &= \frac{(n_i - n_j)^2}{(n_i + n_j)^2}\frac{(n_i + n_j)}{2n_i} + \frac{2n_j}{(n_i + n_j)} = \frac{(n_i - n_j)^2 + 4n_in_j}{2n_i(n_i + n_j)} \\ &= \frac{n_i^2 + n_j^2 - 2n_in_j + 4n_in_j}{2n_i(n_i + n_j)} = \frac{(n_i + n_j)^2}{2n_i(n_i + n_j)} = \frac{(n_i + n_j)}{2n_i} = \frac{1}{t_{ij}}. \end{aligned} \tag{6.9}$$

Therefore, we have

$$a_i = \frac{1}{t_{ij}} a_j + \frac{r_{ij}}{t_{ij}} b_j, \tag{6.10}$$

$$b_i = \frac{r_{ij}}{t_{ij}} a_j + \frac{1}{t_{ij}} b_j. \tag{6.11}$$

Finally, comparing with Eq. 6.2 we obtain

$$S_{ij} = \frac{1}{t_{ij}} \begin{pmatrix} 1 & r_{ij} \\ r_{ij} & 1 \end{pmatrix}. \tag{6.12}$$

6.2.1.3 Propagation Matrix

In order to take into account the effect of the propagation of the wave through medium *i*, we need to establish a matrix relation between the right-going and left-going wave amplitudes at the start ($a_{i,1}$, $b_{i,1}$) and the end ($a_{i,2}$, $b_{i,2}$) of a layer:

$$a_{i,2} = e^{+i\Phi_i} a_{i,1}, \quad b_{i,1} = e^{+i\Phi_i} b_{i,2}, \tag{6.13}$$

where we have defined

$$\Phi_i = 2\pi d_i n_i / \lambda_0, \tag{6.14}$$

being d_i the thickness of the layer, n_i its complex refractive index and λ_0 the illumination wavelength $\lambda_0 = 2\pi c/\omega$, where c is the speed of light in vacuum and ω is the wave frequency.

Therefore, we can define the propagation matrix P_j as

$$\begin{bmatrix} a_1 \\ b_1 \end{bmatrix} = P_j \times \begin{bmatrix} a_2 \\ b_2 \end{bmatrix} = \begin{bmatrix} e^{-i\Phi_i} & 0 \\ 0 & e^{+i\Phi_i} \end{bmatrix} \begin{bmatrix} a_2 \\ b_2 \end{bmatrix} \tag{6.15}$$

6.2.1.4 Transfer Matrix for the Stack

Finally, for the whole stack to be taken into account we need to obtain the transfer matrix M defined in Eq. 6.1. For a stack of N layers, the right going (a) and left going (b) wave amplitudes of the incoming and outgoing waves will be related by

$$\begin{bmatrix} a_{\text{in}} \\ b_{\text{in}} \end{bmatrix} = S_{01} P_1 S_{12} P_2 \cdots S_{N-1,N} P_N S_{N,N+1} \begin{bmatrix} a_{\text{out}} \\ b_{\text{out}} \end{bmatrix}. \tag{6.16}$$

Therefore, the transfer matrix is simply the ordered product of the successive propagation and scattering matrices along the different media and interfaces of the stack.

For the case of MoS_2 flakes on SiO_2–Si wafers there are only four media: air (0), MoS_2 (1), SiO_2 (2), and Si (3), and air and Si are considered semi-infinite:

$$M_{\text{flake}} = S_{01}P_1S_{12}P_2S_{23}. \tag{6.17}$$

In the regions where there's no flake deposited the transfer matrix is given by

$$M_{\text{substrate}} = S_{02}P_2S_{23}. \tag{6.18}$$

It must be noted that the transfer matrices depend only on the illumination wavelength and the thickness and refractive index of the different media in the stack.

6.2.1.5 Optical Contrast Calculation

In order to simplify the optical identification process it is interesting to measure quantity that does not depend on the illumination intensity I_0. The optical contrast C, also named Michelson contrast in the literature, between two objects (the bare substrate and the substrate with a MoS_2 flake on top) is defined as

$$C = \frac{I_{\text{flake}} - I_{\text{substrate}}}{I_{\text{flake}} - I_{\text{substrate}}}, \tag{6.19}$$

where I_{flake} and $I_{\text{substrate}}$ are the reflected intensities from the flake and the substrate. This magnitude is easily measurable and does not dependent on I_0. We can use the TMM to calculate C. The reflected intensity coefficient R for a given media stack is

$$R = \left|\frac{b_{\text{in}}}{a_{\text{in}}}\right|^2; \quad I_{\text{reflected}} = R \cdot I_0, \tag{6.20}$$

being b_{in} the amplitude of the reflected (right going) wave and a_{in} the amplitude of the incident (left going) wave.

If the medium $j = N + 1$ is semi-infinite then $b_{\text{out}} = 0$. Using Eq. 6.1 with $M = M_{\text{flake}}$we find

$$\frac{1}{R} = \left|\frac{a_{\text{in}}}{b_{\text{in}}}\right|^2 = \left|\frac{m_{11}a_{\text{out}}}{m_{21}b_{\text{out}}}\right|^2, \tag{6.21}$$

which gives finally

$$R_{\text{flake}} = \frac{I_{\text{flake}}}{I_0} = \left|\frac{m_{21}}{m_{11}}\right|^2. \tag{6.22}$$

For the regions without a flake deposited, the expression 6.22 is also valid with $M = M_{\text{substrate}}$. The optical contrast, Eq. 6.19, can be now written as

$$C = \frac{I_{\text{flake}} - I_{\text{substrate}}}{I_{\text{flake}} + I_{\text{substrate}}} = \frac{I_0(R_{\text{flake}} - R_{\text{substrate}})}{I_0(R_{\text{flake}} + R_{\text{substrate}})} = \frac{(R_{\text{flake}} - R_{\text{substrate}})}{(R_{\text{flake}} + R_{\text{substrate}})}, \tag{6.23}$$

which does not depend on I_0. An expression for the contrast as a function of the thicknesses and refractive indices of the different media can be obtained substituting R_{flake} and $R_{\text{substrate}}$ with the results derived in Eq. 6.22:

$$R_{flake} = \left| \frac{r_{01}e^{i(\Phi_1+\Phi_2)} + r_{12}e^{-i(\Phi_1-\Phi_2)} + r_{23}e^{-i(\Phi_1+\Phi_2)} + r_{01}r_{12}r_{23}e^{i(\Phi_1-\Phi_2)}}{e^{i(\Phi_1+\Phi_2)} + r_{01}r_{12}e^{-i(\Phi_1-\Phi_2)} + r_{01}r_{23}e^{-i(\Phi_1+\Phi_2)} + r_{12}r_{23}e^{i(\Phi_1-\Phi_2)}} \right|^2, \tag{6.24}$$

$$R_{\text{substrate}} = \left| \frac{r_{02} + r_{23}e^{-2i\Phi_2}}{1 + r_{02}r_{23}e^{-2i\Phi_2}} \right|^2. \tag{6.25}$$

6.2.2 *Optical Contrast of MoS_2 Crystals*

In order to illustrate how optical identification can be used to discriminate flakes of different thicknesses the optical contrast of MoS_2 flakes deposited on two different SiO_2/Si substrates with 90 nm and 290 nm SiO_2 thicknesses, measured as a function of their thickness under different illumination wavelengths, is presented in this section. The quantitative study of the optical contrast of atomically thin MoS_2 was first reported in Ref. [23] in 2010 and subsequent studies [38, 39] have reproduced these results in 2011 and 2012.

The thickness of the studied flakes ranges from a single layer to 200 layers. The samples have been observed under different illumination wavelengths in the visible spectrum using different narrow-band filters [40]. Figure 6.3 shows the measured contrast versus thickness values for MoS_2 flakes transferred onto a 90 nm SiO_2 surface at four different illumination wavelengths, as well as the calculated curve determined using Eq. 6.23. Notice that for a given flake thickness the contrast strongly depends on the illumination wavelength and thicker flakes can yield even lower contrast than thinner ones in some cases. As can be seen in the figure, there is a remarkable good agreement between the experimental results and the Fresnel-based model for both $NbSe_2$ and MoS_2 flakes.

The refractive index for both $NbSe_2$ and MoS_2 obtained at different illumination wavelengths is summarized in Fig. 6.4a and b, respectively. By using these values in combination with Eq. 6.23, it is possible to convert contrast maps, like the ones in Fig. 6.4e–h, to thickness maps with uncertainties lower than 0.3 nm in thickness. Note that in some cases the function thickness versus contrast is multivaluated (see Fig. 6.3c and d) and contrast maps acquired at two or three different incident wavelengths could be necessary to unambiguously determine the thickness of some flake regions [41].

Once the refractive index of MoS_2 is known, the SiO_2 layer thickness that simultaneously maximizes the contrast of a single layer and two layer thick flake can be calculated. By doing so, we find that the optimal SiO_2 substrate thicknesses for the identification of MoS_2 monolayers are 55 and 220 nm, both yielding a contrast of up to -60 %, obtained under an illumination wavelength of 500 nm (Fig. 6.4).

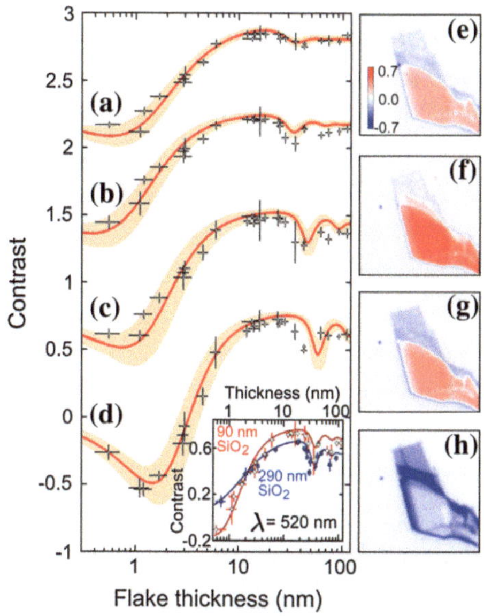

Fig. 6.3 (*Color online*) Measured optical contrast (*symbols*) of MoS_2 flakes onto a 90 nm SiO_2 substrate as a function of their thickness under four different illumination wavelengths λ: 450 nm (**a**), 500 nm (**b**), 546 nm (**c**), and 600 nm (**d**). *Solid red lines* and *shaded area* are the contrast and its uncertainty obtained by fitting Eq. (6.23). For clarity **a**–**c** have been vertically displaced by 2.25, 1.5, and 0.75 units, respectively. The inset below **d** shows the fit to Eq. (6.23) for flakes deposited on two different SiO_2 substrate thicknesses using the same refractive index for both cases. Contrast maps **e**–**h** have been obtained for the same MoS_2 flake shown in Fig. 6.2b at the same four illumination wavelengths. The *color bar* shared by the four maps is shown in the inset in (**e**). Reprinted from Ref. [23] with permission of the American Institute of Physics

6.3 Electric Field Screening in Atomically Thin Layers of MoS_2

The semiconducting properties of MoS_2 nanolayers have motivated the interest in the development of single-layer MoS_2, heterostructures, and combined graphene-MoS_2 devices, [4, 22, 42–45]. Electric field screening is one of the most relevant properties for the engineering of such MoS_2-based devices as well as for understanding the role of their layered structure in their electrical response [46]. Moreover, the role of the interlayer coupling in the screening (negligible for other layered materials such as few-layer graphene) has been considered.

The study of the electrostatic screening in MoS_2 crystals can be carried out by the combination of electrostatic force microscopy and a nonlinear Thomas–Fermi model. A continuum model of decoupled layers, such as the one applied satisfactorily to the electrostatic screening in graphene and graphite, cannot be extended to explain the experimental observations for MoS_2 nanolayers. Instead, a three-dimensional model with an interlayer hopping parameter has to be employed

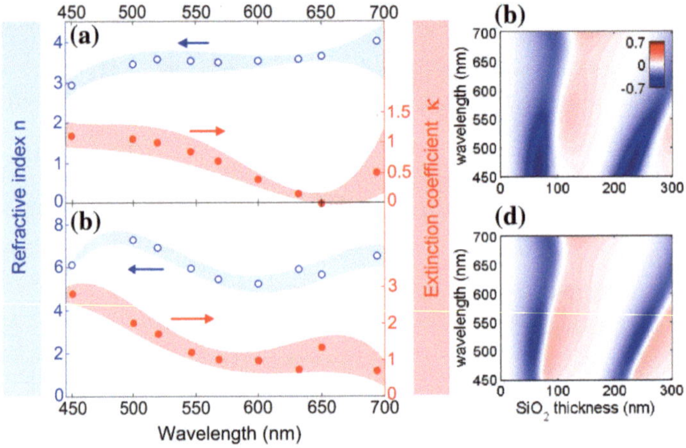

Fig. 6.4 Refractive index (*blue open circles, left axis*) and extinction coefficient (*red filled circles, right axis*) for ultrathin $NbSe_2$ (**a**) and MoS_2 (**b**) thin crystals as a function of the incident illumination wavelength. Reprinted from Ref. [23] with permission of the American Institute of Physics. (**c**) Calculated optical contrast color map for a single layer of MoS_2. (**d**) Same calculation for the contrast between a single layer and a bilayer of MoS_2. Different colors correspond to different values of the optical contrast according to the color scale. Optimal substrate thicknesses for the identification of a single MoS_2 layer are 55 and 220 nm

in order to successfully account for the observed electric field screening, making clear that, in contrast with what happens in graphene, interlayer coupling plays an important role in the screening of MoS_2 films.

6.3.1 Electrostatic Force Microscopy

In order to quantify the electric field screening effect due to the presence of a MoS_2 thin layer one can use electrostatic force microscopy (EFM) to measure the electric field caused by charge impurities present at the MoS_2–substrate interface [25, 49, 50]. The EFM measurements are typically conducted as follows: the AFM tip is placed 20 nm over the surface of the flake (previously characterized by optical microscopy, contact mode AFM and Raman spectroscopy, as shown in Fig. 6.5, to determine the number of layers) and a voltage ramp is applied between the tip and the Si substrate while measuring the oscillation amplitude of the cantilever, which changes depending on the tip-sample electrostatic force [51].

The relation between the applied voltage, V_{tip}, and the electrostatic force, F, is given by [52]

$$F = \frac{1}{2}\frac{\delta C}{\delta z}(V_{tip} - V_s)^2 \tag{6.26}$$

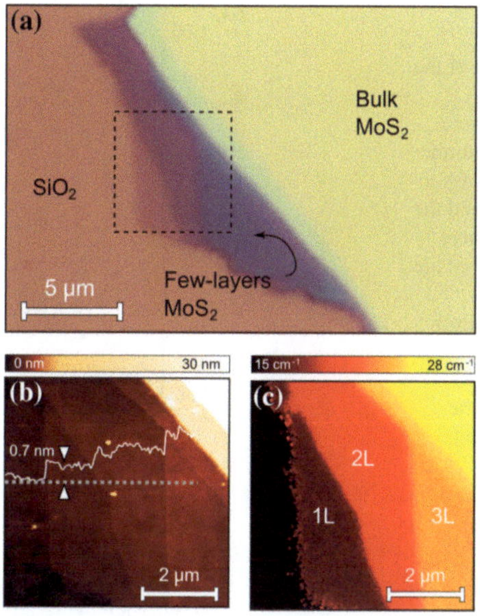

Fig. 6.5 (**a**) Optical micrograph of a multilayered MoS_2 flake deposited on a Si/SiO_2 substrate. As discussed in Sect. 6.2, the regions with different apparent color under *white light* illumination correspond to different thicknesses of the MoS_2 flake and, therefore, to distinct number of atomic layers. Figure 6.5b shows a contact mode atomic force microscopy (AFM) image of the region marked by a *dashed square* in Fig. 6.5a. The thickness of the thinnest region of the flake is 0.7 nm, which is compatible with the expected thickness of a MoS_2 single layer. Figure 6.5c shows a spatially resolved map of the frequency difference between the E^1_{2g} and A_{1g} Raman modes in the same region. Such difference increases monotonically with the number of MoS_2 layers [47, 48] and can be used as an alternative method to confirm the thickness measurements in the different regions of the flake. Reprinted from Ref. [25] with permission of Wiley

where C is the tip-sample capacitance, z the tip-sample distance, and V_s the electrostatic potential at the surface of the flake.

The oscillation amplitude shows a parabolic dependence on the applied voltage, with the vertex positioned at $V_{tip} = V_s$, as shown in Fig. 6.6a. For bulk crystals, the electrostatic field is completely screened by the MoS_2 flake and V_s is simply the tip-sample contact potential difference V_{CPD}, caused by the matching of their chemical potentials, which depends on the work-function difference between the tip and the sample. However, for atomically thin MoS_2 samples the electric field induced by the presence of charged impurities at the interface between SiO_2 and MoS_2 can only be partially screened. This yields a shift in the position of the parabola vertex whose value is dependent on the sample thickness. By repeating the experiment at different values of z between 20 nm and 40 nm one can check that the measured value of V_s does not depend on the tip-sample distance in this range. It has also been observed that the parabola vertex position neither depends

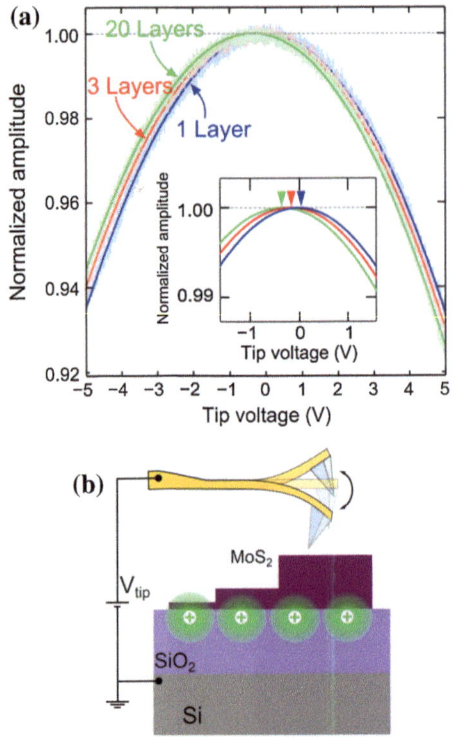

Fig. 6.6 **a** Normalized cantilever oscillation amplitude as a function of the applied tip-sample bias voltage, measured in three regions with different number of layers. The inset shows a zoom in the region around the parabola maximum, where only the fits to the *curves* are shown to facilitate the identification of the *apexes* (indicated by triangles). **b** Schematic of the AFM setup used to perform the measurements. Reprinted from Ref. [25] with permission of Wiley

on the free-oscillation amplitude of the cantilever within the range employed in the experiment (from 2 to 5 nm).

The dependence of the surface potential V_s on the MoS_2 thickness is shown in Fig. 6.7. The difference between the surface potential V_s and its asymptotic limit the value for a thick flake (>30 nm). As expected, the electric field screening grows for increasingly thick MoS_2 flakes, and the surface potential V_s asymptotically approaches its bulk value [53, 54]. For small thicknesses, instead, the sign and magnitude of the deviation from the bulk surface potential $\Delta V_s = V_s - V_s^{bulk}$, due to the partial penetration of the electric field through the MoS_2 layer, is related to the nature and density of the charge carriers present in the flake. In particular, from the positive sign of ΔV_s (meaning that $V_s > V_s^{bulk}$ for thin flakes) for thin flakes one infers the presence of *n*-type extrinsic carriers (electrons), compatible with the presence of positively charged impurities in the SiO_2 substrate. Such *n*-type behavior is commonly observed in MoS_2-based field-effect transistors [4, 14, 55–58] implemented on SiO_2 substrates.

Remarkably, the weak dependence of ΔV_s on the flake thickness indicates a screening as poor as that observed in graphene [54], which is surprising taking into account that the origin of such poor screening in the case of graphene is related to its linear density of states $N(\varepsilon) \sim \varepsilon$, while in MoS_2 the conduction band can be described by a conventional parabolic dispersion and $N(\varepsilon) \sim \varepsilon^{1/2}$. This low

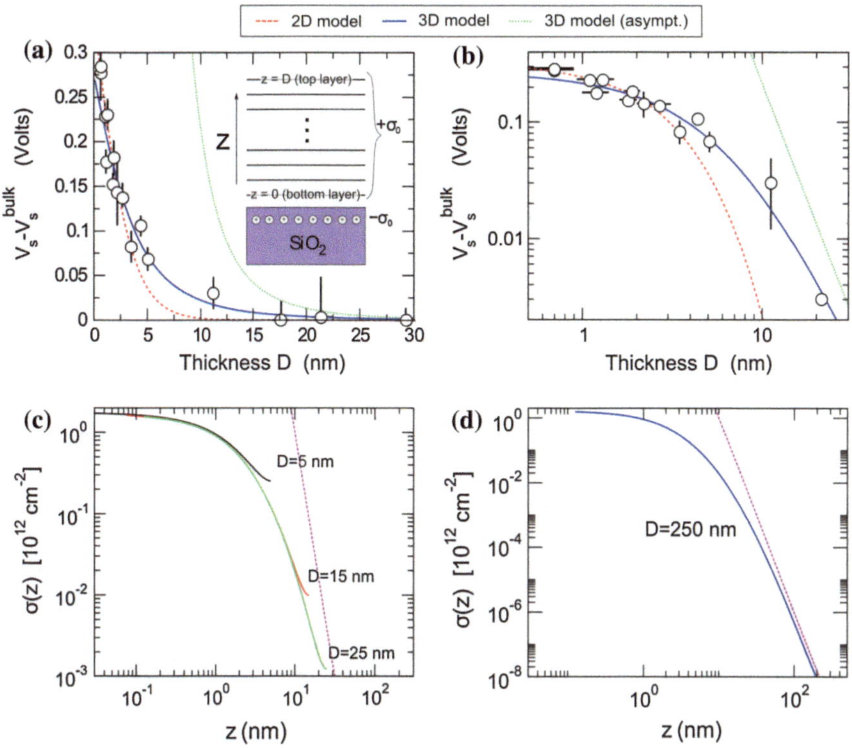

Fig. 6.7 a Dependence of the deviation of V_s from its bulk value as a function of the flake thickness. Theoretical results are shown for a nonlinear Thomas–Fermi 2D model of uncoupled MoS_2 layers (*dashed red line*), for a 3D model with z-axis dispersion (*solid blue line*), and for the asymptotic behavior $\Delta V(D) \sim 1/D^4$of this 3D model. A sketch of the microscopic model is superimposed to the graph. There, *vertical lines* represent MoS_2 layers and the variable $0 < z < D$ represents the vertical position of a given layer in the flake. **b** Same representation with the axes in logarithmic scale to facilitate the comparison between the experimental data and the predictions of the different theoretical models. **c** Calculated charge distribution $\sigma(z)$ according to the 3D model for flakes with different thicknesses D. The *dashed line* represents the strong coupling asymptotic behavior $\sigma(z) \approx z^{-6}$. **d** The strong coupling regime is reached at around 100 nm. Reprinted from Ref. [25] with permission of Wiley

screening can be understood considering a weak interlayer hopping parameter $t_\perp$ in the theoretical model described below.

6.3.2 Thomas–Fermi Model

In order to quantitatively evaluate the results, F. Guinea, E. Cappelluti, and R. Roldán, have developed a nonlinear Thomas–Fermi model [25] similar to that used by Datta et al. to study the electric field screening of graphene [54] and previously used to study graphite [59, 60]. One assumes that the charge transfer takes place between the MoS_2 flakes and the SiO_2 substrate, leaving a charge density $-\sigma_0 e$ (being σ_0 the numeric density of charge carriers) in the substrate surface and a total charge density $+\sigma_0 e$ in the MoS_2 flake. In this situation, the chemical potential μ must be equated in every layer and the free energy of the MoS_2–SiO_2 system can be written as $\Omega = U + E - \mu N_c$, being U and E the total potential and kinetic energy of the charge carriers in the sample and N_c the total number of carriers. We have for this situation (see inset in Fig. 6.10a)

$$\begin{aligned} U &= U_{\text{layer-substrate}} + U_{\text{layer-layer}} \\ &= \left(\sum_{i=1}^{N_L} -\frac{z_i e^2 \sigma_0}{2\varepsilon_0 k_\perp} \sigma_i \right) + \left(\frac{1}{2} \sum_{i=1}^{N_L} \sum_{j=1}^{N_L} -\frac{|z_i - z_j| e^2}{2\varepsilon_0 k_\perp} \sigma_i \sigma_j \right), \end{aligned} \tag{6.27}$$

where $k_\perp$ is the dielectric constant along the Z-axis ($k_\perp = 7.4$ according to the literature [61]), z_i the vertical position of the ith layer, and σ_i the density of charge carriers in the ith layer. The first term in the right side of the equation gives account of the interaction between the N_L layers of MoS_2 and the charged SiO_2 substrate, while the second term accounts for the interaction between the different MoS_2 layers. Therefore, the free energy Ω is

$$\Omega = \left(\sum_{i=1}^{N_L} -\frac{z_i e^2 \sigma_0}{2\varepsilon_0 k_\perp} \sigma_i \right) + \left(\frac{1}{2} \sum_{i=1}^{N_L} \sum_{j=1}^{N_L} -\frac{|z_i - z_j| e^2}{2\varepsilon_0 k_\perp} \sigma_i \sigma_j \right) + \left(\sum_{i=1}^{N_L} E(\sigma_i) - \mu \sigma_i \right), \tag{6.28}$$

or, in the continuum limit [54],

$$\Omega = \int_0^D \frac{dz}{d} \left\{ E[\sigma(z)] - \mu\sigma(z) - \frac{z e^2 \sigma_0}{2\varepsilon_0 k_\perp} \sigma(z) \right\} - \frac{e^2}{4\varepsilon_0 k_\perp} \int_0^D \frac{dz}{d} \int_0^D \frac{dz'}{d} \sigma(z) |z - z'| \sigma(z'), \tag{6.29}$$

where $E[\sigma(z)]$ is the functional of the kinetic energy, D the total thickness of the crystal, and d the interlayer distance, which for MoS_2 takes the value of $d = 0.64$ Å.

Uncoupled MoS_2 layers ($\boldsymbol{t}_\perp = 0$)

If one assumes that the interlayer hopping $t_\perp$ is negligible, the situation can be described by a parabolic conduction band located at the K point and therefore one gets

$$E[\sigma(z)] = A_0 \sigma^2(z); \quad A_0 = \frac{\pi h^2}{N_s N_v m_{||}}, \tag{6.30}$$

where $N_s = 2$ is the spin degeneracy, N_v the valley degeneracy ($N_v = 2$ in this case [19, 62]) and $m_{||}$ the in-plane effective mass of the carriers. By minimizing the free energy of the system, Eq. (6.29), with respect to $\sigma(z)$ one obtains the differential equation

$$\left[\frac{d\sigma(z)}{dz}\right]^2 - \frac{2\beta_0}{d}\sigma^2(z) = \frac{2\beta_0}{d}\sigma^2(D), \tag{6.31}$$

with the boundary conditions

$$\left.\frac{d\sigma(z)}{dz}\right|_{z=0} = 2\beta_0\sigma_0, \quad \left.\frac{d\sigma(z)}{dz}\right|_{z=D} = 0. \tag{6.32}$$

Here we have defined $\beta_0 = e^2/(4\varepsilon_0 k_\perp A_0)$. One can rewrite Eq. (6.31) in the form

$$\int_{r_D}^{1} \frac{du}{(u^2 - r_D^2)^{\frac{1}{2}}} = \sqrt{\frac{2\beta_0}{d}} D, \tag{6.33}$$

where $r_D = \sigma(D)/\sigma(0)$. Therefore, the potential difference $\Delta V(D)$ between the two ends of a sample of thickness D is

$$\Delta V(D) = 2\mathrm{A}_0\sigma_0\sqrt{2\beta_0 d}\frac{1 - r_D}{\left(1 - r_D^2\right)^{1/2}}. \tag{6.34}$$

It is easy to see that in the $\Delta V(D)$ approaches exponentially to its asymptotic bulk value $\lim_{D\to\infty} \Delta V(D) = \left(2\pi\hbar^2\sigma_0/N_s N_v m||\right)\sqrt{2\beta_0 d}$. An analytical expression for the charge density $\sigma(z)$ can also be found for a given thickness D:

$$\sigma(z) = -\sigma_0\sqrt{2\beta_0 d}\frac{\cosh\left(\sqrt{\frac{2\beta_0}{d}}(z - D)\right)}{\sinh\left(\sqrt{\frac{2\beta_0}{d}}D\right)}. \tag{6.35}$$

Coupled MoS_2 layers ($\boldsymbol{t}_\perp \neq 0$)

The two most significant effects of considering an interlayer coupling are a change in the value of the valley degeneracy, which becomes $N_v = 6$ [19, 63–65] and a modification of the functional form of the kinetic energy, that now can be written as:

$$E[\sigma(z)] = A_\perp \sigma^{5/3}(z); \quad A_\perp = \left(\frac{6\pi^2 h^3}{N_s N_v d m_{||}\sqrt{m_\perp}}\right)^{2/3}. \tag{6.36}$$

In this case both the out-of-plane effective mass $m_\perp = \hbar^2 k_\perp^2/(2|t_\perp|d^2)$ and dielectric constant $k_\perp$ become relevant for the problem. Again, by minimizing the free energy one gets

$$\left(\frac{df(z)}{dz}\right)^2 - \frac{2\beta_\perp}{d} f^{\frac{5}{2}}(z) = -\frac{2\beta_\perp}{d} f^{\frac{5}{2}}(D), \tag{6.37}$$

where $f(z) = \sigma^{2/3}(z)$ and $\beta_\perp = 6e^2/(25A_\perp \varepsilon_0 k_\perp)$. The boundary conditions are in this case

$$\left.\frac{\mathrm{d}f(z)}{\mathrm{d}z}\right|_{z=0} = \frac{5}{2}\beta_\perp \sigma_0, \quad \left.\frac{\mathrm{d}f(z)}{\mathrm{d}z}\right|_{z=D} = 0. \tag{6.38}$$

If we define $r_D = f(D)/f(0)$ the integral form of (6.37) with the conditions (6.38) can be written as

$$\left[\frac{25\beta_\perp d\sigma_0^2}{8\left(1 - r_D^{5/2}\right)^{1/2}}\right]^{-1/10} \int_{r_D}^{1} \frac{du}{\left(u^{5/2} - r_D^{5/2}\right)^{1/2}} = \sqrt{\frac{2\beta_\perp}{d}} z, \tag{6.39}$$

and the potential difference $\Delta V(D)$ between the two ends of the sample is

$$\Delta V(D) = \frac{5A_\perp}{3}\left(\frac{25\beta_\perp d\sigma_0^2}{8}\right)^{2/5} \frac{1 - r_D}{\left(1 - r_D^{5/2}\right)^{2/5}} \tag{6.40}$$

In the asymptotic limit $r_D \ll 1$, one can see that $\Delta V(D) \propto D^{-4}$.

In the coupled layers case, it is not possible to obtain an explicit solution for $\sigma(z)$, but an implicit solution can be found as

$$\begin{aligned} \frac{1}{f^{5/2}(D)}\Bigg\{ & f(0)\sqrt{f^{\frac{5}{2}}(0) - f^{\frac{5}{2}}(D)}\,{}_2F_1\left[1, \frac{9}{10}, \frac{7}{5}, \frac{f^{\frac{5}{2}}(0)}{f^{\frac{5}{2}}(D)}\right] \\ & - f(z)\sqrt{f^{\frac{5}{2}}(z) - f^{\frac{5}{2}}(D)}\,{}_2F_1\left[1, \frac{9}{10}, \frac{7}{5}, \frac{f^{\frac{5}{2}}(z)}{f^{\frac{5}{2}}(D)}\right]\Bigg\} = \sqrt{\frac{2\beta_\perp}{d}} z, \end{aligned} \tag{6.41}$$

where ${}_2F_1$ is the Hypergeometric function. In the strong coupling regime $f(0) \gg f(z) \gg f(D)$ the Eq. (6.41) can be expanded to obtain $f(z) \propto z^{-4}$ or, equivalently, $\sigma(z) \propto z^{-6}$.

6.3.3 Electric Field Screening

The predictions of the Thomas–Fermi model are shown in Fig. 6.7 both for the uncoupled 2D model (dashed line) and for the coupled 3D model (solid line). The calculations were done taking $k_\perp = 7.4$ [61] and $m_{||} = 0.01\, m_e$ [66–68]. In order to optimize the agreement with the experiment, the substrate charge density value is $\sigma_0 = 8 \times 10^{12}\,\mathrm{cm}^{-2}$ for the uncoupled model and at a slightly smaller charge density value, $\sigma_0 = 5 \times 10^{12}\ \mathrm{cm}^{-2}$, in the coupled case. Note that the σ_0 value taken in the 3D case is very similar to that estimated for graphene, suggesting that it is an intrinsic property (charged impurities) of the substrate. It can be seen that for a sample with a small thickness the experimental results are satisfactorily reproduced by both models, but as the thickness of the flakes increases, the 2D model screening decreases too fast due to its exponential behavior. However, the experimental results can be explained by means of the coupled 3D model, Eq. (6.40), with a small interlayer hopping. We estimate $t_\perp \approx -0.2\,\mathrm{eV}$ from the splitting of the minimum of the conduction band as a function of the number or layers [20], and we get hence $m_\perp \approx 0.5 m_e$.

The main source of the reduced screening in MoS_2 is the shift toward higher thickness of the transition between the weak and strong coupling regimes. For instance, this crossover takes place for the 2D model at a thickness of 1 nm, whereas for the 3D model it occurs at a thickness above 30 nm. Therefore, the experimentally studied MoS_2 flakes are essentially in the weak coupling regime. The behavior of the charge density between the weak and strong coupling regime is shown in Fig. 6.7d where the strong coupling regime is recovered for very thick samples ($D > 100$ nm). This is quite different from the graphene case, where the strong coupling regime is reached already for $z > 2$ nm.

6.4 Mechanical Properties of Freely Suspended MoS_2 Nanosheets

In this section, the elastic properties of MoS_2 nanocrystals are addressed in order to determine the applicability of such semiconducting crystals in next-generation flexible electronic devices. The measurement of the elastic properties of freely suspended MoS_2 nanosheets with thicknesses ranging from 5 to 25 layers can be done by means of a bending test experiment performed using the tip of an atomic force microscope AFM. These measurements allow for a simultaneous determination of the Young's modulus (E) of the flakes and their pre-tension (T), which depends on the deposition process of the flakes.

6.4.1 Elasticity AFM Measurements

Figure 6.8 shows a contact mode AFM topography image of a 5–7 layers thick MoS_2 flake on a prepatterned SiO_2–Si substrate with holes. It is interesting to notice that the substrate pattern can be observed in the image even in the region

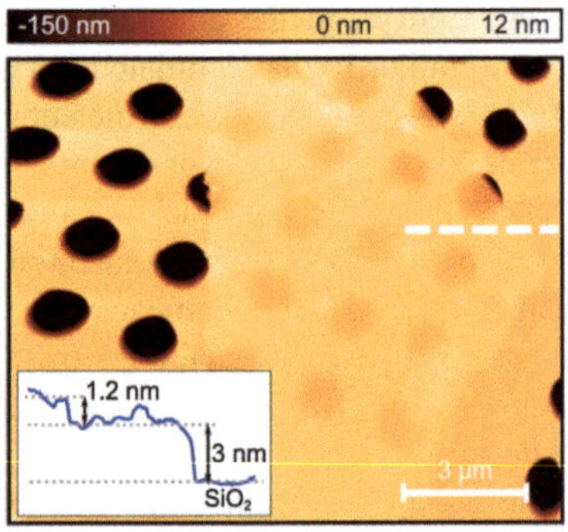

Fig. 6.8 (**a**) Contact mode AFM topography image of a 3–4.2 nm (5–7 layers) thick MoS_2 flake deposited on a 285 nm SiO_2/Si substrate prepatterned with an array of holes with a diameter of 1.1 μm. Different colors correspond to different heights according to the *color bar* on the top of the figure. The topography line profile acquired along the *dashed line* is shown superimposed to the image. Reprinted from Ref. [11] with permission of Wiley

covered by the MoS_2 flakes, indicating a deformation of the flake due to the absence of a solid substrate underneath. Performing a quantitative analysis of such deformation, the elastic properties of the nanosheet can be determined.

By approaching the tip of the AFM to the freely suspended region of the sample one can apply a load at its center while measuring its elastic deformation δ (see Fig. 6.9a). The relation between the deformation of the flake, the deflection of the AFM cantilever Δz_c, and the vertical displacement of the scanning piezotube of the AFM Δz_{piezo} is given by:

$$\delta = \Delta z_{\text{piezo}} - \Delta z_c \tag{6.42}$$

The force F exerted by the cantilever is given by $F = k_c \Delta z_c$ where $k_c = (0.88 \pm 0.20)$ N/m is the spring constant of the cantilever [69].

Figure 6.9b shows the relation between the applied force and the deformation of the MoS_2 membrane for different thicknesses. Clearly, the elastic behavior of the flakes depends on their thickness: the thickest flakes show linear force versus deformation dependence $F(\delta)$ while the thinnest ones show strongly nonlinear traces. Such behavior can be explained by a competition between membrane (bending-dominated) and plate (stretching-dominated) deformation behavior.

Considering a simplified elasticity continuum model, at the center of the MoS_2 freely suspended region, the deformation δ is related with the force by [70, 71]

$$F = F_{\text{plate}} + F_{\text{membrane}} + F_{\text{stiffening}} = \left[\frac{4\pi E}{3(1-\nu^2)} \cdot \left(\frac{t^3}{R^2}\right)\right]\delta + (\pi T)\delta + \left(\frac{q^3 E t}{R^2}\right)\delta^3 \tag{6.43}$$

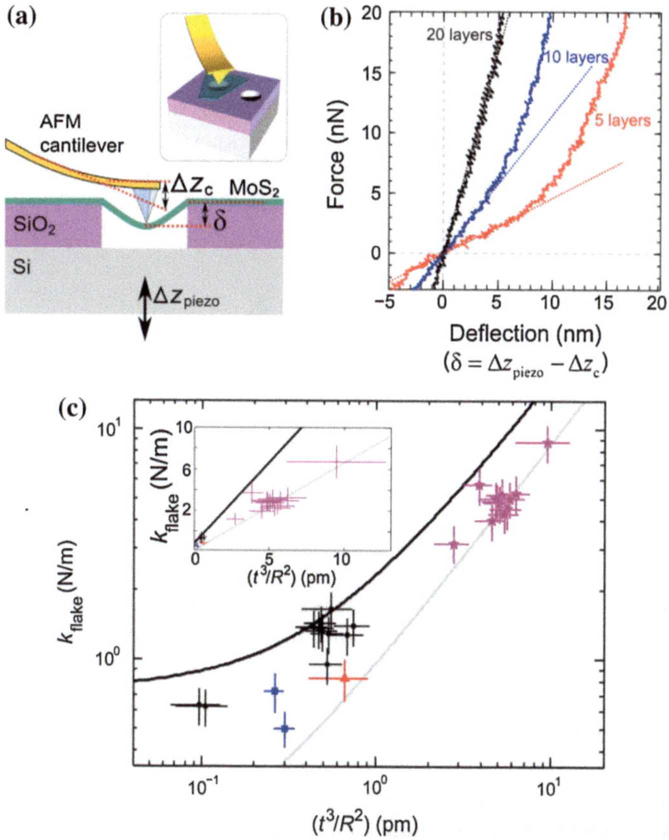

Fig. 6.9 **a** Schematic diagram of the nanoscopic bending test experiment carried out on a freely suspended MoS_2 nanosheet. **b** Force versus deflection traces measured at the center of the suspended part of MoS_2 nanosheets with 5, 10, and 20 layers in thickness. The slope of the traces around zero deflection is marked by a *dotted line*. **c** Elastic constant versus t^3R^{-2} measured for 26 MoS_2 suspended nanosheets with thickness ranging from 25 down to 5 layers. Data points sharing color and symbol correspond to suspended nanosheets of the same MoS_2 flake. The relationship k_{flake} versus t^3R^{-2} calculated with expression (6.44) for $E = 0.21$ TPa and $T = 0.03$ N/m (*gray solid line*) and $E = 0.37$ TPa and $T = 0.23$ N/m (*black solid line*). The inset in (**c**) shows the same graph on a linear scale. Reprinted from Ref. [11] with permission of Wiley

where E is the Young's modulus, ν the Poisson's ratio (0.125 according to the literature [72]), t is the thickness of the nanolayer, R the radius of the suspended region, T its pre-tension and $q = 1/(1.05 - 0.15\nu - 0.16\nu^2)$ is a dimensionless parameter. The first term in the right side, F_{plate}, corresponds to the mechanical response of a plate with a certain bending rigidity [70, 71], while the second term, $F_{membrane}$, corresponds to the behavior of a stretched membrane [73]. The third term accounts for the stiffening of the layer, also called displacement-induced tension, which is responsible of the nonlinear behavior of F for the thinnest flakes.

As the term F_{plate} has a cubic dependence on the thickness t, it dominates for the thickest flakes while it becomes negligible for the thinnest ones compared to $F_{membrane}$ and $F_{stiffening}$. This explains the observed nonlinear regime for small thicknesses. Note that a possible effect of interlayer shearing stresses can been neglected due to the low ratio between plate's thickness and radius ($t/R < 0.02$ for the studied flakes).

At the limit of low deformation, the stiffening term in Eq. 6.43 becomes negligible and the force F becomes linear with the deformation, $F = -k_{flake}\delta$ with a spring constant

$$k_{flake} = \frac{4\pi E}{3(1-\nu^2)} \cdot \left(\frac{t^3}{R^2}\right) + \pi T, \tag{6.44}$$

which scales as t^3/R^2.

6.4.2 Young's Modulus and Pre-tension

By representing the estimated k_{flake} versus t^3 the Young's modulus can be determined through the slope of such relationship while the interception at the origin would provide an estimate for the pre-tension (see Fig. 6.9c).

Using Eq. 6.44, one can evaluate the pre-tension T and Young modulus E values. We find that all our measurements have pre-tension T between 0.03 and 0.23 $N \cdot m^{-1}$ and Young's modulus E values between 0.21 and 0.37 TPa. In the case of thin flakes with an arbitrary deformation, we obtain a nonlinear $F(\delta)$ relationship. Nevertheless, the values of E and T can still be determined by fitting Eq. 6.43, as shown in Fig. 6.10a. For an eight-layer thick MoS_2 nanosheet (the one in the figure), we obtain $E = 0.35 \pm 0.03$ TPa and $T = 0.05 \pm 0.02$ N/m. Proceeding as previously reported in the literature [74], one can also obtain both the Young's modulus and the initial pre-tension of thicker flakes with linear $F(\delta)$. Figure 6.10b shows the compliance $1/(k_{flake})$ of the 8 layers thick MoS_2 flake, studied in Fig. 6.10a, as a function of the radius. The best fit is obtained in this case for $E = 0.40 \pm 0.03$ TPa and $T = 0.02 \pm 0.02$ N/m, which are compatible with the results obtained for nonlinear force versus deformation traces.

It is possible to determine E and T for each MoS_2 flake separately by fitting the experimental force versus deformation traces to Eq. 6.43. The dispersion observed in both properties is attributed to differences both in the density of defects of the flakes and the adhesion forces with the SiO_2 substrate. To quantitatively characterize this dispersion, a histogram of the pre-tension and Young's modulus values has been calculated, as shown in Fig. 6.10b and c. A total of 13 MoS_2 membranes have been studied, with thicknesses ranging between 5 and 10 layers.

The mean Young's modulus and pre-tension obtained for the MoS_2 flakes are $E = 0.33 \pm 0.07$ TPa and $T = 0.13 \pm 0.10$ N/m. Interestingly, this Young's modulus is only about one-third lower than that obtained for exfoliated graphene

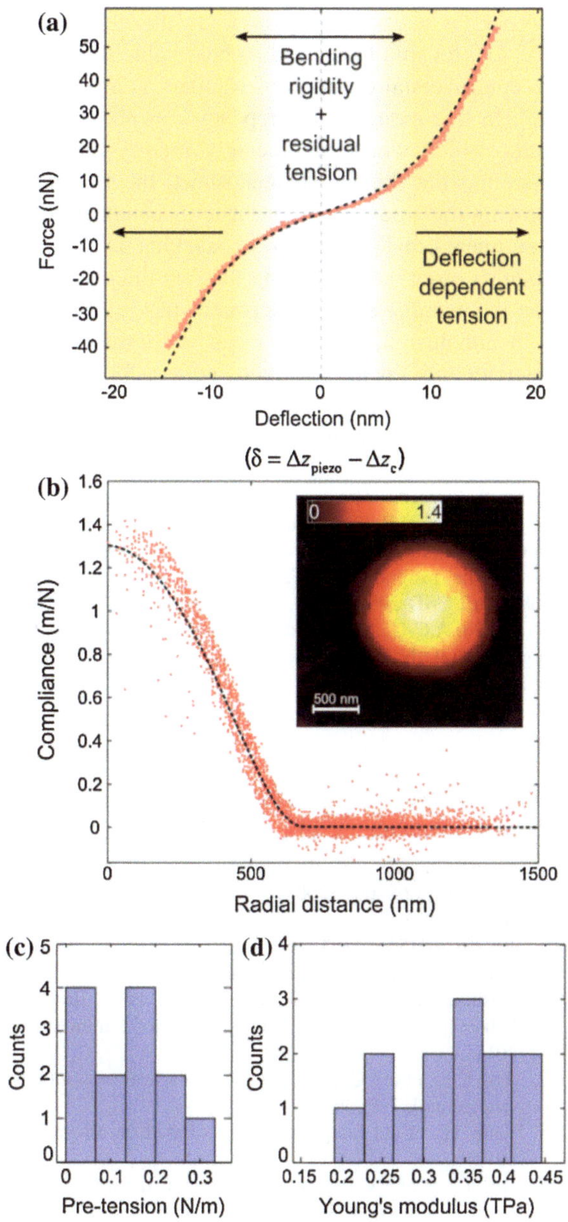

Fig. 6.10 **a** Force *versus* deflection traces obtained on a flake eight layers thick suspended over a hole 1.1 μm in diameter. The *dotted black trace* is the fit to expression (6.43), employed to obtain the Young modulus $E = 0.35 \pm 0.02$ TPa and the pre-tension $T = 0.05 \pm 0.02$ N/m of this nanolayer. **b** Force-volume measurement showing a colormap of the compliance (inset) and its radially averaged profile of the sheet as (**a**). The *solid line* shows a fit using the model from Ref. [74] by H. van der Zant, M. Poot, and co-workers. **b** Histogram of the initial pre-tension obtained from the fit to expression (6.43) for 13 sheets 5–10 layers thick. **c** Histogram of the Young's modulus obtained from fitting *F(d) curves* to expression (6.43)for the same 13 sheets plotted in panel (**b**). Reprinted from Ref. [11] with permission of Wiley

[73, 75] (which is one of the stiffest known materials), being much higher than the values obtained for other 2D materials such as graphene oxide (0.2 TPa) [76, 77], hexagonal boron nitride (0.25 TPa) [78], muscovite mica (200 GPa) [28] carbon nanosheets (10–50 GPa) [79], or some 2D clays (22 GPa) [80].

There is a discrepancy between the obtained Young's modulus $E = 0.33 \pm 0.07$ TPa for thin flakes and its bulk value $E_{bulk} = 0.33 \pm 0.07$ TPa. This leads to the controversial question of whether Young's modulus is size-dependent or not [81]. In this case, the discrepancy can be attributed to the presence of stacking faults, which would significantly modify the mechanical properties the flakes. Therefore the thinner the nanosheet, the lower the presence of stacking faults, allowing the study of the intrinsic mechanical properties of the material. In these experiments, a low density of stacking faults would explain the high Young's modulus observed, indicating that the thickness of our MoS_2 nanolayers is lower than the average distance between stacking faults.

In conclusion, the low mean pre-tension ($T = 0.13 \pm 0.10$ N/m) and high elasticity and Young's modulus ($E = 0.33 \pm 0.07$ TPa) obtained for these MoS_2 crystals make them an attractive substitute or alternative to graphene in applications requiring flexible semiconductor materials [82–84].

Acknowledgments The authors wish to acknowledge fruitful discussion with Emmanuele Cappelluti, Rafael Roldán, Francisco Guinea, Menno Poot, Gary A. Steele, and Herre S.J. van der Zant. J.Q., N.A., and G.R-B. acknowledge the financial support by MICINN/MINECO (Spain) through the programs MAT2011-25046 and CONSOLIDER-INGENIO-2010 "Nanociencia Molecular" CSD-2007-00010, Comunidad de Madrid through program Nanobiomagnet S2009/MAT-1726. A.C-G. acknowledges the financial support by the European Union (FP7) through the program RODIN and Marie Curie project FP7-People-2011-IEF (program no. 300802, 'STRENGTHNANO').

References

1. Novoselov, K.S., et al.: Electric field effect in atomically thin carbon films. Science **306**(5696), 666–669 (2004)
2. Novoselov, K.S., et al.: Two-dimensional atomic crystals. Proc. Natl. Acad. Sci. USA. **102**(30), 10451–10453 (2005)
3. Ayari, A., et al.: Realization and electrical characterization of ultrathin crystals of layered transition-metal dichalcogenides. J. Appl. Phys. **101**(1), 014507 (2007)
4. Radisavljevic, B., et al.: Single-layer MoS2 transistors. Nat. Nanotechnol. **6**(3), 147–150 (2011)
5. Mak, K.F., et al.: Control of valley polarization in monolayer MoS2 by optical helicity. Nat. Nanotechnol. **7**(8), 494–498 (2012)
6. Wang, H., et al.: Integrated circuits based on bilayer MoS2 transistors. Nano Lett. **2**, 4674 (2012)
7. Cao, T., et al.: Valley-selective circular dichroism of monolayer molybdenum disulphide. Nature Commun. **3**, 887 (2012)
8. Feng, J., et al.: Strain-engineered artificial atom as a broad-spectrum solar energy funnel. Nat. Photonics **6**(12), 866–872 (2012)
9. Kim, S., et al.: High-mobility and low-power thin-film transistors based on multilayer MoS2 crystals. Nature Commun. **3**, 1011 (2012)
10. Mak, K.F., et al.: Tightly bound trions in monolayer MoS2. Nat. Mater. **12**(3), 207–211 (2013)
11. Castellanos-Gomez, A., et al.: Elastic properties of freely suspended MoS2 nanosheets. Adv. Mater. **24**(6), 772–775 (2012)

12. Ross, J.S., et al.: Electrical control of neutral and charged excitons in a monolayer semiconductor. Nature Commun. **4**, 1474 (2013)
13. Wu, S., et al.: Electrical tuning of valley magnetic moment through symmetry control in bilayer MoS2. Nat. Phys. **9**(3), 149–153 (2013)
14. Li, H., et al.: Fabrication of single- and multilayer MoS2 film-based field-effect transistors for sensing NO at room temperature. Small **8**(1), 63–67 (2012)
15. Wang, H., et al.: Integrated circuits based on bilayer MoS2 transistors. Nano Lett. **12**(9), 4674–4680 (2012)
16. Perkins, F.K., et al.: Chemical vapor sensing with monolayer MoS2. Nano Lett. **13**(2), 668–673 (2013)
17. Buscema, M., et al.: Large and tunable photothermoelectric effect in single-layer MoS2. Nano Lett. **13**(2), 358–363 (2013)
18. Sundaram, R.S., et al.: Electroluminescence in single layer MoS2. Nano Lett. **13**(4), 1416–1421 (2013)
19. Mak, K.F., et al.: Atomically thin MoS_{2}: a new direct-gap semiconductor. Phys. Rev. Lett. **105**(13), 136805 (2010)
20. Splendiani, A., et al.: Emerging photoluminescence in monolayer MoS2. Nano Lett. **10**(4), 1271–1275 (2010)
21. Fuhrer, M.S., Hone, J.: Measurement of mobility in dual-gated MoS2 transistors. Nat. Nanotechnol. **8**(3), 146–147 (2013)
22. Bao, W., et al.: High mobility ambipolar MoS 2 field-effect transistors: substrate and dielectric effects. Appl. Phys. Lett. **102**(4), 042104–042104-4 (2013)
23. Castellanos-Gomez, A., Agrait, N., Rubio-Bollinger, G.: Optical identification of atomically thin dichalcogenide crystals. Appl. Phys. Lett. **96**(21), 213116 (2010)
24. Castellanos-Gomez, A., et al.: Fast and reliable identification of atomically thin layers of TaSe2 crystals. Nano Research **6**,1–9 (2013)
25. Castellanos-Gomez, A., et al.: Electric-field screening in atomically thin layers of MoS2: the role of interlayer coupling. Adv. Mater. **25**(6), 899–903 (2013)
26. Bonaccorso, F., et al.: Production and processing of graphene and 2D crystals. Mater. Today **15**(12), 564–589 (2012)
27. Castellanos-Gomez, A., et al.: Atomically thin mica flakes and their application as ultrathin insulating substrates for graphene. Small **7**(17), 2491–2497 (2011)
28. Castellanos-Gomez, A., et al.: Mechanical properties of freely suspended atomically thin dielectric layers of mica. Nano Research **5**(8), 550–557 (2012)
29. Gao, J., et al.: Layer-by-layer removal of insulating few-layer mica flakes for asymmetric ultra-thin nanopore fabrication. Nano Research **5**(2), 99–108 (2012)
30. Frindt, R.F.: Superconductivity in ultrathin NbSe_ 2 layers. Phys. Rev. Lett. **28**(5), 299–301 (1972)
31. Novoselov, K.S., et al.: Two-dimensional atomic crystals. Proc. Natl. Acad. Sci. **102**(30), 10451 (2005)
32. Staley, N.E., et al.: Electric field effect on superconductivity in atomically thin flakes of NbSe[sub 2]. Phy. Rev. B **80**(18), 184505–184506 (2009)
33. Moreno–Moreno, M., et al.: Ultralong natural graphene nanoribbons and their electrical conductivity. Small **5**(8), 924–927 (2009)
34. Nemes-Incze, P., et al.: Anomalies in thickness measurements of graphene and few layer graphite crystals by tapping mode atomic force microscopy. Carbon **46**(11), 1435–1442 (2008)
35. Fan, J., et al.: Investigation of the influence on graphene by using electron-beam and photo-lithography. Solid State Commun. **151**(21), 1574–1578 (2011)
36. Blake, P., et al.: Making graphene visible. Appl. Phys. Lett. **91**(6), 063124–063124-3 (2007)
37. Blake, P., et al.: Making graphene visible. Appl. Phys. Lett. **91**(6), 063124 (2007)
38. Benameur, M., et al.: Visibility of dichalcogenide nanolayers. Nanotechnology **22**(12), 125706 (2011)

39. Li, H., et al.: Optical identification of single-and few-layer MoS2 sheets. Small **8**(5), 682–686 (2012)
40. Narrow band-pass filters (10 nm bandwidth at half height) with central frequencies: 450 nm, 500 nm, 520 nm, 546 nm, 568 nm, 600 nm, 632 nm, 650 nm and 694 nm. From Edmund optics
41. Nolen, C.M., et al.: High-throughput large-area automated identification and quality control of Graphene and few-layer Graphene films. ACS Nano **5**(2), 914–922 (2011)
42. Kośmider, K., Fernández-Rossier, J.: Electronic properties of the MoS_{2}-WS_{2} heterojunction. Phys. Rev. B **87**(7), 075451 (2013)
43. Yu, W.J., et al.: Vertically stacked multi-heterostructures of layered materials for logic transistors and complementary inverters. Nat. Mater. **12**(3), 246–252 (2013)
44. Britnell, L., et al.: Strong light-matter interactions in heterostructures of atomically thin films. Science **340**, 1311 (2013)
45. Sup Choi, M., et al.: Controlled charge trapping by molybdenum disulphide and graphene in ultrathin heterostructured memory devices. Nat. Commun. **4**, 1624 (2013)
46. Das, S., Appenzeller, J.: Screening and interlayer coupling in multilayer MoS2. physica status solidi (RRL)-Rapid Res. Lett., **7**, 268–273 (2013)
47. Lee, C., et al.: Anomalous lattice vibrations of single-and few-layer MoS2. ACS Nano **4**(5), 2695–2700 (2010)
48. Li, H., et al.: From bulk to monolayer MoS2: evolution of Raman scattering. Adv. Funct. Mater. **22**(7), 1385–1390 (2012)
49. Li, Y., Xu, C.-Y., Zhen, L.: Surface potential and interlayer screening effects of few-layer MoS 2 nanoflakes. Appl. Phys. Lett. **102**(14), 143110–143110-4 (2013)
50. Hao, G., et al.: Electrostatic properties of few-layer MoS2 films. AIP Advances **3**(4), 042125–042125-6 (2013)
51. Castellanos-Gomez, A., et al.: Spatially resolved electronic inhomogeneities of graphene due to subsurface charges. Carbon **50**(3), 932–938 (2012)
52. Glatzel, T., Sadewasser, S., Lux-Steiner, M.: Amplitude or frequency modulation-detection in Kelvin probe force microscopy. Appl. Surf. Sci. **210**(1–2), 84–89 (2003)
53. Lee, N.J., et al.: The interlayer screening effect of graphene sheets investigated by Kelvin probe force microscopy. Appl. Phys. Lett. **95**(22), 222107 (2009)
54. Datta, S.S., et al.: Surface potentials and layer charge distributions in few-layer Graphene films. Nano Lett. **9**(1), 7–11 (2009)
55. Ghatak, S., Pal, A.N., Ghosh, A.: The nature of electronic states in atomically thin MoS 2 field effect transistors. ACS Nano **5**(10), 7707–7712 (2011)
56. Yin, Z., et al.: Single-layer MoS2 phototransistors. ACS Nano **6**(1), 74–80 (2012)
57. Castellanos-Gomez, A., et al.: Laser-thinning of MoS2: on demand generation of a single-layer semiconductor. Nano Lett. **12**(6), 3187–3192 (2012)
58. Choi, W., et al.: High-detectivity multilayer MoS2 phototransistors with spectral response from ultraviolet to infrared. Adv. Mater. (2012)
59. Safran, S., Hamann, D.: Self-consistent charge densities, band structures, and staging energies of graphite intercalation compounds. Phys. Rev. B **23**(2), 565 (1981)
60. Pietronero, L., et al.: Charge distribution in c direction in Lamellar Graphite acceptor intercalation compounds. Phys. Rev. Lett. **41**(11), 763–767 (1978)
61. Molina-Sánchez, A., Wirtz, L.: Phonons in single-layer and few-layer MoS_ 2 and WS_ {2}. Phys. Rev. B **84**(15), 155413 (2011)
62. Lebègue, S., Eriksson, O.: Electronic structure of two-dimensional crystals from ab initio theory. Phys. Rev. B **79**(11), 115409 (2009)
63. Bromley, R.A., Murray, R.B., Yoffe, A.D.: The band structures of some transition metal dichalcogenides. III. Group VIA: trigonal prism materials. J. Phys. C: Solid State Phys. **5**(7), 759 (1972)
64. Coehoorn, R., et al.: Electronic structure of MoSe_{2}, MoS_{2}, and WSe_{2}. I. Band-structure calculations and photoelectron spectroscopy. Phys. Rev. B, **35**(12), 6195–6202 (1987)

65. Mattheiss, L.F.: Band structures of transition-metal-Dichalcogenide layer compounds. Phys. Rev. B **8**(8), 3719–3740 (1973)
66. Evans, B., Young, P.: Exciton spectra in thin crystals: the diamagnetic effect. Proc. Phys. Soc. **91**(2), 475 (1967)
67. Baldereschi, A., Diaz, M.: Anisotropy of excitons in semiconductors. Il Nuovo Cimento B (1965–1970), **68**(2), 217–229 (1970)
68. Harper, P.: Anomalous effective mass in MoS2. J. Phys. C: Solid State Phys. **7**(7), 1247 (1974)
69. Sader, J.E., Chon, J.W.M., Mulvaney, P.: Calibration of rectangular atomic force microscope cantilevers. Rev. Sci. Instrum. **70**, 3967 (1999)
70. Timoshenko, S., Woinowsky-Krieger, S., Woinowsky, S.: Theory of plates and shells. Vol. 2, McGraw-Hill, New York (1959)
71. Landau, L., et al., (eds.).: Theory of elasticity. 3rd edn. Theoretical Physics1959, Butterworth-Heinemann. 195 (1959)
72. Lovell, M., Khonsari, M., Marangoni, R.: A finite element analysis of the frictional forces between a cylindrical bearing element and MoS2 coated and uncoated surfaces. Wear **194**(1–2), 60–70 (1996)
73. Lee, C., et al.: Measurement of the elastic properties and intrinsic strength of monolayer graphene. Science **321**(5887), 385 (2008)
74. Poot, M., Van der Zant, H.S.J.: Nanomechanical properties of few-layer graphene membranes. Appl. Phys. Lett. **92**(6), 063111 (2008)
75. Li, P., et al.: Graphene fixed-end beam arrays based on mechanical exfoliation. Appl. Phys. Lett. **98**, 253105 (2011)
76. Gómez-Navarro, C., Burghard, M., Kern, K.: Elastic properties of chemically derived single graphene sheets. Nano Lett. **8**(7), 2045–2049 (2008)
77. Suk, J.W., et al.: Mechanical Properties of Monolayer Graphene Oxide. ACS Nano **4**(11), 6557–6564 (2010)
78. Song, L., et al.: Large scale growth and characterization of atomic Hexagonal Boron Nitride layers. Nano Lett. **10**(8), 3209–3215 (2010)
79. Turchanin, A., et al.: One nanometer thin carbon nanosheets with tunable conductivity and stiffness. Adv. Mater. **21**(12), 1233–1237 (2009)
80. Kunz, D., et al.: Deformation measurements on thin clay Tactoids. Small **5**(16), 1816–1820 (2009)
81. Xiong, Q., et al.: Force-deflection spectroscopy: a new method to determine the Young's Modulus of nanofilaments. Nano Lett. **6**(9), 1904–1909 (2006)
82. Liu, J., et al.: Preparation of MoS2-Polyvinylpyrrolidone nanocomposites for flexible nonvolatile rewritable memory devices with reduced Graphene oxide electrodes. Small **8**(22), 3517–3522 (2012)
83. Pu, J., et al.: Highly flexible MoS2 thin-film transistors with ion gel dielectrics. Nano Lett. **12**(8), 4013–4017 (2012)
84. He, Q., et al.: Fabrication of flexible MoS2 thin-film transistor arrays for practical gas-sensing applications. Small **8**(19), 2994–2999 (2012)

Chapter 7
Insights into Vibrational and Electronic Properties of MoS_2 Using Raman, Photoluminescence, and Transport Studies

Achintya Bera and A. K. Sood

Abstract We review the relevant vibrational and electronic properties of a single and a few layer MoS_2 to understand their resonant and nonresonant Raman scattering results. In particular, the optical modes and low frequency shear and layer breathing modes show significant dependence on the number of MoS_2 layers. Further, the electron doping of the MoS_2 single layer achieved using top-gating in a field effect transistor renormalizes the two optical modes A_{1g} and E^1_{2g} differently due to symmetry-dependent electron–phonon coupling. The issues related to carrier mobility, the Schottky barrier at the MoS_2–metal contact pads and the modifications of the dielectric environment are addressed. The direct optical transitions for single-layer MoS_2 involve two excitons at *K*-point in the Brillouin zone and their stability with temperature and pressure is reviewed. Finally, the Fermi level dependence of spectral shift for a quasiparticle, called trion, is discussed.

7.1 Introduction

Scientific challenges and potential applications of a single and a few layer graphene in the last few years [1–6] have given impetus to look for similar opportunities in other layered materials like BN, MoS_2, $MoSe_2$, and WS_2. Field-effect transistors made of two-dimensional semi-conducting materials are of particular interest for technological applications. In the recent years, the enormously studied single-layer graphene has made its importance in sensors and high-frequency devices [7, 8] because of the high carrier mobility ($\sim 10^5$ for suspended graphene [9]) and the response to the surrounding charges, although in logical devices it is not very useful due to the vanishing electronic bandgap at the *K*-points

A. Bera · A. K. Sood (✉)
Department of Physics, Indian Institute of Science, Bangalore 560 012, India
e-mail: asood@physics.iisc.ernet.in

Z. M. Wang (ed.), *MoS_2*, Lecture Notes in Nanoscale Science and Technology 21, DOI: 10.1007/978-3-319-02850-7_7, of the Brillouin zone (BZ). Recently, a single-layer transition dichalcogenide material Molybdenum disulfide (MoS_2) field-effect transistor was demonstrated at room temperature to have a high on–off ratio $\sim 10^8$ [10], whereas ~ 10 nm-thick MoS_2 device gives a phonon limited mobility value as high as ~ 700 cm^2/Vs [11]. Production of single-layer MoS_2 is not limited to the mechanical exfoliation method pioneered by Geim et al. for graphene; it is produced by the chemical vapor deposition (CVD) [12], high temperature annealing method [13], solution method [14–16], intercalation method [17–19], and many other techniques [20–23]. CVD-based MoS_2 is used with graphene to form a heterojunction which produces photoresponsitivity value $\sim 10^7$ A/W [12]. MoS_2 nanoparticles deposited on reduced graphene oxide act as an advanced catalyst for the hydrogen evolution reaction [24]. The diamagnetic single-layer MoS_2 shows a magnetic ordering at room temperature upon exposure of proton of 2 MeV energy, which is attributed to the formation of various defects and edge states [25, 26].

The photoluminescence (PL) spectra confirms the gradual transition from an indirect bandgap of bulk-MoS_2 to the direct bandgap of single-layer MoS_2 [27]. The presence of the spin-filtered valence band at the K-point valleys of BZ and the time reversal symmetry guarantees the strong couplings of the spin and the valley index of the charge carriers and hence consequent optical selection rules [28]. It has been shown with the circularly polarized light that the observed chirality of the PL emission has almost the same degree of polarization as the incident one at low temperatures [29]. The voltage-dependent transfer of the spectral weight [30] of the exciton complexes ~ 1.9 eV in PL spectra is associated with the observed metal–insulator transition [31] in the two-dimensional MoS_2 at low temperature. The measured ratio of the Coulomb potential energy to the kinetic energy is ≈ 60 at a charge density of $\sim 1 \times 10^{11}$ cm^{-2} for the electron gas [30]. Such high electron–electron interactions set the single-layer MoS_2 as a platform to observe Wigner crystallization in the insulating regime, yet to be experimentally verified. Moreover, the existence of nonzero Berry curvature predicts the Hall effect in presence of circularly polarized light without applying magnetic field [32]. The single-layer MoS_2 shows an electric-field induced superconductivity in the metallic phase with transition temperature of $T_c \sim 9.4$ K [33]. This observed superconductivity is attributed to the electron–electron interactions and the weak electron–phonon interactions [34]. In this review, we discuss various aspects of phonon-assisted phenomena and their probing by Raman spectroscopy as a function of temperature, pressure, carrier doping and the layer numbers, and the dependence of carrier's mobility on various factors. Finally, the PL spectra at different gate voltages are discussed leading to the discovery of negatively charged excitons or trions in single-layer MoS_2.

7.2 Crystal Lattice and Electronic Band Structure of MoS_2: Indirect to Direct Bandgap

Bulk MoS_2 is a layered semiconducting material with each layer having hexagonal symmetry as graphene. Unlike graphene which consists of only carbon atoms, each MoS_2 layer unit is formed by stacking three hexagonal planes of S–Mo–S atoms in order along the *c*-axis [35] as shown in Fig. 7.1. Within the stable unit (S–Mo–S), atoms are bonded in a prismatic configuration through ionic-covalent interactions and each such stable units are connected via van der Waals interactions along the *c*-axis to form the bulk MoS_2 crystal. Two units, forming the unit cell of bulk MoS_2, are stacked vertically in such a way that the Mo atom of one layer sits on top of the S atoms of the other layer as shown in Fig. 7.1. There are two polytypes [36] of MoS_2: (i) 2H–MoS_2 (shown in Fig. 7.1) is more commonly available in nature as molybdenite salt and it has two layers per unit cell. Our discussions are related to this type of structure; and (ii) 3R-MoS_2 has three layers per unit cell which are stacked in rhombohedral symmetry.

Like Si, bulk MoS_2 has an indirect bandgap of $\sim$1.2 eV [27, 37, 38] with valence band maximum at Γ point and conduction band minimum in between the Γ–K direction as shown in Fig. 7.2a. Now if we cleave the bulk MoS_2 one by one

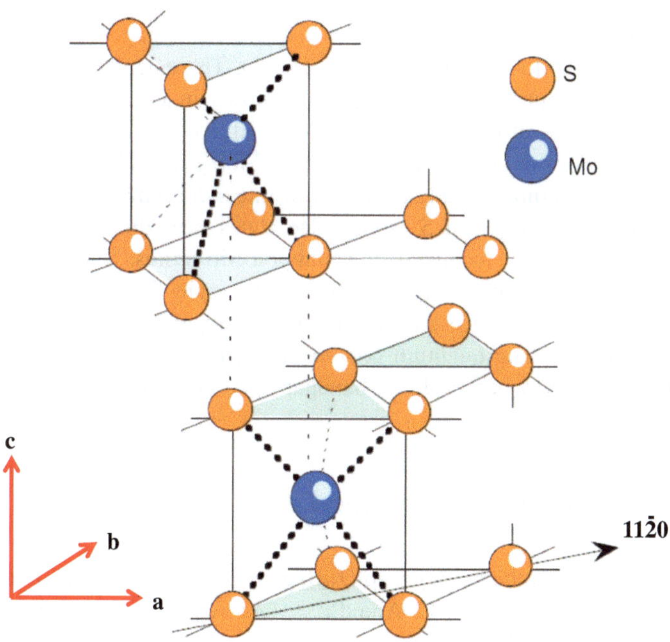

Fig. 7.1 The trigonal prismatic configurations of Mo atoms are shown for the structure of 2H-MoS_2 type. The (*a*, *b*) axis is in the *x*–*y* plane and the *c*-axis is along the *z*-direction. Along *c*-axis S–Mo–S units are stacked via van der Waals interactions. Taken from Ref. [35]

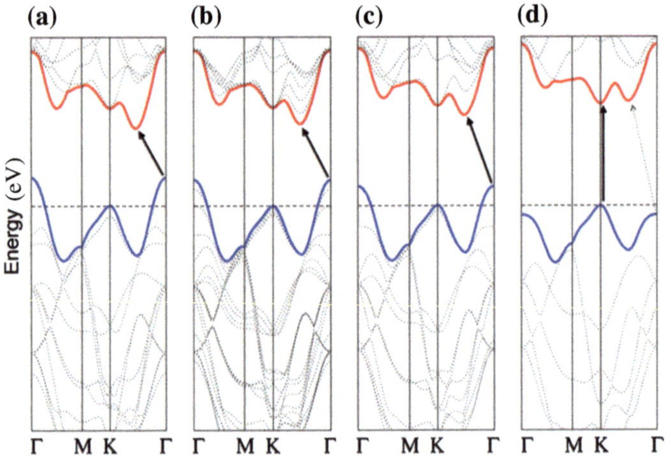

Fig. 7.2 Calculated electronic energy band-diagrams are shown for **a** bulk-MoS_2, **b** 4-layer MoS_2, **c** 2-layer MoS_2, and **d** single-layer MoS_2. The *solid arrow lines* shown in Fig. **a, b,** and **c** indicate the indirect transitions from valence band (at Γ point) to the conduction band minimum (along Γ-*K* direction), whereas in **d** it indicates the direct electronic transition at the *K*-point. Taken from Ref. [39]

stable unit (S–Mo–S) and examine the electronic band structure, we will see that there is a transition from indirect bandgap to a direct bandgap ~1.9 eV at *K*- point for single-layer (S–Mo–S unit) MoS_2 (SL-MoS_2). The calculated electronic band structure with the number of layers [39] is shown in Fig. 7.3. The photoluminescence spectra [27, 39] confirms the direct bandgap of SL-MoS_2, which will be discussed later. For the bulk or a few layers of MoS_2, it has been shown [40] that the states near the indirect bandgap consist of linear combinations of *d* orbitals on Mo atoms and p_z orbitals on S atoms. The states near the conduction band at *K*- point consist of only localized *d* orbitals on Mo atoms. Since Mo atoms are sandwiched between two planes of S atoms in single S–Mo–S unit, the direct bandgap has less dependence than the indirect bandgap on number of layers (or interlayer coupling). The quantum confinement effect along the *c*-axis increases the indirect bandgap, whereas direct bandgap does not vary much.

7.3 Phonons in Bulk and Single-Layer MoS_2

Bulk MoS_2 belongs to the point group D_{6h} (space group $P3m1$) [41]. Each unit cell with lattice parameters $a = 3.12$ Å and $c = 12.3$ Å [41] contains two molecular units shown in Fig. 7.4 [42] and hence, it has a total of 18 normal modes of vibrations corresponding to $2A_{2u} + 2E_{1u} + 2B_{2g} + 2E_{2g} + A_{1g} + E_{1g} + B_{1g} + E_{2u}$ irreducible representations [43–45]. The calculated values of the frequencies and

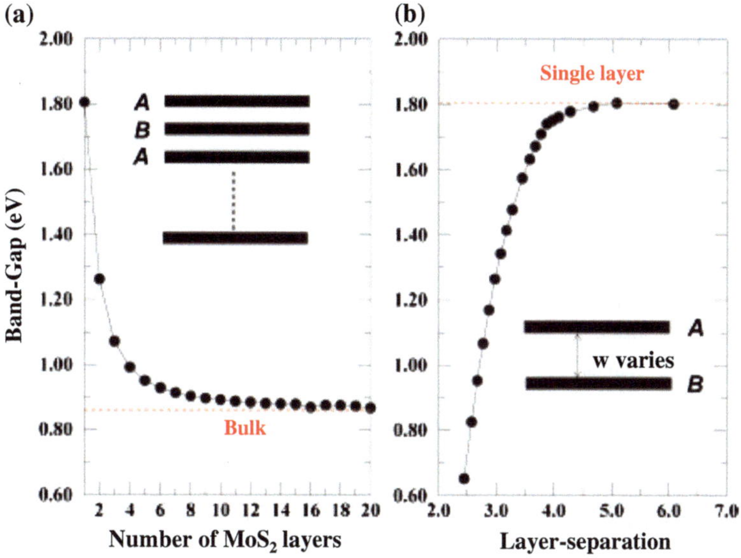

Fig. 7.3 Calculated changes in the bandgap for MoS_2 flakes as a function of **a** number (N) of layers and **b** the separation (w) between two-layers. Here, *A* and *B* denote S-Mo-S unit. Taken from Ref. [40]

optical activity of all the phonon modes are shown in Table 7.1 [41]. A^1_{2u} and E^1_{2u} phonon modes belong to acoustic branches. Since bulk MoS_2 has center of inversion symmetry, the infrared (IR) active phonon modes (A^2_{2u} and E^2_{1u}) and Raman active phonon modes are mutually exclusive. Excluding the inactive

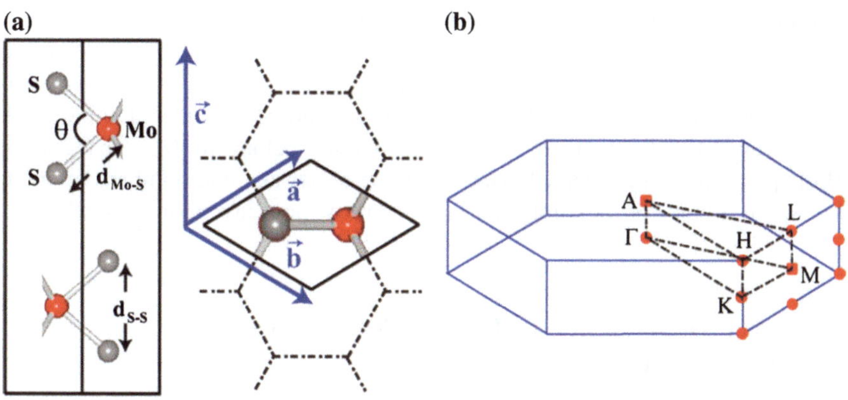

Fig. 7.4 Hexagonal lattice structure of 2H-MoS_2 is shown in **a** for side and top views. The unit cell is marked by solid black lines, which contains one Mo atom (shown by *red balls*) and one S atom (shown by *gray balls*). The corresponding Brillouin zone is shown in **b**, where *red dots* mark the high-symmetry points and *dashed lines* the corresponding directions. Taken from Ref. [42]

Table 7.1 List of high-frequency optical phonon modes for bulk and SL-MoS_2

SL-MoS_2 D_{3h}(S)	Bulk D_{6h}(B)	Character	Direction	Atoms Involved	$\omega(cm^{-1})$ Calculated[a]		$\omega(cm^{-1})$ Exp.
A_2''	A_{2u}	Acoustic	Out of plane	Mo + S	0.0 (S)	0.0 (B)	
	B_{2g}^2	Inactive	Out of plane	Mo + S		55.7 (B)	
	E_{2g}^2	Raman	In plane	Mo + S		35.2 (B)	22 (Bi) 33 (B)[c,d]
A_1'	A_{1g}	Raman	Out of plane	S	410.3 (S)	412.0 (B)	402 (S) 408 (B)[b]
	B_{1u}	Inactive	Out of plane	S		407.8 (B)	
A_2''	A_{2u}	Inactive (E‖c)	Out of plane	Mo + S	476.0 (S)	469.4 (B)	
	B_{2g}^1	Inactive	Out of plane	Mo + S		473.2 (B)	
E_2'	E_{2g}^1	Raman	In plane	Mo + S	391.7 (S)	387.8 (B)	382 (S) 380 (B)[b]
	E_{1u}	Infrared (E⊥c)	In plane	Mo + S		391.2 (B)	
E_2''	E_{1g}	Raman	In plane	S	289.2 (S)	288.7 (B)	
	E_{2u}	Inactive	In plane	S		287.1 (B)	

[a] Ref. [41]. [b] Ref. [46]. [c] Ref. [50]. [d] Ref. [54]. Here, 'Bi' indicates Bi-layer

phonon modes $2B_{2g}$, $B_{1u}$and E_{2u}, the four Raman-active optical phonon modes are A_{1g}, E_{1g} and $2E_{2g}$. The vibrational configurations for all the phonon modes are shown in Fig. 7.5a and the phonon dispersion curves for bulk MoS_2 are shown in Fig. 7.5b [42].

Single-layer MoS_2 has a point group symmetry of D_{3h} (space group $P6m2$) [41]. Since SL-MoS_2 has no inversion symmetry, the labelings for the phonon modes get changed. The total number of phonon branches is reduced to nine and the modes at Γ-point are given by $2A_2'' + 2E' + A_1' + E''$ [45]. Here A_2'' and E' are acoustic phonons; another E' mode is both IR and Raman active. Second A_2'' mode is IR active. The total Raman active phonon modes are A_1', E' and E'' corresponding to bulk Raman modes A_{1g}, E_{2g} and E_{1g}, respectively. The eigenvectors of the phonon modes and the calculated dispersions are shown in Fig. 7.6a and b, respectively [42]. Even number of S–Mo–S units belong to D_{6h} (space group $P3m1$) point group symmetry having inversion center and odd numbers belong to D_{3h} (space group $P6m2$) point group symmetry without inversion center [41].

7.4 Layer Dependence of Optical Raman Modes: Off-Resonance

With the ability of making single and bilayer samples by mechanical exfoliation technique, the layered materials have potential technological applications in nanodevices and logic circuits. The nondestructive Raman spectroscopy has been

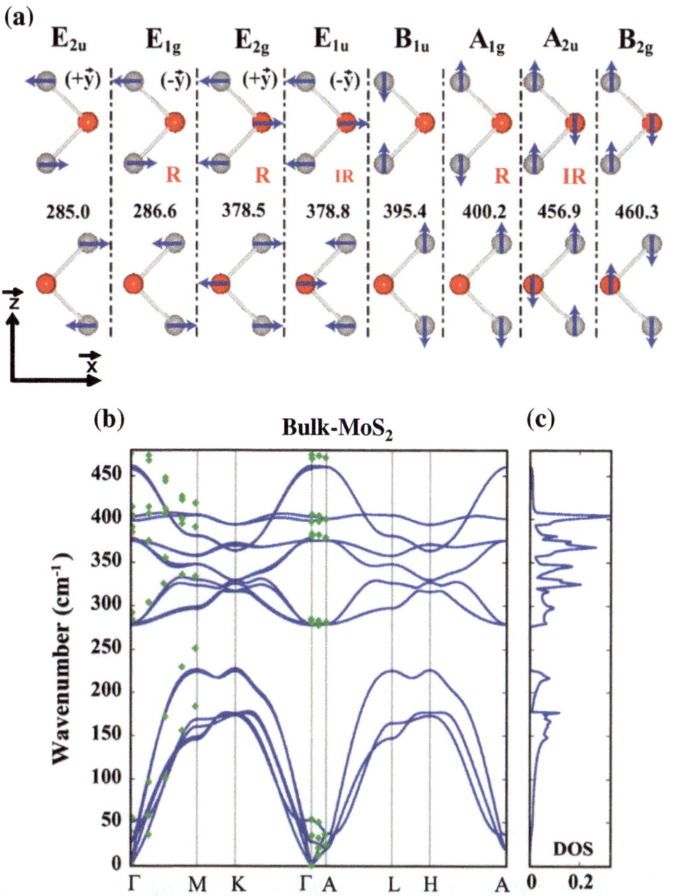

Fig. 7.5 **a** Different optical phonon modes at the Γ point with the corresponding symmetries and vibrational configurations for bulk-MoS_2. The numbers are the corresponding vibrational frequencies in cm^{-1}. The phonon dispersion curves along different symmetry directions and the corresponding density of states are shown in **b** and **c**, respectively. Taken from Ref. [42]

widely used to measure the number of layers precisely for single, bilayer, and multilayer graphene systems via the signatures of the 2D phonon band near $\sim$2,700 cm^{-1} (two-phonon double resonance feature) [48]. Similarly, different number of MoS_2 layers have been identified through the mode frequencies. In this section, we discuss the off-resonance Raman spectra (laser excitation energy 2.41 and 2.33 eV as compared to the direct bandgap of 1.9 eV).

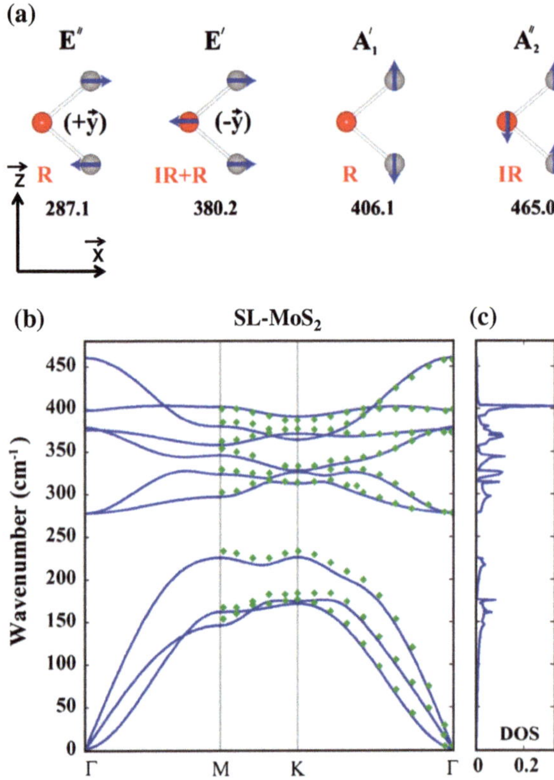

Fig. 7.6 **a** Different optical phonon modes at the Γ point with the corresponding symmetries and vibrational configurations for single-layer MoS_2. The numbers are the corresponding vibrational frequencies in cm^{-1}. **b** Phonon dispersion curves along different symmetry directions and **c** the density of states. Taken from Ref. [42]

7.4.1 High Frequency Optical Modes (Around 400 cm^{-1}): Opposite Trends of A_{1g} and E^1_{2g} Modes as a Function of Number of Layers

The single and few layers of MoS_2 deposited on SiO_2/Si wafer are shown in Fig. 7.7a taken by optical microscope and the thicknesses of the corresponding flakes measured by atomic force microscopy (AFM) are shown in Fig. 7.7b. The measured thickness of a single-layer MoS_2 is about 0.6–0.7 nm. The Raman spectra for bulk and few layers MoS_2 are shown in Fig. 7.8a and the observed frequencies of the two modes are shown in Fig. 7.8b [47]. The A_{1g} Raman mode hardens with increasing the number of layers by $\sim$6 cm^{-1}, where as E^1_{2g} softens by $\sim$2 cm^{-1}. Above four layers, both modes, show saturation values with the bulk values $\sim$408 and 382 cm^{-1}, respectively. The opposite trend of the two Raman modes leads to increase in frequency difference (shown in Fig. 7.8c) which can be

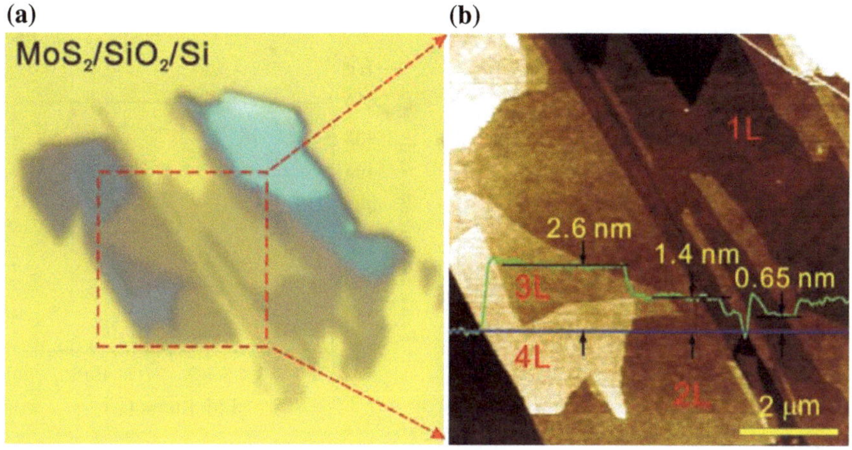

Fig. 7.7 **a** Optical image of the MoS_2 flake deposited on SiO_2/Si wafer. The dashed box indicates the area of $8 \times 8\ \mu\ m^2$ for which the AFM image has been taken and shown in **b**. The different layers and the corresponding height profiles are shown in AFM image. Taken from Ref. [46]

used to probe the number of layers of MoS_2 system. The hardening behavior of out-of-plane mode (A_{1g}) can be understood in terms of increasing the restoring force constant perpendicular to the basal plane through van der Waals interactions as we increase the number of layers. In comparison, the in-plane vibrational mode $\left(E^1_{2g}\right)$ shows anomaly. The difference between these two vibrational modes is that E^1_{2g} involves vibration of Mo and S atoms in the basal plane; whereas in A_{1g} vibrational configuration, Mo atoms remain fixed and only S atoms vibrate along c-axis. It has been shown [41] that the long-range coulomb interaction part of the self-interaction remains negligible for S atoms, whereas for Mo atoms it decreases considerably as we increase the number of stacking layers. Because of increased screening and consequent increase in dielectric tensor with the number of layers, the in-plane mode E^1_{2g} softens. The qualitative agreement between the theory [41] and the experiment [46] for the observed two Raman modes is shown in Fig. 7.9.

The integrated intensity variation of the A_{1g} and E^1_{2g} modes with the number of layers is shown in Fig. 7.10a. The Raman intensity increases linearly up to four layers and then decreases to the bulk value. The increase in intensity for both the Raman modes can be attributed to the optical field enhancement due to multiple reflections of Raman field within the sample layers, and consequent interference effects between these reflected lights and those coming from the thin layer of SiO_2 (300 nm) deposited on Si substrate. The optical field enhancement and interference phenomena also lead to increase in intensity of Raman modes for graphene systems and show a maximum of up to nine layers [49]. It has been

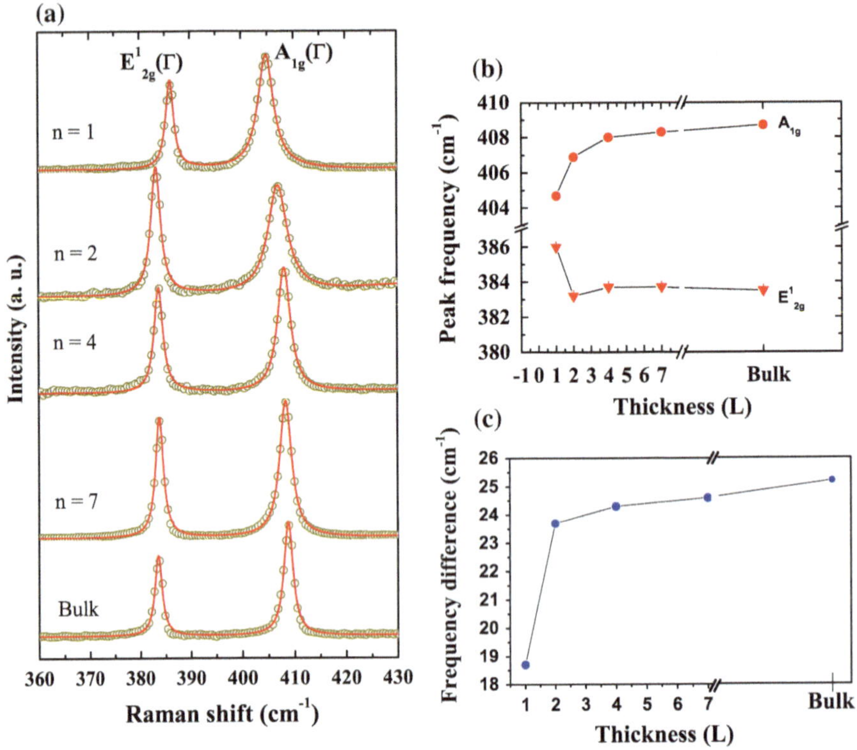

Fig. 7.8 **a** Raman spectra of a few layers and bulk MoS_2. The laser excitation energy used is 2.41 eV (514 nm) [47]. *Solid lines* represent the Lorentzian fits to the experimental data (*circles*). **b** Raman shifts of these two modes and **c** the frequency difference $\left(\omega_{A_{1g}} - \omega_{E^1_{2g}}\right)$ as a function of thickness of the sample. Taken from Ref. [47]

observed [46] that SL-MoS_2 deposited on quartz substrate has Raman intensity of ~20 % of that of the SL-MoS_2 deposited on SiO_2/Si substrate. The integrated intensity ratio of the two modes, $I\left(E^1_{2g}\right)/I(A_{1g})$, is also shown in Fig. 7.10a. Since the interference enhancement phenomena affects both the modes equally, the abrupt increase of the integrated intensity ratio from single to bilayer is not clear, and it may be due to the differences in underlying crystal symmetries between them. The full width at half-maximum (FWHM) of A_{1g} and E^1_{2g} modes are shown in Fig. 7.10b [47]. The FWHM of A_{1g} Raman mode increases from 5 to 7 cm^{-1} as the layer number decreases from seven to two layers. The E^1_{2g} mode shows no thickness-dependent variations (1 cm^{-1}) of the FWHM. The FWHM dependence of both the modes A_{1g} and E^1_{2g} needs further understanding. Lee at el [46] showed that the broadening of the A_{1g} phonon is also observed for the suspended MoS_2 layers and hence cannot be attributed to the interaction with the substrate.

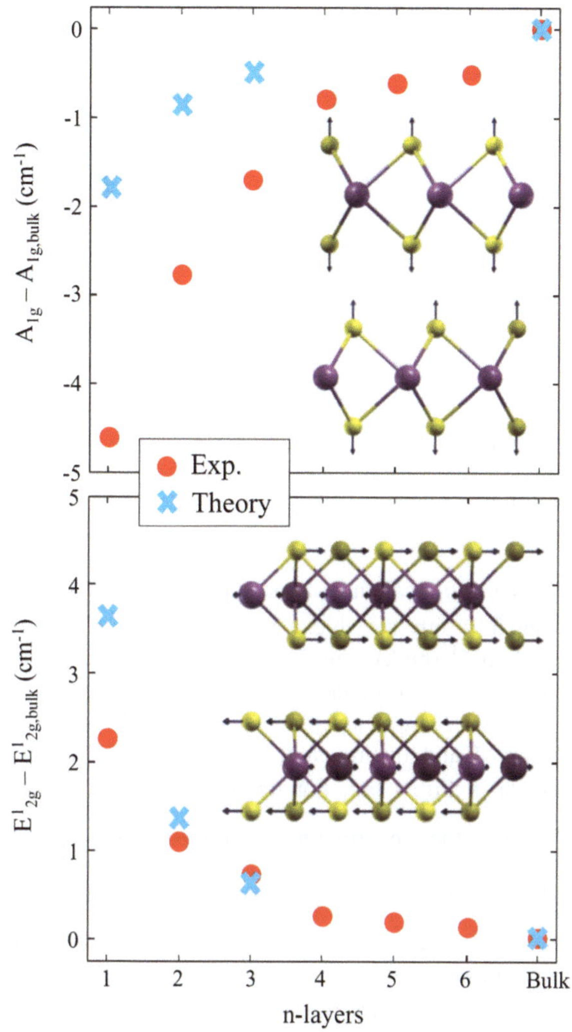

Fig. 7.9 The phonon frequency differences between n-layers and bulk for A_{1g} (*top panel*) and E^1_{2g} (*bottom panel*). The calculated data from Moli et al. [41] are shown with blue crossed symbols and the experimental data from Lee et al. [46] are shown with red circles. Taken from Ref. [41]

7.4.2 Low Frequency Optical Modes (Below 50 cm^{-1}): Shear $\left(E^2_{2g}\right)$ and Compressional Modes

In the E^2_{2g} optical phonon mode, which is an in-plane rigid layer vibration (Fig. 7.11a), the atoms in each layer vibrate in phase such that each layer rigidly oscillates against the adjacent layers and the only restoring force is the weak van der Waals interactions acting between the layers. This mode, also known as a shear mode, exists in all layered compound materials with low frequency ($\sim$30–40 cm^{-1}). Raman spectra [50] of a few layers and bulk MoS_2 for the low

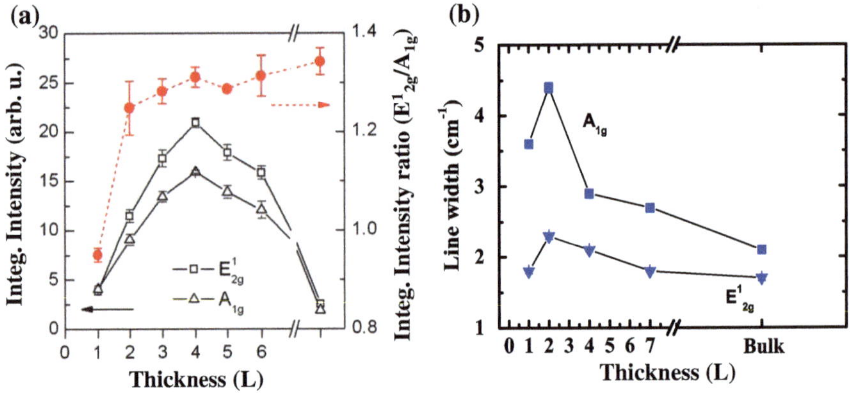

Fig. 7.10 **a** The integrated intensity of A_{1g} and E^1_{2g} modes (corresponding to left vertical scale) and their ratio (corresponding to right vertical scale). Taken from Ref. [46]. **b** The line width (FWHM) of these two modes. Taken from Ref. [47]

frequency optical phonons are shown in Fig. 7.11b. The E^2_{2g} mode shows stiffening behavior with the thickness of the sample from $\sim 22\ \text{cm}^{-1}$ (two layers) to $\sim 33\ \text{cm}^{-1}$ (bulk) (Fig. 7.11c). Another low frequency mode with broad feature attributed to the compressional mode (C-mode) shows softening behavior from $\sim 42\ \text{cm}^{-1}$ (two layers) to $\sim 15\ \text{cm}^{-1}$ (seven layers). The experimentally observed and theoretically calculated [50] frequencies for both the modes are shown in Fig. 7.11c. For the shear E^2_{2g} mode, the fitted frequency (Fig. 7.11c) follows the expression (obtained from the linear-chain model) $\omega = \frac{1}{\sqrt{2}\pi c}\sqrt{\frac{\alpha}{\mu}}\sqrt{1+\cos\left(\frac{\pi}{N}\right)}$ [51]. Here ω denotes the peak position of the shear mode; N, the number of layers; c, the speed of light in cm/s; α, the interlayer force constant per unit area; and μ, the unit layer mass per unit area. For multilayer graphene [51], the low frequency shear modes also follow the same trends, i.e., with the increasing layer numbers, the restoring force constant acting between the layers increases.

The thickness dependence of the C-mode shows $\frac{1}{N}$ dependence instead of $\frac{1}{\sqrt{N}}$. The $\frac{1}{N}$ behavior also observed for the longitudinal acoustic phonons on Na films [deposited on Cu (001)] [52] is due to open standing waves in multilayer (or thin films) systems [53]. A strong coupling between the substrate and the atomic layer of the sample does not allow the atoms close to the substrate vibrate in the C-mode and creates a node there, while there are no restrictions for the atoms in the top layer; thus creating open standing waves (kind of organ-pipe modes) and hence $\frac{1}{N}$ dependence.

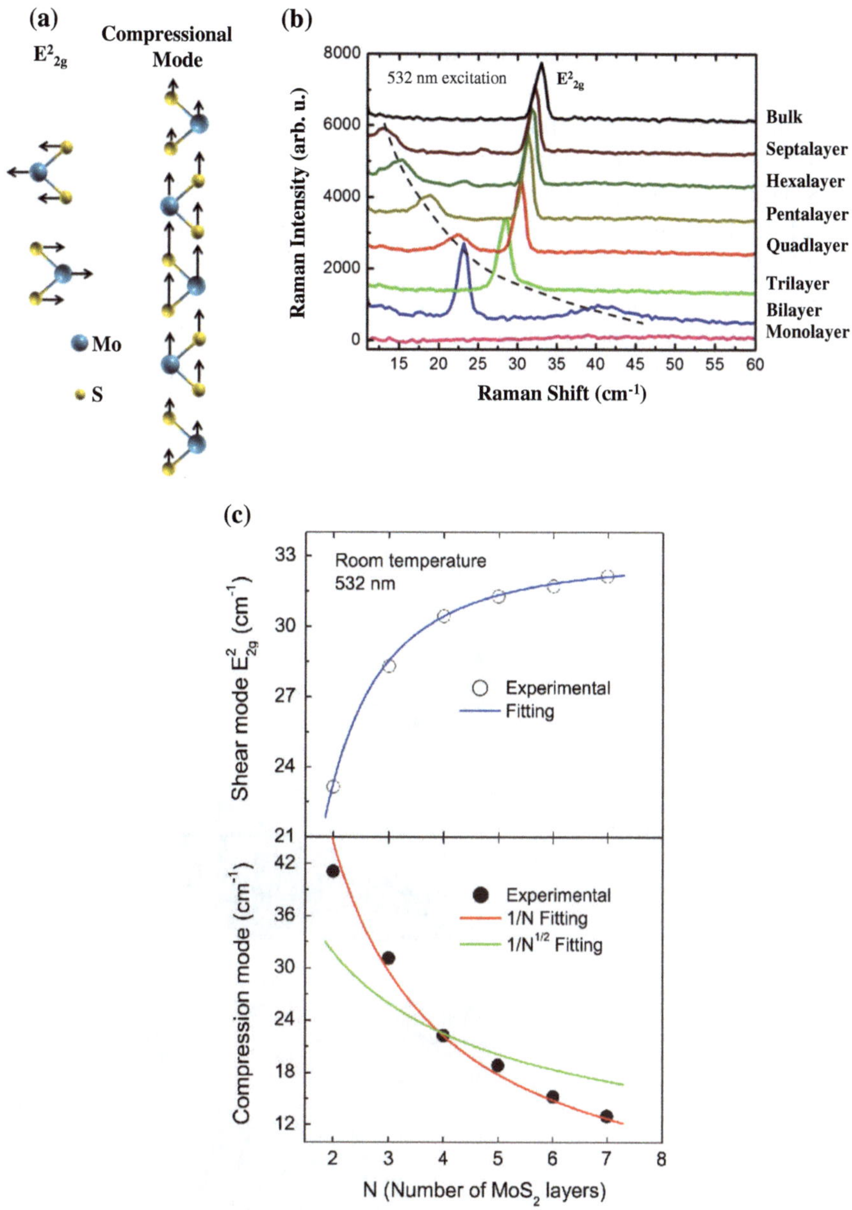

Fig. 7.11 **a** The vibrational configurations of the E_{2g}^2 and the compressional modes. **b** Raman spectra of two modes for different layers and bulk MoS_2 deposited on SiO_2/Si substrate. The *dashed line* is the guide to the eye. **c** The peak positions as a function of thickness of the samples. The interlayer force constant per unit area (α) and the unit layer mass per unit area (μ) used for the least square fitting of E_{2g}^2 mode are 2.9×10^{19} Nm^{-3} and 30.3 kg $\AA^{-2}$, respectively. Taken from Ref. [50]

7.4.3 Thickness Monitoring of MoS_2 Flakes via Raman Imaging: Two Pairs of Raman Modes

From the above two subsections, we have seen that the frequencies of the four Raman active modes (two pairs) change differently by sensing only the number of MoS_2 layers. The two pairs of optical phonons are {A_{1g} and E^1_{2g}} and {E^2_{2g} and C-mode}. Each member of the first pair shows an opposite trend with the result that the difference $\omega_{A_{1g}} - \omega_{E^1_{2g}}$ shows a sizable change, which helps in Raman imaging. Since, the C-mode shows softening behavior and has frequency lower (15 cm^{-1} for seven layers) than the E^2_{2g} mode (32 cm^{-1} for seven layers), the shear E^2_{2g} mode is the feasible one from the experimental point of view. Moreover, the shear mode shows a ~50 % change in peak positions going form bilayer (22 cm^{-1}) to bulk

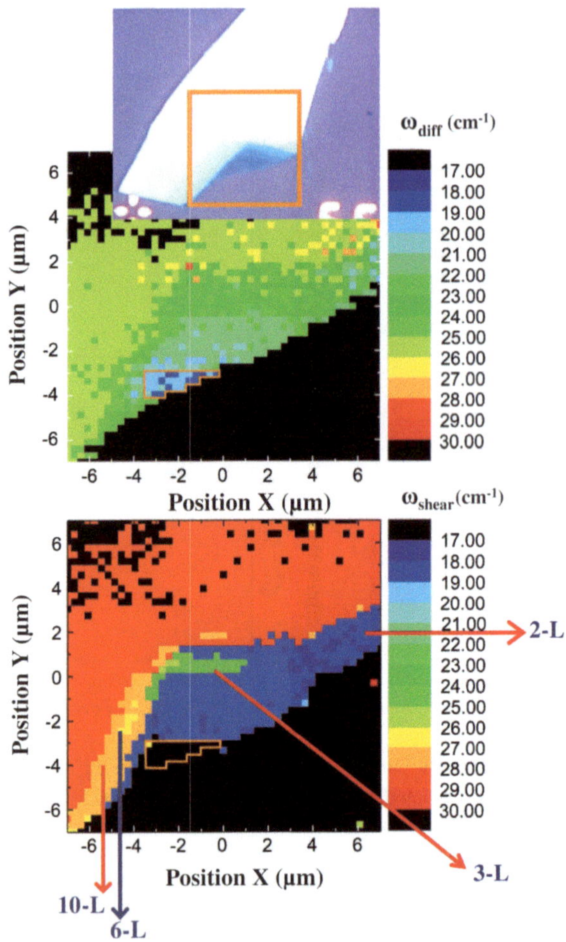

Fig. 7.12 **a** The optical image of different layers of MoS_2. The boxed area used for Raman imaging. **b** The Raman image taken by locking the $\omega_{diff} = \omega_{A_{1g}} - \omega_{E^1_{2g}}$. Single layer is marked by a boxed area. **c** Raman image recorded by locking the $\omega_{shear} = \omega_{E^2_{2g}}$. Here the color scales used for ω_{diff} and ω_{shear} are arbitrary. Taken from Ref. [54]

value (33 cm^{-1}). Plechinger et al. [54] recorded (shown in Fig. 7.12) Raman images of a few layers of MoS_2 by locking the frequency difference of two high frequency Raman modes $\left(\omega_{A_{1g}} - \omega_{E^1_{2g}}\right)$ and a low frequency shear mode $\left(E^2_{2g}\right)$. The Raman imaging of MoS_2 flakes (rectangular area marked in Fig. 7.12a) is shown in Fig. 7.12b for the frequency difference of $\omega_{A_{1g}} - \omega_{E^1_{2g}}$. The single-layer MoS_2 for which, $\omega_{A_{1g}} - \omega_{E^1_{2g}} = 18\,cm^{-1}$ (from Fig. 7.8b), is shown by the marked area with false color in Fig. 7.12b. The signature of E^2_{2g} mode on single layer is zero and is shown in Fig. 7.12c by a closed marked region. The bilayer (2L), trilayer (3L), hexalayer (6L), and 10-layers (10L) and their boundaries are clearly seen by tracking the shear mode (in Fig. 7.12c), whereas the counterpart for the frequency difference is not clear as shown in Fig. 7.12b.

7.4.4 *Shear and Layer Breathing Modes of* N-*Layers of* MoS_2 $\{N = 1 \rightarrow 19\}$

For the rigid layer vibrations of N-layer MoS_2, there are $N-1$ shear modes parallel to the basal plane (perpendicular to the c-axis) and $N-1$ layer breathing modes (LBMs) along the c-axis. For even N, LMBs are not Raman active and $\frac{N}{2}$ shear modes are Raman active (doubly degenerate). For odd, $\frac{N-1}{2}$ LBMs and $N-1$ shear modes are Raman active [45]. Figure 7.13a shows the newly observed low frequency shear and LBMs Raman modes for 5L and 6L with different polarizations [45]. As stated earlier, the Raman active modes A'_1 and E' for odd layers (ONL-MoS_2) are the counterpart of the A_{1g} and E^2_{2g} for even layers (ENL-MoS_2), respectively. The eigenmodes of A'_1 and B^2_{2g} are shown in Fig. 7.13b. The B^2_{2g} mode is Raman inactive and the reason of observation is discussed later. All the observed low frequency modes are listed in Table 7.2.

To understand the low-frequency rigid layer modes, the calculated [45] frequencies with the diatomic chain model (DCM) and the experimentally observed ones for the ENL-MoS_2 and ONL-MoS_2 are shown in Fig. 7.14, separately. For the ONL-MoS_2, the shear mode E' (28 cm^{-1} for 3L) stiffens and the one LBM A'_1 (29 cm^{-1} for 3L) softens with increasing the thickness of the sample from 3L to 19L (see Fig. 7.14b, d). According to DCM calculations, a new mode appears as N reaches a value of $4\,N + 3$, where $N = 1,2,3,\ldots$ And that new mode splits into two branches such that one branch softens and the other stiffens until N reaches a value of $4\,N + 3$ and so on. The similar softening and stiffening behavior of a shear mode E^2_{2g} (23 cm^{-1} for 2L) and an LBM mode B^2_{2g} (40 cm^{-1} for 2L) for ENL-MoS_2 are shown in Fig. 7.14a, b respectively. For ENL-MoS_2, the appearance and splitting of a new mode into two branches follow according to $4\,N + 2$. The DCM calculated value matches well with the experimentally observed Raman inactive B^2_{2g} mode and this is the reason, along with the polarization behavior, that

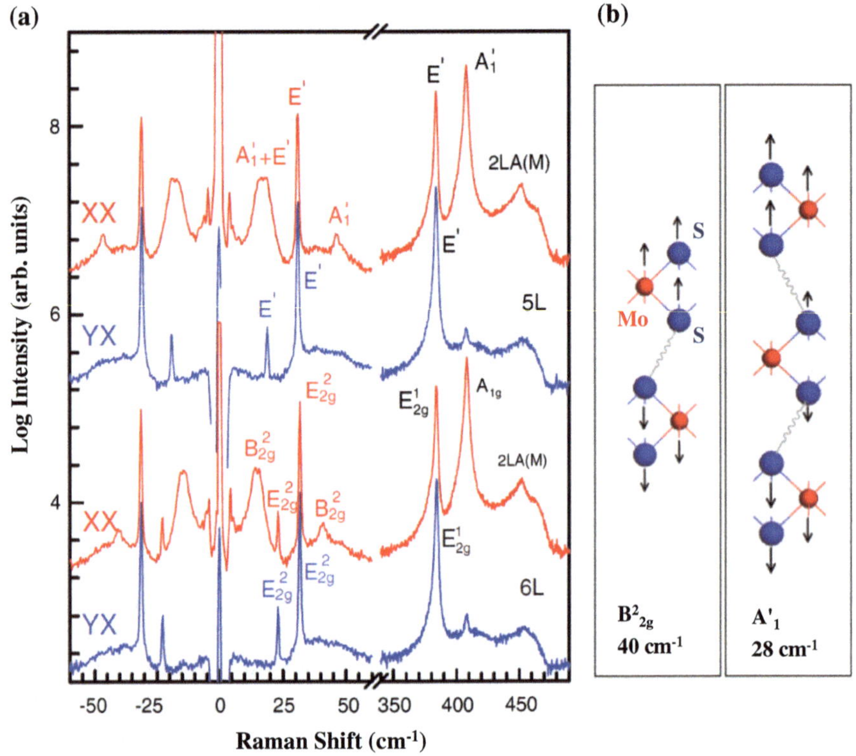

Fig. 7.13 **a** Polarized Raman spectra for 5L and 6L of MoS_2. YX represents the direction of incident laser polarization (*Y*) and the direction of analyzer's polarization (*X*); both lie in the basal plane. **b** The vibrational configurations of the atoms in A'_1 and B^2_{2g} modes. Taken from Ref. [45]

Table 7.2 List of low-frequency optical phonon modes for 5L and 6L of MoS_2 [45]

5 Layers (5L)				6 Layers (6L)			
Shear modes	Exp.	LBMs	Exp.	Shear modes	Exp.	LBMs	Exp.
$2E'$ (R, IR)	19 and 33[a]	$2A'_1$ (R)	17[a] and 47[b]	$3E^2_{2g}$ (R)	23[a] and 32[b]	$3B^2_{2g}$ [c]	15 and 41[a]
$\{XX, YX\}$	(in cm^{-1})	$\{XX\}$	(in cm^{-1})	$\{XX, YX\}$	(in cm^{-1})	$\{XX\}$	(in cm^{-1})
$2E''$(R)	–	$2A''_2$(IR)	–	$2E1u$ (IR)	–	$2A^2_{2u}$(IR)	–
$\{XZ, YZ\}$	–	–	–	–	–	–	–

[a] Ref. [45]. [b] Ref. [50]. [c] This mode is a silent mode; it is neither Raman nor infrared (IR) active

Zhang et al. [45] named it as a E^1_{2g} mode. Although the B^2_{2g} for ENL-MoS_2 follows exactly the same polarization behavior as A'_1 for ONL-MoS_2 (see Fig. 7.13a), the deeper understanding of its appearance, polarization behavior, and matching with the DCM calculated value needs further work. The most striking features from

Fig. 7.14 are that the experimentally observed shear modes coming from the upper branch of cone-like curve with stiffening character and LBMs are from the lower branch with softening feature for both the ENL-MoS_2 and ONL-MoS_2. The ratio of the shear E^2_{2g} mode for bulk to 2L is given by $\omega_{\text{Bulk}}/\omega_{2\text{L}} = 32.7/22.6 = 1.447$ [45], which is very close to the value of graphene $\left(\omega_{\text{Bulk}}/\omega_{2\text{L}} = \sqrt{2}\right)$ [51].

In general, in order to know the exact trend of softening and stiffening features of low-frequency rigid layer modes of layered materials, a reduced monatomic chain model (MCM) has been considered [45]. The DCM calculated values showed that the relative displacements between Mo atoms and S atoms are less ($\sim$0.6 % for 2L) [45] and decreases with increasing N. Therefore, for rigid layer modes, the relative displacements can be ignored and hence, in MCM model the additive mass ($M_{\text{add}} = M_{\text{MO}} + 2M_{\text{S}}$ instead of a reduced mass $\frac{1}{\mu} = \frac{1}{M_{\text{MO}}} + \frac{1}{2M_{\text{S}}}$) is taken. All the observed shear modes for ENL-MoS_2 $\left(E^2_{2g}\right)$ and ONL-MoS_2 (E') from Fig. 7.14a, b are plotted together in Fig. 7.15a with the DCM and MCM calculated values [45], and Fig. 7.15b is for all the LBMs. The interesting feature from Fig. 7.15 is that all the experimentally observed low-frequency modes originate from ENL-MoS_2, i.e., 2, 4, 6, and 10L. The black solid lines and the gray

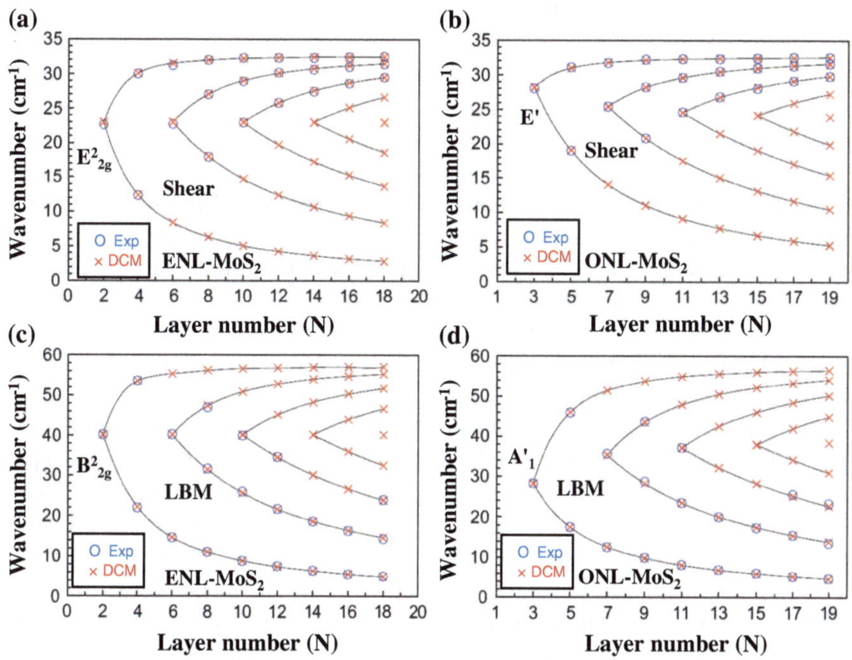

Fig. 7.14 Calculated (*red cross*) according to the diatomic chain model (DCM) and the experimentally observed frequencies or wavenumbers (*blue circle*) **a** with the even (ENL) and **b** odd number (ONL) of layers for the shear modes, and for the layer breathing modes (LBM) in **c** and **d**. Here the connected solid lines are guide to the eye. Taken from Ref. [45]

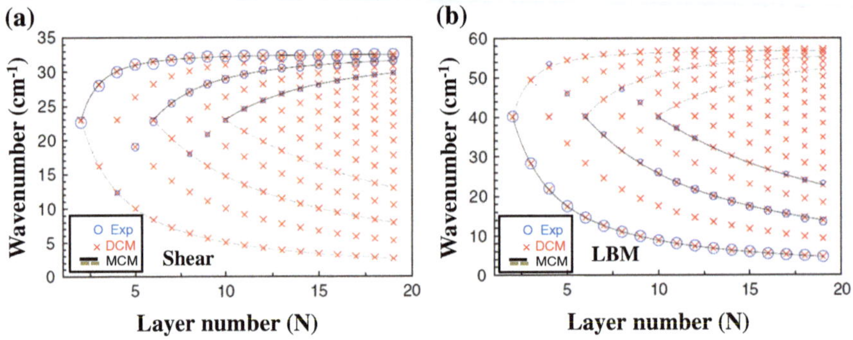

Fig. 7.15 Calculated (*red cross*) according to the diatomic chain model (DCM) and the experimentally observed frequencies or wavenumbers (*blue circle*) of all the layers together for the shear modes in **a** and the LBM modes in **b**. Size of the circle represents the observed Raman intensity. The *black solid lines* (in **a** and **b**) are fitted curves according to the monatomic chain model (MCM) for those modes generating from 2L ($N_0 = 1$), 6L ($N_0 = 3$), and 10L ($N_0 = 5$) branches. Another set of modes (shear and LBMs) according to the MCM model are shown by the gray dashed lines in **a** and **b**. Taken from Ref. [45]

dashed lines are the fitted curves which follow $\omega(N) = \omega(2N_0)\sqrt{1 \pm \cos(N_0\pi/N)}$ (+ and −sign for upper and lower branch, respectively) such that $N \geq 2N_0$ and $N_0 = 1, 2, 3...$ Since $\omega(2N_0)$ is almost equal to $\omega(2)$ for all rigid layer modes, the fitted frequency obeys $\omega(N) = \omega(2)\sqrt{1 \pm \cos(N_0\pi/N)}$ formula according to MCM. Here, the model MCM did not take into account the interactions between one layer of MoS_2 and the supported substrate and explains well the experimentally observed frequencies; whereas in contrast, Zeng et al. [50] included the substrate effect and showed the 1/*N* behavior of one LBM as shown in Fig. 7.11c. It was shown that for suspended multilayer graphene, the scaling of the shear mode with thickness follows the trend predicted by MCM calculations [51]. Recently, suspended 2L-MoS_2 showed no substrate effect on the rigid layer phonon frequencies and their FWHMs [55]. It would therefore be reasonable to assume that the interaction between substrate and MoS_2 is not responsible for the observed 1/*N* scaling with the thickness [45].

Recently, two new low frequency rigid layer modes were observed [55] (Fig. 7.16a): one is LBM (B2) and the other a shear mode (S2). Both modes appear from 4L-MoS_2 such that B2 mode softens from 54 cm^{-1} (4L) to 35 cm^{-1} (7L) and S2 mode stiffens from 13 cm^{-1} (4L) to 30 cm^{-1} (12L). The frequency variation with the thickness of the sample is shown in Fig. 7.16b. The fitted solid curves are according to linear chain model calculations [55]. S2 mode shows softening behavior because the out-of-phase displacement between layers increases with *N* [55]. B2 mode stiffens as in-phase motion increases with the thickness *N*. The appearance of B2 LBM mode for even layers of MoS_2 (4L) needs further studies as, according to Zhang et al. [45], for even layers there should be no Raman active LBM.

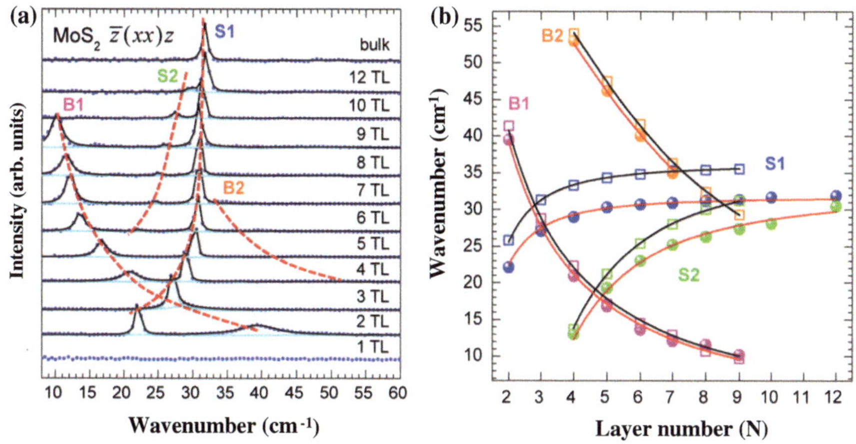

Fig. 7.16 **a** Raman spectra of few layers and bulk MoS_2 for the low frequency shear (S) and layer breathing modes (B). Here 1TL means one S-Mo-S unit, i.e., one layer (1L). The dashed lines are the guide to the eye. **b** The variations of the measured (*solid dots*) and the calculated frequencies (*open squares*) from first principles. *Solid lines* represent the fitting of the data according to the linear chain model. Taken from Ref. [55]

7.5 Layer Dependence of Optical Raman Modes: Resonance

Resonance Raman Scattering (RRS) occurs for a system when the incident laser excitation energy is close to the electronic bandgap. In this section, we will discuss the RRS with the laser energy 1.9 eV (633 nm), which matches with the direct bandgap at the K-point of the BZ. Figure 7.17 shows the resonant Raman spectra [47] of different layers ($N = 1$, 2, 4, and 7L) and bulk MoS_2. Since the RRS enhances the possibility of higher order Raman scattering (multiphonon process) compared to the first-order process in normal Raman scattering, Fig. 7.17 shows many Raman peaks along with the off-resonant Raman modes. Figure 7.18a–c show the variations of frequencies, FWHMs, and integrated intensity ratio of two Raman modes A_{1g} and E^1_{2g}, respectively, for the RRS [47] which show the same trends as for the off-resonant case (see Figs. 7.8b and 7.10). The new Raman bands, which were not observed with the 514 and 533 nm laser lines, are understood as follows.

7.5.1 New Modes in Low of 160–230 cm^{-1}

(i) A Raman band $\sim$179 cm^{-1} is observed and this mode is assigned as a difference combination mode of A_{1g} (M)–LA(M) [56, 57]. The frequency of this

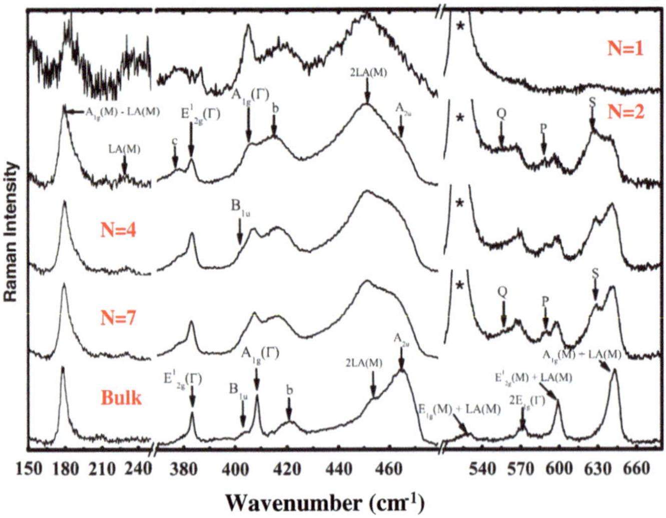

Fig. 7.17 Resonant Raman spectra of few layers and bulk MoS_2 using 633 nm laser line. The strong mode coming from the SiO_2/Si substrate (marked by *) is absent for thick samples (bulk). Here N represents the number of layers. Taken from Ref. [47]

Fig. 7.18 **a** Raman shifts and **b** the line widths (FWHM) of the A_{1g} and the E^1_{2g} modes as a function of the number of layers. **c** The integrated intensity ratio $\mathrm{I}\left(\mathrm{E}^1_{2g}\right)/\mathrm{I}\left(\mathrm{A}_{1g}\right)$. Taken from Ref. [47]

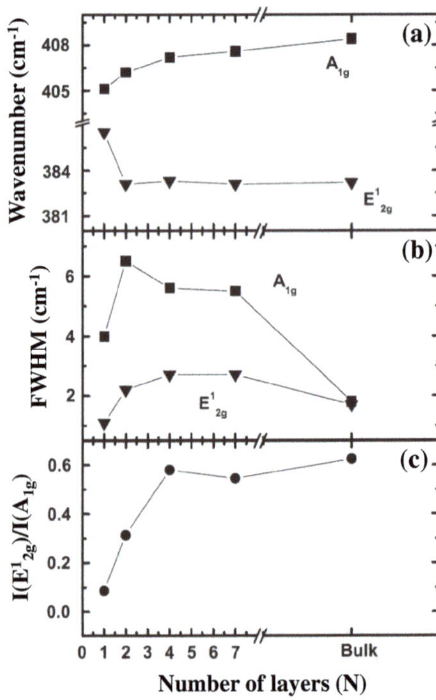

mode does not show layer-dependent feature. The three Raman active $\left(A_{1g}, E_{1g} \text{ and } E^1_{2g}\right)$ phonons at Γ point of BZ are almost dispersionless along the Γ–M direction; the frequency of rigid layer mode E^2_{2g} increases with q (phonon wave-vector) and reaches to a longitudinal acoustic (LA) mode at M (233 cm^{-1}) [57]. At M point of BZ, the dispersion of LA mode shows an inverse parabolic nature, i.e., on either side of M-point, the frequency decreases. The asymmetric nature of the peak on the higher side is due to the involvement of the LA(M) mode as a difference combination band [57]. (ii) Another band at $\sim$233 cm^{-1} is observed for 2, 4, and 7L and is assigned [58] as LA(M). The activation of this first-order mode implies the relaxation of the $q \sim 0$ selection rule. The appearance of the zone boundary phonon (LA) for MoS_2 nanoparticle is attributed to the defect-induced Raman scattering [59]. In the present case, the observed LA(M) mode for 2, 4, and 7L MoS_2 could be due to the structural defects [47].

7.5.2 Frequency Region of 360–430 cm^{-1}

There are three new Raman active bands in the frequency range of 360–430 cm^{-1}. (i) a band named as 'c' $\sim$ 377 cm^{-1} (shown in Fig. 7.17) becomes sharper with increasing the layer from 1 to 7L. Sekine et al. [60] assigned this mode as E^2_{1u} mode. Since the experimentally observed E^2_{1u} mode has higher frequency than the E^1_{2g} mode [43, 61, 62] and these are Davydov pair [60], the observed c mode cannot be assigned as a E^2_{1u} mode [47]. The c mode does not show thickness-dependent frequency variations. The integrated intensity ratio $\mathrm{I(C)}/\mathrm{I}\left(\mathrm{E}^1_{2g}\right)$ decreases with the number of layers as shown in Fig. 7.19b. (ii) One peak $\sim$409 cm^{-1} evolves with a number of layers and is assigned as B_{1u} mode [61, 62]. The B_{1u} and A_{1g} mode are the small frequency splittings of Davydov pair [60] and the appearance of the B_{1u} mode is due to the resonance effect. Figure 7.19c shows that the integrated intensity ratio $I(B_{1u})/I(A_{1g})$ increases with the layers of the sample. (iii) The prominent feature at $\sim$420 cm^{-1}, marked as 'b' in Fig. 7.17, is observed under resonant conditions. This mode is assigned as a two-phonon process and the appearance of this mode is related to the A exciton ($\sim$1.9 eV) at the K-point [58–60]. The dispersion curve for polariton–exciton is shown in Fig. 7.20. Here, the intermediate polariton state involves the two phonons under resonant condition. First, the photon-like state from the high energy upper branch (inner branch) scatters to exciton–polariton state (outer branch) by scattering a longitudinal optical phonon (this dispersive quasi-acoustic phonon is the silent B^2_{2g} mode at the Γ point) and this interaction is due to the deformation potentials; second, the polariton state decays to photon-like state by scattering the transverse optical phonon (with E^2_{1u} symmetry) [60]. This mode is a combination band, $\omega_b = \omega_{QA} + \omega_{TO}$ [47].

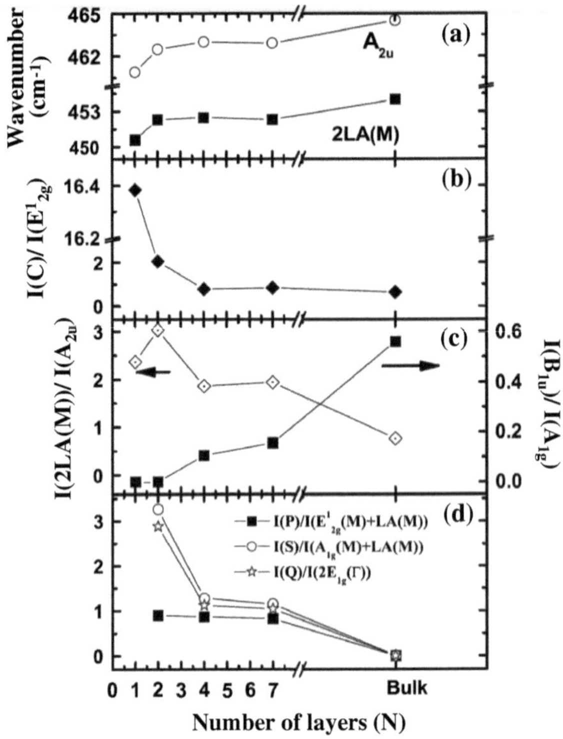

Fig. 7.19 **a** Raman shifts of the 2LA and the A_{2u} modes as a function of the number of layers. **b–d** The changes in the integrated intensity ratios for different modes. Taken from Ref. [47]

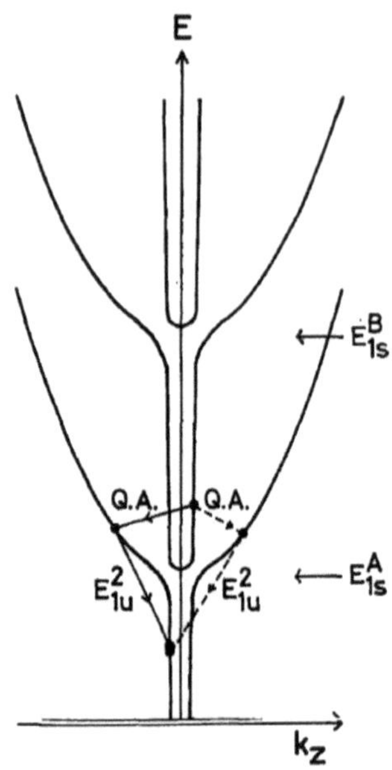

Fig. 7.20 Schematic diagram for the dispersion curve of a polariton and the two-phonon Raman scattering involving a longitudinal quasi-acoustic (Q.A.) mode and the E_{1u}^2 mode. The dashed and the solid lines marked by arrows represent the α and β processes, respectively. Here E_{1s}^A and E_{1s}^B represent the energy of 1 s level for the A and B excitons, respectively. Taken from Ref. [60]

7.5.3 Frequency Region of 440–490 cm^{-1}

(i) The mode $\sim$450 cm^{-1} (for bulk) is attributed to second-order Raman (2LA(*M*)) [57, 59, 60] and (ii) the mode $\sim$466 cm^{-1} (for bulk) is assigned as A_{2u} [59, 63]. The peak positions of the two modes (2LA and A_{2u}) increase with the number of layers (*N*) as shown in Fig. 7.19a. Figure 7.19c shows the decrease in the integrated intensity ratio $I(2LA)/I(A_{2u})$ with the thickness [47, 59].

7.5.4 Frequency Region of 510–630 cm^{-1}

For the bulk MoS_2, (i) the observed mode $\sim$526 cm^{-1} is assigned as $E_{1g}(M)$ + 2LA(*M*), (ii) $\sim$571 cm^{-1} mode as $2E_{1g}(\Gamma)$, (iii) $\sim$599 cm^{-1} as $E^1_{2g}(M) + LA(M)$, and (iv) $\sim$642 cm^{-1} as $A_{1g}(M) + LA(M)$ [59]. In all the multiphonon processes involving LA(*M*) mode, the sum combination band has asymmetric tail at lower side [57]. All the new high frequency bands $\sim$554 cm^{-1} (marked as Q in Fig. 7.17), $\sim$588 cm^{-1} (marked as P), and $\sim$628 cm^{-1} (marked as S) are observed under the resonance condition for 2, 4, and 7L MoS_2 [47] and were not observed for bulk MoS_2 and nanoparticles of MoS_2 [59]. The peak positions of these modes do not show variations with *N*. Figure 7.19d shows the variations in the relative integrated intensity of the bands (*Q*, *P*, and *S*) with respect to the neighboring band.

Figure 7.21 shows another new broad feature $\sim$38 cm^{-1} (5 meV) [50] observed under the resonant condition for the cross-polarization geometry. The peak position and the shape of the mode do not change with the layer numbers. Since the binding energy of the excitons (*A* and *B* with energies 42 and 134 meV, respectively [64]) are much greater than the observed peak with 5 meV, the appearance of this mode is not responsible for the exciton mechanisms. Defects and impurity effects are also excluded due to having energy comparable to the

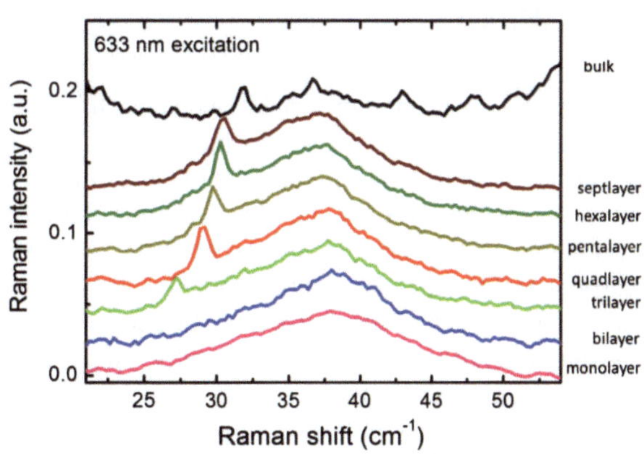

Fig. 7.21 Resonant Raman spectra using 633 nm laser line for samples of different thickness. The new broad feature $\sim$ 38 cm^{-1} appears only under resonant condition. Taken from Ref. [50]

excitons in optical measurements [50]. The oscillating collective modes (plasmons, charge density waves) are also excluded due to the reluctant behavior of the position and shape of the peak, as the long-range Coulomb interaction would be screened with the layer thickness [50]. The spin–orbit coupling is responsible for the splitting near the top of the valence band at the *K*-point of BZ, which leads to the origin of *A* and *B* excitons. Zeng et al. [50] argued that the same spin-orbit coupling mechanisms also split the conduction band at the *K*-point of $\sim$4 meV, which leads to the observed broad feature ($\sim$38 cm^{-1}) associated with the resonant Raman scattering. The understanding of the appearance of all the new Raman modes under resonant condition requires further theoretical work.

7.6 Transport Characteristics as a Function of Temperature and Carrier Concentration: Various Aspects Toward Improvement of Mobility

In technological applications of nanodevices, the two-dimensional (2D) layered materials have a lot of potential due to easy fabrication. The highest mobility observed for the suspended graphene device is larger than 10^5 cm^2/V s [9]. Due to the lack of electronic bandgap in single-layer graphene, the 2D semiconducting materials like MoS_2 (on–off ratio is 10^8 [10]) family compounds have potential applications in digital electronics with low power dissipation. In order to replace silicon-based logic devices, a high current on–off ratio $\sim 10^4$–10^7 [9] and electronic bandgap larger than 400 meV [65] are reasonable. The various scattering mechanisms (intrinsic and extrinsic effects), which are responsible for the momentum relaxations and thereby decreasing the mobility of charge carriers in MoS_2, are discussed in the following sections.

7.6.1 Temperature Dependence of Carrier Mobility in Bulk-MoS_2

Bulk MoS_2 is a layered material in which each individual 2D layer is stacked along the z-direction via van der Waals interactions. To capture this anisotropic feature, a potential model was invoked [66] such that the total potential $V(\boldsymbol{r}) = V(x, y) + V(z)$, where $V(x, y)$ and $V(z)$ represent the potential in basal plane and perpendicular to the basal plane, respectively. According to this model, the excess carriers feel a series of parallel potential wells along the perpendicular direction and between each of them, there existing local energy levels for the carriers which depend upon the local width of the wells. This implies that the localization energy within each layer depends on the layer thickness. The variation of the layer thickness associated with thermal phonons may give a friction to the carriers. The energy of the excess carriers (due to decoupled potential) is given as $E(\mathrm{p}) = \frac{p_x^2}{2m_x} + 2I_z\left(\cos\frac{p_z d_z}{h} - 1\right)$ [67], where m_x is the effective mass in basal plane, d_z the interlayer spacing, I_z small overlap energy between adjacent layers, and the second term of E(**p**) corresponds to the tight binding model. The corresponding density of states (DOS) is given as $D(E) = \frac{m_x N_z}{2\pi^2\hbar^2}\cos^{-1}(1 - E/2I_z)$ for $0 \leq E \leq 4I_z$ and $D(E) = \frac{m_x N_z}{2\pi^2\hbar^2}$ for $E \geq 4I_z$ [67], where N_z denotes the number of layers/length along the z-direction. The DOS is almost constant except for the low energy range, which is the characteristic feature for two-dimensional electron gas.

In most isotropic semiconductors (nonpolar), the high temperature mobility is limited by the quasi-elastic scattering with acoustic phonons via deformation potentials [68]. The acoustic phonon mode corresponds to local dilatations of the lattice. Since between two adjacent layers there is a weak van der Waals force, the change in deformation potentials (CDP) due to the acoustic modes polarized in the basal plane will be more than that by the vertically polarized ones. Therefore, in layered semiconductors, the carriers will be scattered significantly by the horizontally polarized acoustic phonons. The calculated mobility due to the deformation potential scattering by acoustic modes is $\mu_{\mathrm{bulk}}^{\mathrm{acoustic}} \propto T^{-\gamma^{\mathrm{acoustic}}}$ [67] where $\gamma^{\mathrm{acoustic}} = 1$ which is quite different from the well-known value of $\gamma^{\mathrm{acoustic}} = 3/2$ for isotropic case because of peculiar behavior of DOS.

In contrast to the acoustic mode, the optical modes correspond to the distortion of the unit cell. In isotropic materials (nonpolar), the CDP due to optical modes is less and can be neglected. But in layered materials, the deformation potential is highly anisotropic [67]. Those optical modes which can modulate the thickness of the individual layer will interact with the charge carriers as discussed above (because of series of parallel potentials along z). These thickness modulating optical modes which do not produce first-order dipoles, irrespective of having effective charges on each atom, are known as *homopolar* modes. The mobility due to the vertically polarized (along z) homopolar optical modes is $\mu_{\mathrm{bulk}}^{\mathrm{hpolar}} = \mu_0(T/T_0)^{-\gamma^{\mathrm{hpolar}}}$ [67], where $\gamma^{\mathrm{hpolar}} > 1$ and $\mu_0 = \mu_{\mathrm{bulk}}$ at $T = T_0$. In case of polar materials, where long-range Coulomb interaction associated with polar optical modes couples the charge carriers (via Frohlich interaction), it has been shown that $\gamma^{\mathrm{polar}} \cong \gamma^{\mathrm{hpolar}} - 1$ [67]. The experimentally observed Hall mobility of bulk MoS_2 is shown in Fig. 7.22. The fitted solid line corresponds to $\mu_{\mathrm{bulk}}^{e} \cong 100(T/T_0)^{-2.6}$ $\mathrm{cm^2/V\,s}$, where $T_0 = 300$ K. To get a comparison of different scattering mechanisms of charge carriers by thermal phonons, the exponent (γ) for the mobility has been plotted with the phonon energy as shown in Fig. 7.23. Since for bulk MoS_2 the only one homopolar mode [67] corresponds to 60 meV ($A_{1g} \sim 408\ \mathrm{cm^{-1}}$ for bulk), it is clear from Fig. 7.23 that the vertically polarized nonpolar (homopolar mode) mode is responsible for the high-T limited mobility.

Fig. 7.22 Hall mobility versus temperature for the n-type bulk-MoS_2. The solid line represents the fitting to the observed data. At low temperature, the increment of mobility is due to extrinsic properties. Taken from Refs. [66, 67]

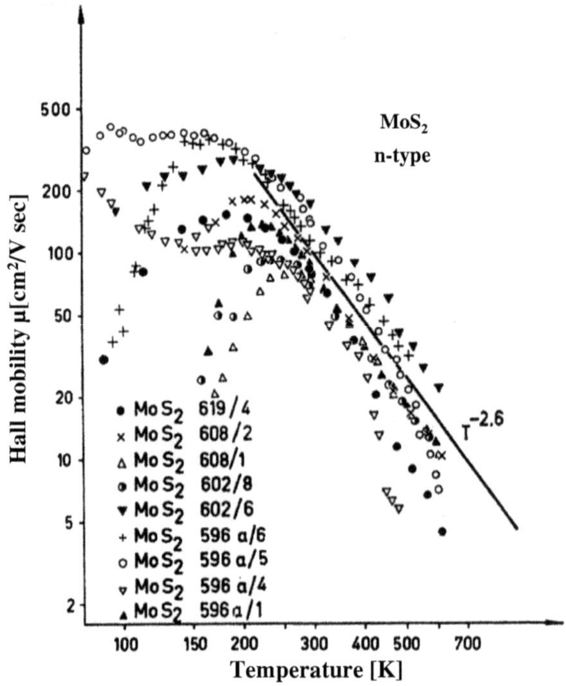

Fig. 7.23 The exponent (γ) for mobility as a function of phonon energies ($\hbar\omega$). Exponent is given by $\mu_{\text{bulk}} = \mu_0(T/T_0)^{-\gamma}$. The graphs a, b, and c correspond to the scattering by homopolar phonon modes, by polar phonon modes, and by acoustic phonon modes, respectively. Here, bulk-MoS_2 corresponds to $\hbar\omega = 60$ meV and $\gamma = 2.6$ (indicated by an *arrow* along the vertical axis). Taken from Ref. [67]

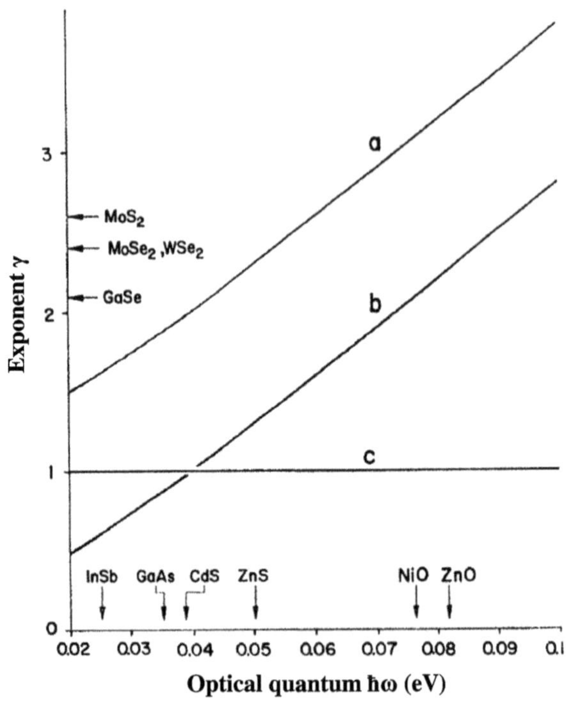

7.6.2 Temperature Dependence of Carrier Mobility in Single-Layer MoS_2

From theoretical point of view. The electrical properties of single-layer MoS_2 transistor have been probed extensively in past few years. The different phonon scattering mechanisms with the charge carriers discussed in the previous section are calculated [69] and plotted at $T = 300$ K as a function of energy for nondegenerate carriers, as shown in Fig. 7.24a. The Brillouin zone for SL- MoS_2 is hexagonal and similar to graphene, two valleys exist at K and K' point. Kaasbjerg et al. [69] calculated both the intravalley and intervalley scattering contributions to the relaxation time of the charge carriers, as these two valleys are occupied by the carriers for *n*-type SL-MoS_2. Here the first-order deformation potential is also shown in addition to other scattering mechanisms. The longitudinal acoustic (LA) and transverse acoustic (TA) modes contribute to quasielastic intravalley scattering process through the acoustic deformation potential (ADP). The contribution from TA is nonzero because of the inclusion of Umklapp process and results in a coupling of highly anisotropic nature [69]. From Fig. 7.24a, it is clear that at low carrier energy the *total* scattering rate is dominated by ADP. At higher energy, the total scattering rate is dominated by zero-order optical deformation potential

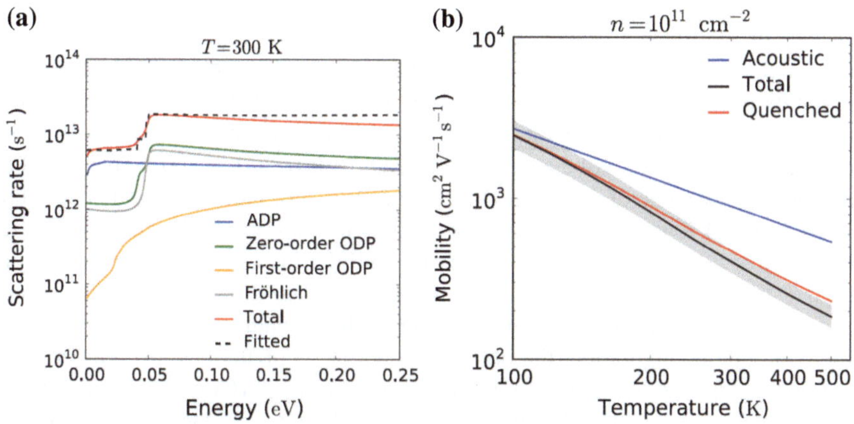

Fig. 7.24 **a** The first-principles calculations of different electron–phonon scattering rates (and the total) at room temperature versus the charge carrier energy. Here ADP = A_λ, Zero-order ODP = D_λ^0 and First-order ODP = D_λ^0. The *dashed line* shows the fitted deformation potential parameters given by $M_{\mathbf{q}\lambda} = A_\lambda q$ for acoustic modes and $M_{\mathbf{q}\lambda} = D_\lambda^0 q$ for all the other modes by choosing corresponding deformation potentials. The kinks in the curves for the optical phonon processes are attributed to the onset of phonon emissions. **b** Calculated mobility for different temperatures. The shaded area denotes the variations of the mobility due to the uncertainty (10 %) in the calculated deformation potentials. For comparison, the mobility because of the presence of the acoustic modes only is plotted, where $\mu \sim 1/T$. The *red solid line* represents the improvement of the mobility due to the quenching of the homopolar modes. Taken from Ref. [69]

(ODP) and the scattering via Frohlich interaction; the sudden jump in the optical scattering rates is because of onset of the emission of optical phonons by the charge carriers. At all energy scales, the first-order ODP scattering is one order of magnitude smaller than the other scattering rates.

Similar to the bulk case, the mobility due to ADP is $\mu_{SL}^{acoustic} \propto T^{-\gamma^{acoustic}}$ [69], where $\gamma^{acoustic} = 1$. The calculated temperature-dependent mobility for SL-MoS_2 is shown in Fig. 7.24b. At lower temperature ($\sim$100 K) the value of $\gamma = 1$. At higher temperatures the mobility follows according to $\mu_{SL}^{optical} \propto T^{-\gamma}$, where $\gamma = 1.69$ due to zero-order ODP and Frohlich scattering [69]. The calculated mobility at room temperature is of $\sim$400 cm^2/V s. If we exclude the zero-order ODP (homopolar mode), the resulting mobility is increased by $\sim$70 cm^2/V s and the reduced $\gamma = 1.52$, as shown in Fig. 7.24b for the quenched case, although the effect of quenching on the mobility is not much.

Comparison with the experimentally observed mobility. The experimentally observed (at room temperature) low mobility in the range of 1–50 cm^2/V s [10, 70, 71] reported earlier is due to the charge impurity scattering and increased up to 200 cm^2/V s [10] by depositing a high dielectric material (HfO_2) on top of the SL-MoS_2. This enhancement of the mobility is partially attributed to the strong damping of Coulombic scattering from charge impurities [72]. The enhancement of mobility is also achieved by polymer (PEO + $LiClO_4$) top gating from $\sim$0.1 to $\sim$150 cm^2/V s [73]. The difference between the calculated value ($\sim$410 cm^2/V s) and the experimentally observed ($\sim$200 cm^2/V s) may be due to the defects and surface optical phonon scattering [69]. The temperature-dependent mobility with the single-gate configuration is shown in Fig. 7.25. The field effect mobility is given by $\mu = [d\sigma/dV_{bg}][L/WC_{bg}]$, where σ denotes conductance, V_{bg} applied back gate voltage, L channel length, W the width of the channel, and C_{bg} represents back-gate capacitance associated with the SiO_2/Si layer. At $\sim$200 K, the mobility reaches a peak value of $\sim$18 cm^2/V s. The decrease of mobility with temperature from 200 to 4 K is attributed to the well-known scattering from charged impurities. Above 200 K, the mobility decreases with the increase in temperature. This high-T limited mobility due to phonons is fitted according to $\mu_{SL} \sim T^{-\gamma}$, where $\gamma \approx 1.4$. This is in good agreement with the predicted value of $\gamma \approx 1.69$ [69]. With the dual-gate configuration, the observed mobility is plotted against temperature, as shown in Fig. 7.26. The mobility varies from $\sim$168 cm^2/V s at 4 K to $\sim$60 cm^2/V s at 240 K. The mobility with dual-gate configuration is quite different from that with the single-gate configuration, where the mobility monotonously decreases with decreasing temperature. This distinct feature with dual-gate device is attributed to the strong damping of charged impurity scattering in presence of dielectric media and to the metallic top gate which also changes the dielectric environment for the SL-MoS_2 [31]. In the high temperature (100–300 K), the fitted value of mobility due to the phonon scattering shows $\gamma \approx 0.73$. All the other dual-gate devices of SL-MoS_2 show the variations of γ from 0.3 to 0.75 [31]. These observed values are much less than the predicted $\gamma \approx 1.52$ [69]. In addition to the quenching of homopolar mode due to the

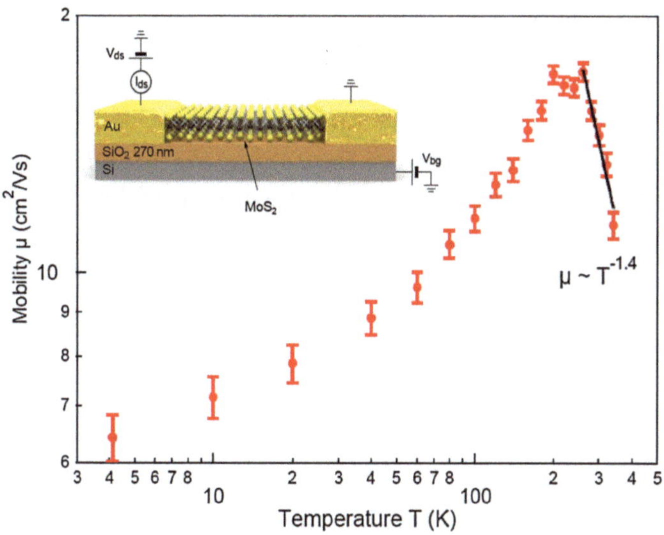

Fig. 7.25 The temperature-dependent field effect mobility for single-layer MoS_2 deposited on SiO_2/Si substrate. The back gate voltage (V_{bg}) is in the range of 30-40 V to extract the mobility. In the high-T range, the fitted value of the exponent (γ) is 1.4. Taken from Ref. [31]

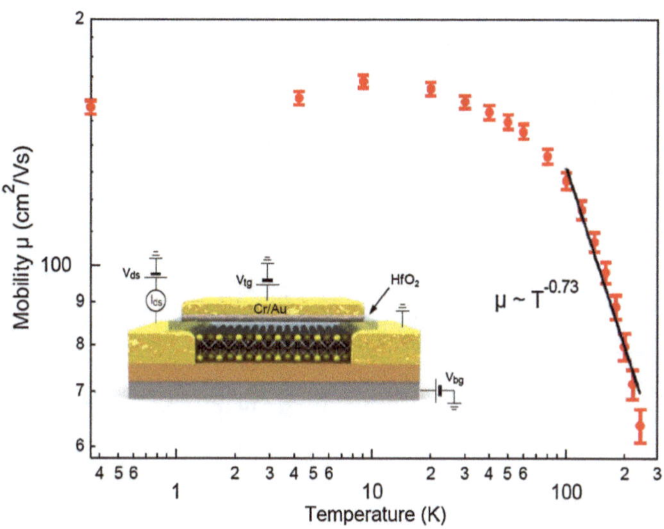

Fig. 7.26 The temperature-dependent field effect mobility $\left(\mu = [\mathrm{d}\sigma/\mathrm{dV}_{bg}][\mathrm{L}/\mathrm{WC}_{tg}]\right)$ for single-layer MoS_2 deposited on SiO_2/Si substrate. Applied top gate voltage (V_{tg}) is in the range of 3-5 V to extract the mobility. In the high-T range, the fitted value of the exponent (γ) is 0.73. Taken from Ref. [31]

dielectric environment, other phonon scattering mechanisms are also reduced which needs further theoretical work to clarify this discrepancy.

7.6.3 Substrate Effect Toward Improving the Mobility of MoS_2: Suppression of Short Range and Long Range Interactions

It has been established that the presence of the top gate material on MoS_2 enhances the mobility of the charge carriers [10, 31, 73]. Before putting the top-gate material the SL-MoS_2 deposited on bare SiO_2/Si substrate shows a less amount of mobility in the range of 1–50 cm^2/V s [10, 70, 71]. In order to investigate the substrate effect on the measured mobility, two different kinds of devices are considered for the various thickness (1–80 nm) of MoS_2: (i) the MoS_2 flakes are deposited on SiO_2/Si substrate (shown in Fig. 7.27a) and (ii) those deposited on polymethyl methacrylate (PMMA) (shown in Fig. 7.27b). Figure 7.27c shows the observed room temperature conductivity (σ) with the applied back gate voltage (V_{bg}) for

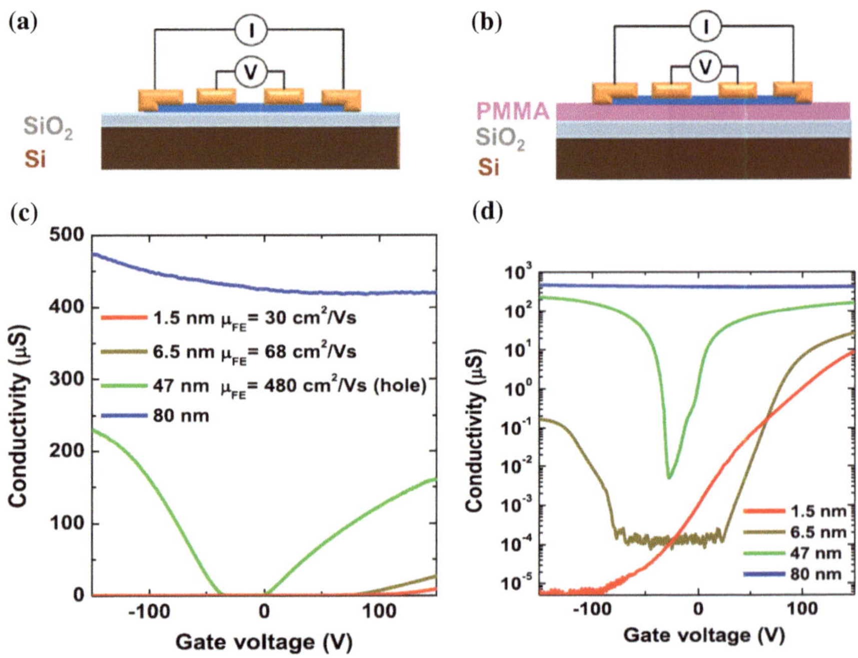

Fig. 7.27 Schematic diagrams of the four-probe devices (contacts made of Ti/Al electrodes) **a** without PMMA and **b** with PMMA (300 nm) deposited on 300 nm SiO_2/Si substrate. The measured conductivity (σ) of four MoS_2-devices having PMMA with different thicknesses (1.5, 6.5, 47, and 80 nm) **c** with linear scale and **d** with the semi-log scale. Taken from Ref. [74]

four different thicknesses of sample deposited on 300 nm PMMA [74]. Figure 7.27d shows the semi-logarithmic behavior of the conductivity. While the 6.5 and 47 nm thick MoS_2 show the ambipolar behavior, 1.5 nm thick (2L-MoS_2) shows unipolar nature with mobility 30 cm^2/V s. The mobilities for the 6.5 nm thick sample are given by $\mu_e(6.5) \sim 68 cm^2/V$ s and $\mu_h(6.5) \sim 1 cm^2/V$ s for electrons and holes, respectively; for 47 nm thick MoS_2, these are given by $\mu_e(47) \sim 270$ cm^2/V s and $\mu_h(47) \sim 480$ cm^2/V s. It is clear from Fig. 7.27d that the region (off-state) separating electron and hole conductance decreases with the increasing thickness of the sample; the off-state conductance also increases with the thickness. For an 80-nm-thick sample, there is no off-state indicating the dominance of conductance from the bulk [74].

The thickness-dependent mobility at room temperature for two different kinds of device configurations is shown in Fig. 7.28. The ambipolar nature has been marked by a dashed line connecting the hole and electron mobility. From Fig. 7.28, all the devices can be grouped into two categories: (a) unipolar type-shown by all the devices deposited on SiO_2/Si substrate and by all thin samples (1–30 nm) supported by PMMA; (b) ambipolar type—observed for all the thick sample of width in the range of 40–70 nm deposited on PMMA [74]. The observed mobility peaks are for ~ 50 nm thick MoS_2 samples of ambipolar type and are given by $\mu_e(50) \sim 470$ and $\mu_h(47) \sim 480$ in cm^2/V s. From Fig. 7.28 it can be inferred that for most of the devices of ambipolar type, the μ_h is larger than the μ_e. Hence, the multilayer MoS_2 for the *p*-type operation can be chosen for faster electronics. The thickness-independent mobility shown by unipolar-type devices deposited on SiO_2/Si substrate is in the range of 30–60 cm^2/V s. The observed higher mobility at room temperature ($\mu_e(50) \sim 470$ and $\mu_h(47) \sim 480$ in cm^2/V s)

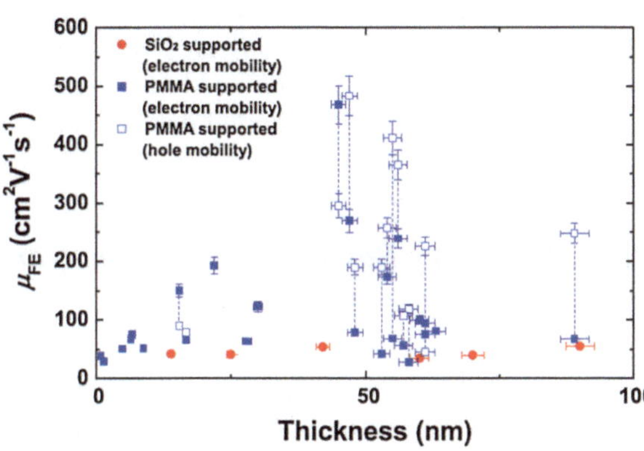

Fig. 7.28 The field effect mobility (at 300 K) versus thickness of the MoS_2 flakes for 25 devices with PMMA and 6 devices without PMMA. For PMMA supported devices (range of $V_{bg} \pm 150$ V), electron (solid *squares*) and hole (*empty squares*) mobilities are connected with dashed lines. For the SiO_2 supported devices (range of $V_{bg} \pm 75$ V), only electron mobility (*red circle*) is shown. Taken from Ref. [74]

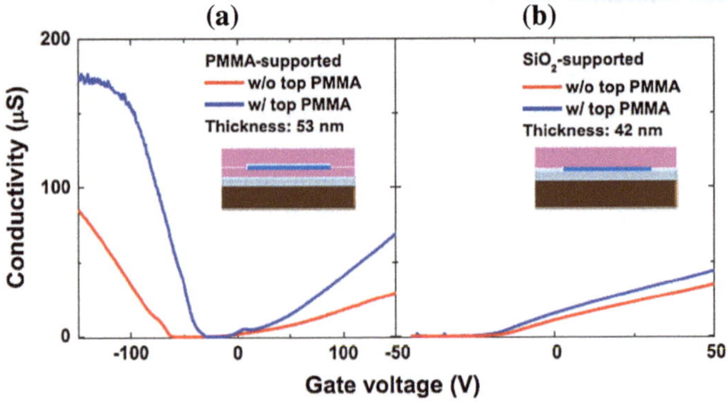

Fig. 7.29 The conductivity with back gate voltages **a** for PMMA supported device of 53 nm thick MoS_2 and **b** for SiO_2 supported device of 42 nm thick MoS_2. The *red and blue curves* show the conductivity before and after PMMA coating (from top) on MoS_2 flakes. The insets show the schematics of the devices. Taken from Ref. [74]

for the multilayer MoS_2 is larger than the theoretically predicted phonon limited mobility for SL-MoS_2 [69] and for the thick layer of MoS_2 (~ 100 cm^2/V s) [75].

In case of MoS_2 devices, the charge carriers are confined close to the dielectric interface (within few nm) used for the back gating [75]. The thickness-dependent enhancement of the carriers mobility can be attributed to the role of the additional MoS_2 layers to screen the long range disorder [74]. To check the screening effect, PMMA coating has been done on the top of the devices for the two different types of configurations (see Fig. 7.27a, b). The resulting measured conductivity is shown in Fig. 7.29a, b. While the PMMA coating helps to screen the long range scattering for the PMMA supported devices, the effect is less for SiO_2/Si supported devices. The maximum improvement mobility observed with the PMMA coating is $\sim$more than 300 % for 60 nm thick sample [74]. The observed variation of the mobility on the thickness or to the PMMA coating is attributed to the presence of short range disorder (rough surface) for the SiO_2/Si supported devices [74].

7.6.4 Reduction of Schottky Barrier at Metal–Semiconductor Contact to Improve the Mobility of MoS_2

In previous studies, attention has been paid to the interface physics, i.e., the interface between gate dielectric and the sample, the interface between the substrate and the sample to improve the mobility, and hence device performance. However, the presence of Schottky barrier between metal contacts and semiconductor can affect the mobility significantly. In order to understand the Schottky barrier contacts and their role in improving the mobility, four different metals scandium (work function $\Phi_M = 3.5$ eV), titanium ($\Phi_M = 4.3$ eV), nickel

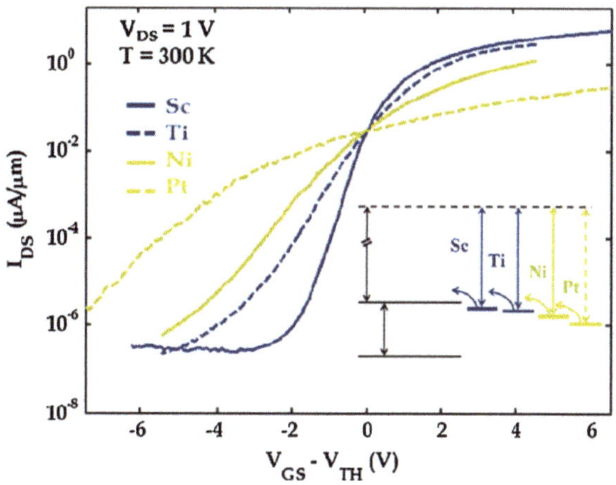

Fig. 7.30 The drain-source current versus back gate voltage for 6-nm-thick MoS_2. The threshold voltage, $V_{Th} = -6.0, -1.0, 1.5$ and 4.0 V for Sc, Ti, Ni, and Pt contact devices, respectively. The *inset* shows the alignment of the Fermi levels of four metals with the conduction band edge (irrespective of different work functions of the metals) according to the observed n-type device characteristics. Taken from Ref. [11]

$(\Phi_M = 5.0\ \text{eV})$, and platinum $(\Phi_M = 5.9\ \text{eV})$ are used to make contacts with the multilayer MoS_2 (electron affinity = 4.0 eV [76]) for the devices using a back gate. Figure 7.30 shows the experimentally measured transfer characteristics for the 6-nm-thick MoS_2 device [11]. All the four devices show the unipolar *n*-type operation, irrespective of the different work functions of the four metals. On the positive side of V_{GS}–V_{TH}, there is a clear decrease in on-current from Sc to Pt contacts, and this is consistent with the formation of Schottky barrier at the metal to channel interface.

In order to estimate the true height of the Schottky barrier (Φ_{SB}) quantitatively, the temperature dependence (100–400 K) transfer characteristics are measured for the Ni contact as shown in Fig. 7.31a. In the subthreshold region, the number of charge carriers are less and hence diffusive temperature-dependent thermionic current $(I_{\text{thermionic}})$ and tunneling current $(I_{\text{tunneling}})$ become important. The temperature-dependent current (I_{DS}) is then plotted for various back gate voltages, as shown in Fig. 7.31b. The thermionic emission current is fitted to extract the effective barrier height (Φ_B) with the well-known formula $I_{DS} = I_{\text{thermionic}} = AT^2 \exp\left[\frac{e\Phi_B}{K_B T}\right]\left[1 - \exp\left(\frac{eV_{DS}}{K_B T}\right)\right]$, where A is the Richardson's constant, K_B the Boltzmann constant, *e* the electronic charge. Figure 7.31c shows the variation of the extracted Φ_B with the applied V_{GS}. The effective barrier height $\Phi_B = \Phi_{SB} - C_{\text{eff}}(V_{GS} - V_{FB})$ [11], for $V_{GS} \leq V_{FB}$ and $\Phi_B = \Phi_{SB} - \Delta$, for $V_{GS} > V_{FB}$. The factor $C_{\text{eff}} = 1\Big/\left(1 + \frac{C_{it}+C_{ch}}{C_{ox}}\right)$ represents the band movement of the channel with V_{GS}, where C_{it} and C_{ox} denote the capacitance for the interface trap charges

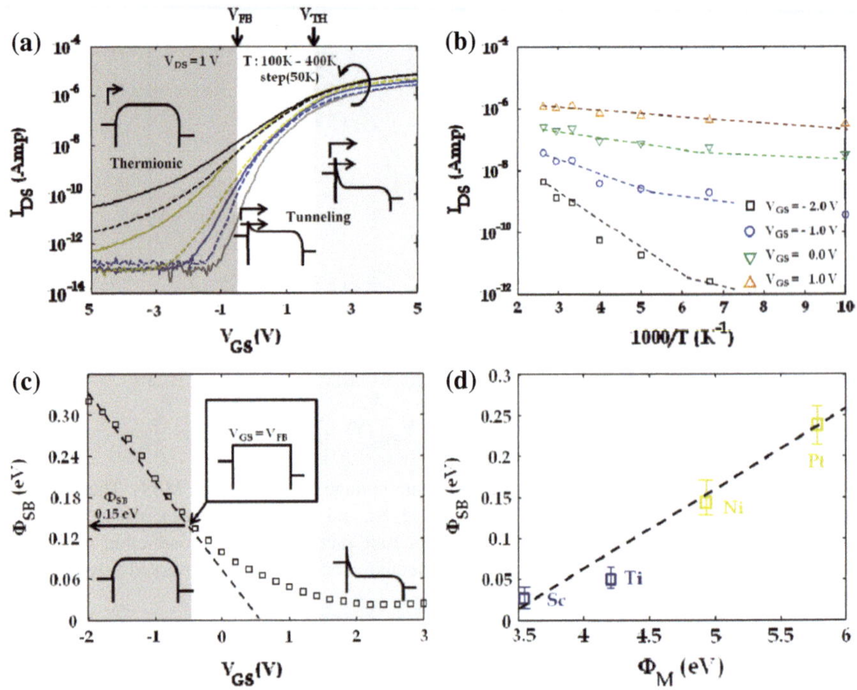

Fig. 7.31 **a** The drain-source current versus back gate voltage at different temperatures showing three distinct regions of operations: below flat-band voltage (V_{FB}), below threshold voltage, and above threshold voltage. The insets show the energy-band diagrams in the corresponding regions. **b** The drain-source current versus the inverse-temperature (1/T) (taken from a). **c** The effective barrier height (Φ_B from **b**) as a function of gate voltage and the extraction of true Schottky barrier (Φ_{SB}) for Ni-contact device. **d** All Φ_{SB} versus work functions for all the different metal contact (Sc, Ti, Ni and Pt) devices. Taken from Ref. [11]

and oxide layers, respectively. Δ is a positive quantity but not constant. V_{FB} is the flat band voltage. The $I_{\mathrm{thermionic}}$ is the only current through the channel for the negative $V_{\mathrm{GS}} < V_{\mathrm{FB}}$ (see Fig. 7.31a) and hence, the linear relationship between Φ_{B} and Φ_{SB} as shown in Fig. 7.31c. When $V_{\mathrm{GS}} \geq V_{\mathrm{FB}}$, the $I_{\mathrm{tunneling}}$ starts to flow through the channel and this current is not included in the fitted equation for $I_{\mathrm{thermionic}}$. Because of this fact, the linear relationship between Φ_{B} and Φ_{SB} does not hold and starts to deviate after V_{FB}, as shown in Fig. 7.31c. The extracted value of Φ_{SB} is 0.15 eV fo Ni contact. It must be noteworthy that this is the tunneling current which is responsible for getting ohmic contact even for the higher work function metals including gold ($\Phi_M = 5.4$ eV) [11]. Figure 7.31d represents the measured Φ_{SB} for different metals, which are 30, 50, 150, and 230 meV for Sc, Ti, Ni and Pt, respectively. The fitted dashed line gives the slope $\mathrm{d}\Phi_{\mathrm{SB}}/\mathrm{d}\Phi_M \sim 0.1$ (0.27 for Si [77]) implying that there is a strong pining of Fermi level at the semiconductor interface before the metal contact is made. This pining of Fermi level near neutral level is related to the charge impurities present on the

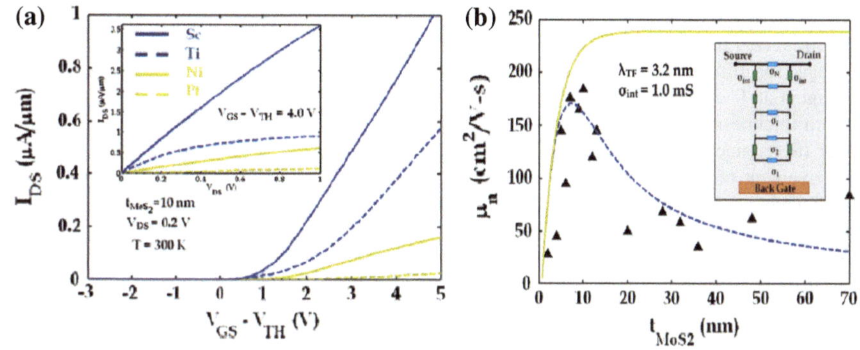

Fig. 7.32 **a** The drain-source current versus back gate voltage (T = 300 K) for Sc, Ti, Ni, and Pt-contact devices. Here $V_{GS}-V_{TH} = 4$ V corresponds to $\sim 1.4 \times 10^{12}/cm^2$. The value of the back gate capacitance is $\sim 3 \times 10^{-4}$ F/m^2. **b** The extracted field effect mobility for different thicknesses of the sample with Sc-contact devices. The *dashed line* shows the fitting of the data according to the resistor network model, where two model parameters are Thomas–Fermi screening length ($\lambda_{TH} \sim 3$ nm) and the interlayer conductivity ($\sigma_{int} = 1$ mS). The *solid line* represents the calculated mobility without taking σ_{int}. The *inset* shows the schematic of the resistor network model. Taken from Ref. [11]

semiconductor-metal interface, which is usually the case with the covalent semiconductors [77]. This explains clearly why only *n*-type device characteristics are observed irrespective of high work functions of the metals used for making contacts.

To understand the effect of measured Schottky barrier on mobility, the output and transfer characteristics for 10-nm-thin MoS_2 are shown in Fig. 7.32a. The exponential rise (see the inset) of I_{DS} with V_{DS} for the Pt contact is in accordance with the Schottky barrier of 230 meV. For Ni-contacted devices having $\Phi_{SB} = 150$ meV, it shows linearity. Since at 300 K, the temperature assisted tunneling current is responsible for this observed linear behavior between the current and drain voltage, the issue of getting ohmic contact is irrelevant [11]. The measured field-effect mobilities from transfer characteristics are 21, 36, 125, and 184 in cm^2/V s for the devices having Pt, Ni, Ti, and Sc contacts, respectively [11]. The measured low mobility [10, 70, 71, 73] (<50 cm^2/V s) using Au or Ti/Au contacts without top gate material is in agreement with the effect of Schottky barrier.

Figure 7.32b shows the variations of the mobility with the thickness of the sample in the range of 2–70 nm for the device having Sc contacts. The maximum mobility observed corresponds to the ~10-nm-thin MoS_2 device. The observed mobility is fitted with the resistor network model (see inset of Fig. 7.32b). For thicker samples, the access to the lower layer involves the interlayer resistance. For thinner samples, the screening of the substrate is much lesses than the thicker one. The observed peak is therefore related to the competition between the interlayer conductivity (σ_{int}) and the screening of the substrate [11]. Figure 7.33

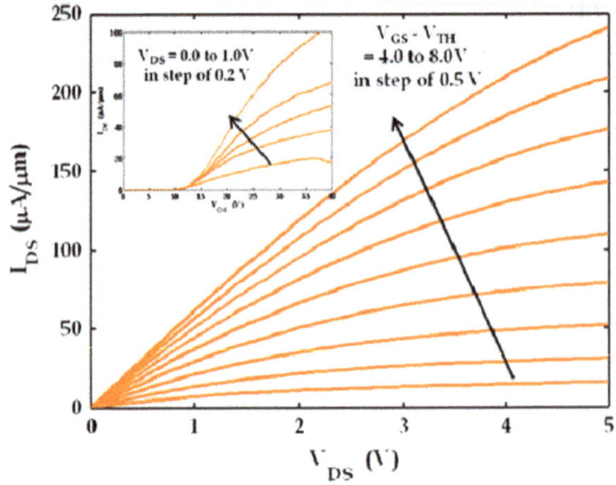

Fig. 7.33 The drain-source current versus drain-source voltage (for different top gate voltages) at T = 300 K for 10 nm thickness of the MoS_2 layer (Sc-contact). The improvement of the transport coefficients is achieved by depositing a top-gate material (Al_2O_3) of 15 nm thickness. The *inset* shows the corresponding transfer characteristics for different drain-source voltages. Taken from Ref. [11]

shows the output and transfer characteristics for 10-nm-thin MoS_2 with the addition of 15-nm dielectric material (Al_2O_3) on the top of the sample. The field effect mobility due to the screening of long range scattering is now enhanced to the value of ~700 cm^2/V s from 184 cm^2/V s for Sc contacted device. This huge increment is attributed to the (i) use of low Schottky barrier $\left(\Phi_{SB}^{SC} = 30\text{ meV}\right)$, (ii) choosing of optimum value of layer thickness (10 nm) which corresponds to the optimization of suppression of substrate effect as well as the interlayer resistor, and (iii) screening of the long range Coulomb interaction. Finally, a high saturation current density ~240 μA/μm at the carrier density of ~$2.8 \times 10^{12}/cm^2$ and the higher transconductance ~4.7 μS/μm at $V_{DS} = 1.0$ V are measured for a 5 μm channel length. In order to get a higher mobility (≥700 cm^2/V s), it is desirable to perform experiment on a device having Sc contact deposited on a high dielectric material instead of SiO_2/Si substrate (maybe PMMA [74]) with high dielectric top gating material by varying only the thickness of the sample.

7.7 Phonon Renormalization in a Single-Layer MoS_2 Transistor: Aspects of Symmetry

It has been established that the high temperature (~200–300 K) phonon limited mobility in single and bulk-MoS_2 (homopolar mode A_{1g} ~ 408 cm^{-1}) is because of dominant scattering of charge carriers by the optical phonons [31, 67]. Electron–phonon scattering not only affects carrier momentum relaxation, but also affects the frequency and the full width at half maximum (FWHM) of phonons via renormalization of their self-energy; the real part of the self-energy is related to the frequency shift and the imaginary part to the FWHM [68]. The nondestructive Raman spectroscopy is exploited extensively to probe the electron–phonon coupling (EPC) quantitatively for single-layer graphene [79–81] and bi-layer graphene [82–84]. To quantify the EPC for single-layer MoS_2, in situ gate voltage-dependent Raman experiments have been reported [78]. The schematic diagram for the device is shown in Fig. 7.34a. A mixture of $LiClO_4$ and polyethylene oxide (PEO) in 1:8 weight ratio is used as a top gating material. Figure 7.34b shows the atomic force microscopy (AFM) image in contact mode to measure the height (0.7 nm) of the MoS_2. The output characteristics of the device is shown in Fig. 7.34c. The observed slight nonlinearity between I_{DS} and V_{DS} is due to the Schottky barrier for Au contacts [11] (as discussed earlier). The transfer characteristics in semi-log scale are shown in Fig. 7.34d. The *n*-type operation was attributed to the Fermi level pinning at the semiconductor interface due to interface charges [11]. An extracted mobility was $\sim 50\ cm^2/V\ s$ and the on–off ratio was 10^5 (maximum observed $\sim 10^8$ [10]).

The observed Raman spectra with different top gate voltages is shown in Fig. 7.35a. In the spectral range of 350–450 cm^{-1}, two Raman modes are observed. From Table 7.1, these two modes are A_1' ($\sim 402\ cm^{-1}$) and E_2^1 ($\sim 382\ cm^{-1}$) corresponding to the bulk A_{1g} and E_{2g}^1 modes, respectively. We will

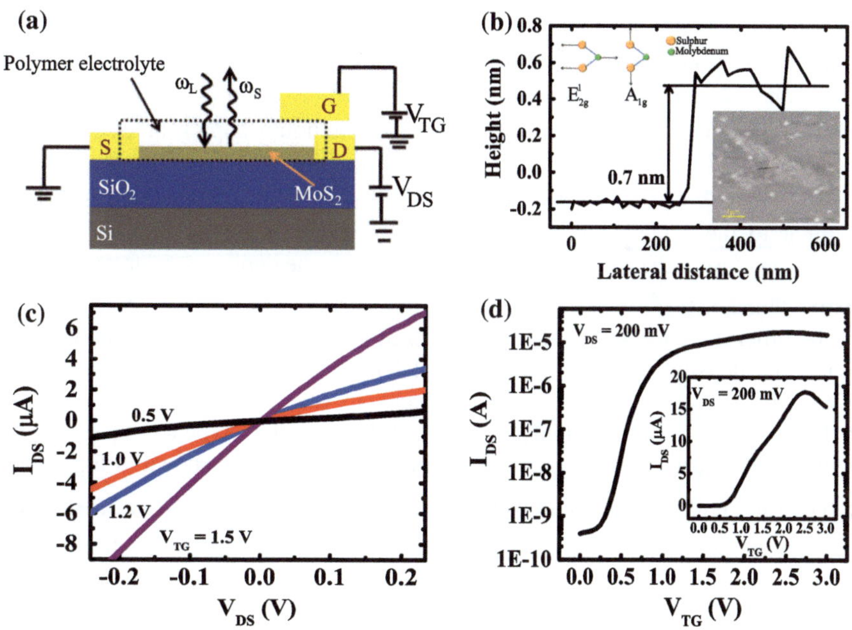

Fig. 7.34 **a** Schematic of the single-layer MoS_2 device (50 nm thick Au contact used). Here ω_L and ω_s represent the incoming and scattered photons, respectively. The length of the device is $\sim$ 2.5 μm and width is $\sim$ 1.5 μm. **b** The AFM height profile of the single layer. The right inset (AFM image) shows the dotted line on the sample along which the AFM height profile has been taken. **c** The output characteristics of the device at different top gate voltages. **d** The transfer characteristics in semi-log scale (the inset showing in linear scale) at $V_{DS} = 200$ mV. It shows the on–off ratio $\sim 10^5$. Taken from Ref. [78]

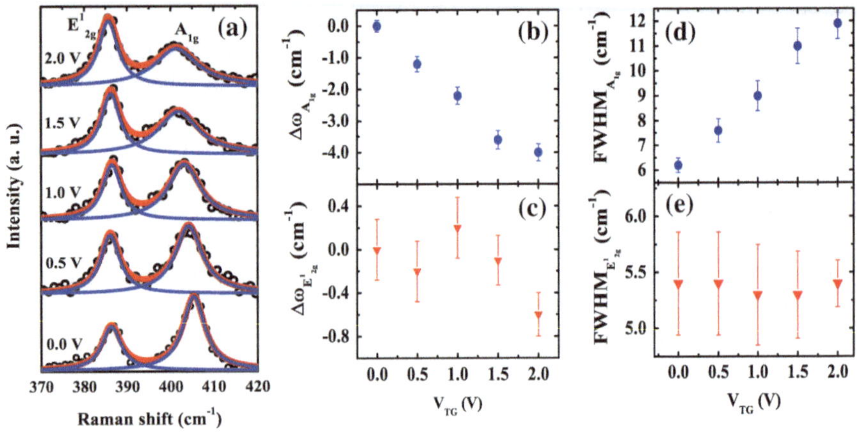

Fig. 7.35 **a** Raman spectra along with Lorentzian fits at different top gate voltages for the single-layer MoS_2. The changes in frequency and FWHM for the A_{1g} **b** and **d** and E^1_{2g} **c** and **e** modes as a function of top gate voltages. Taken from Ref. [78]

continue to use the notation A_{1g} and E^1_{2g} for the observed two modes. The frequency variations of the two phonon modes with the top gating (V_{TG}) are shown in Fig. 7.35b. The A_{1g} mode softens by 4 cm^{-1} and the E^1_{2g} mode shows almost no change ($\sim$0.6 cm^{-1}) in frequency up to the maximum doping of electron concentration $\sim 1.8 \times 10^{13}/cm^2$ ($V_{TG} \sim 2$ V). The FWHM of A_{1g} mode increases by 6 cm^{-1} (see Fig. 7.35c), but the FWHM of E^1_{2g} mode remains constant [78]. Similar to the detection principle of layer numbers of MoS_2 by measuring the $\omega_{A_{1g}} - \omega_{E^1_{2g}}$ [46], the $\Delta\omega_{A_{1g}}$ can be used to estimate the doping concentration of the MoS_2 flake.

In order to understand the renormalization of the A_{1g} phonon mode, first-principles density functional theoretical (DFT) calculations have been carried out [78]. Single-layer MoS_2 has a direct bandgap of 1.9 eV at the K-point. The hybridized conduction band edge consists of dominated d_{z^2}-type Mo states and the hybridized valence band edge states having d_{xy}-type [78], as shown in Fig. 7.36e, f. The square of the electronic wavefunction $\left(|\psi(r)|^2\right)$ or the charge density near the valence band top and the conduction band bottom have the same full symmetry as the single-layer MoS_2 crystal [78]. The EPC for a phonon mode ν with momentum q and frequency $\omega_{q\nu}$ is given by [85] $\lambda_{q\nu} = \frac{2}{\hbar\omega_{q\nu}N(\epsilon_f)}\sum_k \sum_{ij} |g^{q\nu,ij}_{k+q,k}|^2 \times \delta\left(\epsilon_{k+q,i} - \epsilon_f\right) \times \delta\left(\epsilon_{k,j} - \epsilon_f\right)$, such that the EPC matrix is given by $g^{q\nu,\,ij}_{k+q,\,k} = \left(\frac{h}{2M\omega_{q\nu}}\right)^{\frac{1}{2}}\langle\psi_{k+q,\,i}|\Delta V_{q\nu}|\psi_{k,\,j}\rangle$ where $N(\epsilon_f)$ is the density of states at the Fermi energy for electrons, $\epsilon_{k,j}$ the electronic energy with momentum **k** in a band j, $\psi_{k,j}$ corresponding electronic wavefunction. Here $\Delta V_{q\nu}$ denotes the change in the

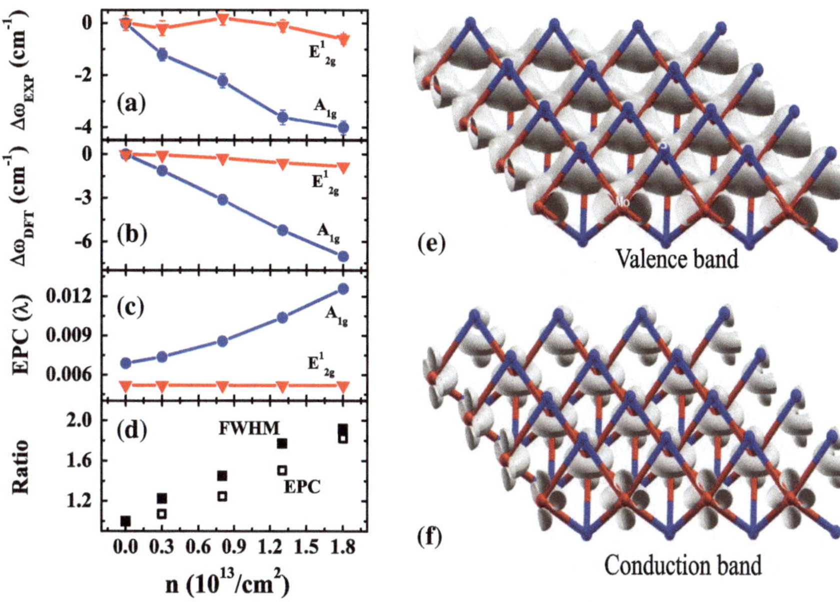

Fig. 7.36 Experimentally observed and the DFT calculated changes in frequency of the **a** A_{1g} and **b** E^1_{2g} modes versus electron concentrations (n). **c** Calculated EPC for the A_{1g} and E^1_{2g} modes. **d** Calculated two ratios for EPC [EPC(n $\neq$ 0)/EPC(n = 0)] and FWHM [FWHM^{total} $(n \neq 0)$/ FWHM^{total} $(n = 0)$] of the A_{1g} mode. The charge densities $\left(|\psi(r)|^2\right)$ for the **e** top of the valence band and **b** bottom of the conduction band. Here, the *dark gray* (*red*) and *black* (*blue*) spheres indicate Mo and sulfur atoms, respectively. Taken from Ref. [78]

potential due to the vibrations of the lattice associated with a phonon (ω_{qv}). In the A_{1g} phonon mode, the atoms vibrate in such a way that the total symmetry of the single-layer MoS_2 does not change, and hence it corresponds to the identity representation. For *n*-type device operations, the electrons are occupying gradually the conduction band edge states and for those electrons $|\psi(r)|^2$ transforms according to the identity representation. Hence, EPC matrix element is nonzero, in which the product of $\Delta V_{A_{1g}}$ and $|\psi(r)|^2$ is involved. The irreducible representation for the E^1_{2g} mode is orthogonal to the A_{1g} mode representation and hence the EPC matrix vanishes leading to no response of this mode to the channel doping [78].

To understand dependence of FWHM on doping, the total FWHM is expressed as $\text{FWHM}^{\text{total}} = \text{FWHM}^{\text{EPC}} + \text{FWHM}^{\text{an}}$; where FWHM^{an} is due to anharmonic effects. The two ratios $\text{FWHM}^{\text{total}}$ $(n \neq 0)$/$\text{FWHM}^{\text{total}}$ $(n = 0)$ and EPC$(n \neq 0)$/ EPC$(n = 0)$ have been calculated for the A_{1g} mode and they show the same trend as shown in Fig. 7.36d. This explains that only EPC is responsible to broaden the FWHM of A_{1g} mode with the doping. It is noteworthy that the top gating by an amount of $\sim 1.8 \times 10^{13}/\text{cm}^2$ leads to hardening of the G phonon mode by $\sim$10 cm^{-1} in case of graphene [81], while the same condition gives rise to softening (by $\sim$4 cm^{-1}) of A_{1g} mode for single-layer MoS_2. In case of graphene adiabatic approximation fails, whereas in case of MoS_2, EPC is within the adiabatic approximation.

7.8 Resonant Raman Scattering of Bulk MoS_2 Tuned by Temperature and Pressure: Stability and Contributions of A and B Excitons to Resonance

Resonant Raman scattering at ambient conditions on single and multilayer MoS_2 have been discussed [47], where 633 nm red laser line was used for excitation. It has been shown that there exist two excitons corresponding to direct transitions at the *K*-point [64]. The exciton *A* having binding energy 42 meV corresponds to $K_4 \rightarrow K_5$ optical transition, whereas B exciton of binding energy 134 meV corresponds to $K_1 \rightarrow K_5$ optical transition. The energy difference between the *A* and *B* excitons is due to the spin–orbit splitting and interlayer interaction [64]. Since the pressure and temperature variations affect the lifetime of the excitons, we will address their stability and the limits of applied pressure and temperature to which the resonance effects can be observed. The appearance of all the Raman active modes and inactive modes $\left(E^2_{1u} \text{ and } B_{1u}\right)$ due to resonance effects is already discussed [47] in Sect. 7.5. Here, our focus is on the 'b' mode attributed to the two-phonon Raman process (see Fig. 7.20) such that it corresponds to the emission of a dispersive quasi-acoustic (QA) longitudinal phonon followed by emission of a dispersionless transverse optical mode (here c-band) [60]. The more interesting feature about the b mode is that the observed frequency for the Stokes $\left(\omega_{\rm b}^{\rm S}\right)$ and anti-Stokes $\left(\omega_{\rm b}^{\rm AS}\right)$ spectra is not the same and they differ by $\sim$4.7 cm^{-1}, as shown in Fig. 7.37.

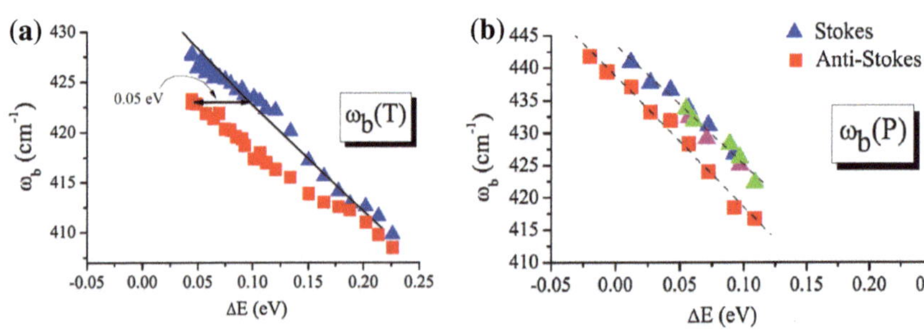

Fig. 7.37 **a** Temperature and **b** pressure-dependent ω_b^S and ω_b^{AS} as a function of the energy difference $\Delta E = E_L - E_{A1}$. Solid line in (**a**) indicates the fitting for ω_b^S. Taken from Ref. [86]

Since, longitudinal phonons are strongly coupled to excitons than the transverse ones, the contribution of QA phonons will be more dominant to the b mode [60]. The rate at which frequency decreases with temperature $(\partial\omega/\partial T)$ for the b mode is $\sim 2.66 \times 10^{-2}$, whereas for all the other modes it is in the range of ~ 1.0–1.6×10^{-2}. Figure 7.37a, b shows the comparison between the shifting of the peak position of the b band for the Stokes and anti-Stokes scattering as a function of temperature and pressure, respectively. Here, the $\omega_{\mathrm{b}}^{\mathrm{AS}}$ is redshifted compared to the $\omega_{\mathrm{b}}^{\mathrm{S}}$ part.

Since the intermediate exciton state plays an important role for the observed b band and from the above experimental findings, three major things need to be addressed: (i) identification of tuned excitonic states with respect to temperature and pressure to observe resonance effects for Stokes as well as anti-Stokes processes, (ii) the stability of the A and B excitons, and (iii) the observed redshifted frequency $\omega_{\mathrm{b}}^{\mathrm{S}} - \omega_{\mathrm{b}}^{\mathrm{AS}} \sim 4.7\ \mathrm{cm}^{-1}$. As already noted [78], the d_{z^2} (this orbital points along the c-axis) character of the K_5 final state predominantly couples the A_{1g} phonon with the excitons than the E_{2g}^1 mode. This means that the A_{1g} phonon mode is much more sensitive to the resonance effect than the E_{2g}^1. Hence the resonance effect will be captured from the intensity ratio by comparing $\left[I_{A_{1g}}/I_{E_{2g}^1}\right]^{\mathrm{reso}}$ with the $\left[I_{A_{1g}}/I_{E_{2g}^1}\right]^{non-reso}$ (i.e., with 514 nm laser line). Here, the measured $\left[I_{A_{1g}}/I_{E_{2g}^1}\right]^{reso}$ implies the Raman scattering cross-section. The intensity ratio $\left[I_{A_{1g}}/I_{E_{2g}^1}\right]^{\mathrm{reso}}$ is denoted by R^S for Stokes process and by R^{AS} for anti-Stokes process.

The probability of the Raman scattering of a phonon by an excitonic 1s state is given by [68, 86]

$$P_{\text{phonon}} \approx (2\pi\hbar)\left|\frac{\langle 0|H_{eR}(\omega_L)|1\rangle\langle 1|H_{e-\text{ion}}|1\rangle\langle 1|H_{eR}(\omega_s)|0\rangle}{(E_i-\hbar\omega_L-i\Gamma_1)(E_i-\hbar\omega_s-i\Gamma_1)}\right|^2.$$

Here, H_{eR} and $\mathrm{H}_{\mathrm{e-ion}}$ represent the electron–photon and electron–ion interaction Hamiltonian, respectively. The states indicated by 0 and 1 correspond to the ground and the exciton states (A_1 and B_1), respectively. The incident and scattered laser frequencies are given by ω_{L} and ω_{s}, respectively. $\mathrm{E}_i = \hbar\omega_{\mathrm{L}}$ or $= \hbar\omega_{\mathrm{s}}$ depending on incoming or outgoing resonance, respectively. It has been shown experimentally that the Raman scattering cross-section for the b-band shows a peak at $\mathrm{E}_{1\mathrm{s}}^{\mathrm{A}} + \hbar\omega_{\mathrm{b}}$ (for the A exciton), but does not correspond to energy of the exciton 1 s state, i.e., at $\mathrm{E}_{1\mathrm{s}}^{\mathrm{A}}$ [60]. Here, the outgoing probability for the Stokes and anti-Stokes Raman scattering of A_{1g} mode is given by $P_{A_{1g}}^{\mathrm{S}}$ and $P_{A_{1g}}^{\mathrm{AS}}$ corresponding to $E_i + \hbar\omega_{A_{1g}}$ and $E_i - \hbar\omega_{A_{1g}}$ energy levels, respectively. Ho et al. [88] measured the temperature dependence of the exciton energies by fitting a formula given by $E_i(T) = E_{iPo} - a_{iP}\left[1 + \frac{2}{\exp(\Theta_{iP}/T)-1}\right]$, where a_{iP} and Θ_{iP} represent the exciton–

phonon interaction strength and the average phonon temperature, respectively. The measured values of E_{iPo}, a_{iP} and Θ_{iP} are for A_1 are 1.976 eV, 46 meV, and 220 K, respectively; and for B_1 they are 2.179 eV, 42 meV, and 200 K, respectively [88] for $T \leq 300$ K. The broadening parameter has also been measured by fitting the equation $\Gamma(T) = \Gamma_0 + \frac{\Gamma_{LO}}{\exp(\Theta_{LO}/T) - 1}$ where $\Gamma_{LO} = 75$ meV, represents the exciton-LO phonon interaction strength and $\Theta_{LO} = 560$ K. The Γ_0 for A_1 and B_1 excitons are 18.0 and 37.4 meV, respectively. Hence, it is clear that the broadening is because of exciton–optical phonon interactions [88].

The temperature-dependent binding energies of the A_1 and B_1 excitons are shown in Fig. 7.38a. The measured intensity ratios R^S and R^{AS}, and the calculated probabilities $P^S_{A_{1g}}$ and $P^{AS}_{A_{1g}}$ are also plotted (blue and red solid lines). The outgoing resonance energies for the A_{1g} mode are also shown for Stokes and anti-Stokes scattering processes of A_1 and B_1 excitons. The interactions of excitons with the acoustic and optical phonons are responsible for the observed temperature-dependent shift, and also for the variations in the lattice parameters. Since the exciton-optical phonon interaction holds for the high temperature, the fitted parameters are extended to 550 K. The agreement between the theory and experiment is reasonable. From Fig. 7.38a, it is clear that at $T = 0$ K the excited

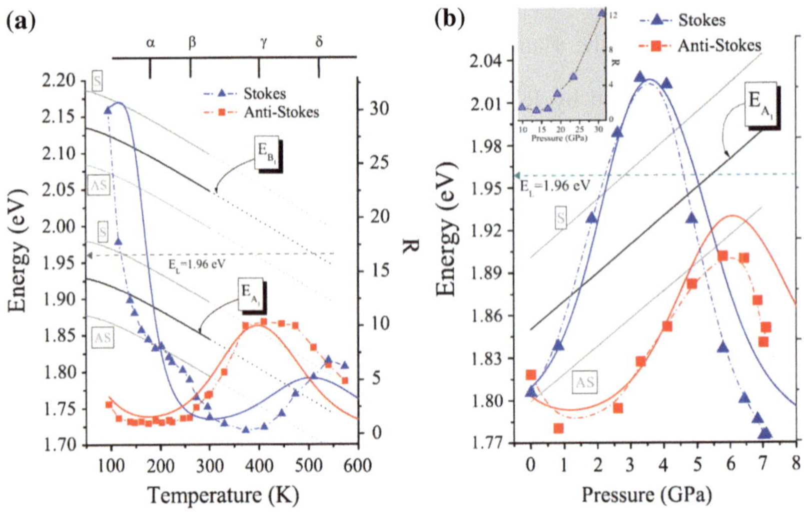

Fig. 7.38 **a** (corresponding to the left scale) Temperature (measured for T ≤ 300 K and then extended to 550 K)-dependent binding energies (E_{A_1} and E_{B_1}) of the A_1 and B_1 excitons. **b** Pressure-dependent binding energy for the A_1 exciton is plotted (corresponding to the left scale). Here, AS and S represent the outgoing resonance energies $E_i + \hbar\omega_{A_{1g}}$ and $E_i - \hbar\omega_{A_{1g}}$ for anti-Stokes and Stokes Raman scattering, respectively. The measured intensity ratios R^S for Stokes process, R^{AS} for anti-Stokes process, and the corresponding calculated probabilities (P^S and P^{AS}) are plotted together (corresponding to the right scale) in **a** and **b**. The inset in **b** shows the R^S for pressures above 10 GPa. The upper scale in **a** showing α-δ represent different temperature regions of operations discussed in the text. Taken from Ref. [86]

laser line ($E_L = 1.96$ eV) lies between E_{A1} and $E_{A1} + \hbar\omega_{A1g}$. As we increase the temperature the $E_{A1} + \hbar\omega_{A1g}$ (Stokes-process) approaches to E_L and around $\approx$ 95 K, outgoing resonance shows a peak for the R^S, i.e., for the Raman cross-section. Now with the further increase in temperature for $T > 175$ K (marked as α on the upper scale) $E_{A1} + \hbar\omega_{A1g}$ goes away from the E_L and $E_{B1} - \hbar\omega_{A1g}$ approaches E_L and continues up to ~260 K (marked as β). In the boundary of 175 K $\leq T \leq$ 260 K, the resonance effects are of mixed type: due to the proximity effects of E_L to the Stokes/anti-Stokes shifted outgoing resonance for the A_1 and B_1 excitons. With increase of T further, R^S goes to a minimum value around 350 K, whereas R^{AS} reaches a maximum value at ~400 K (marked as γ). For $T \geq 520$ K (marked as δ), R^S and R^{AS} follows the nonresonant temperature-dependence behavior of $\exp(-\hbar\omega_{A1g}/kT)$ [86].

Connell et al. have shown experimentally that the binding energy of the A_1 exciton decreases with pressure at the rate of ~ 7 meV/GPa at room temperature [89]. The decrease in the exciton binding energy is associated with the pressure-induced free charge carriers and hence more screening. Therefore, at ~6 GPa and higher values, the A_1 exciton ceases to exist. The pressure variations in the binding energy (~ 0.02eV/GPa [86]), Stokes, and anti-Stokes shifted energies of the A_1 exciton are plotted in Fig. 7.38b along with the measured (R^S and R^{AS}) and calculated $\left(P^S_{A_{1g}} \text{ and } P^{AS}_{A_{1g}}\right)$ resonance Raman scattering cross-sections. The pressure-dependent broadening parameter $\Gamma(P) \approx 35 + 3P$ (in meV) has been taken to calculate the $P^S_{A_{1g}}$ and $P^{AS}_{A_{1g}}$ values. According to calculations, the peak positions for the $P^S_{A_{1g}}$ and $P^{AS}_{A_{1g}}$ are ~3.8 and ~ 6.3 GPa, respectively [86]. Since $E_{A1} + \hbar\omega_{A1g}$ is closer to E_L at low pressure regime, R^S increases and peaks at ~ 3.8 GPa in accordance with the theoretical value. At higher pressure, $E_{B1} - \hbar\omega_{A1g}$ approaches to E_L (not shown in the figure) and hence R^{AS} increases to the peak value at ~ 6 GPa. The deviation of $P^{AS}_{A_{1g}}$ from R^{AS} for the pressure $\geq$ 6 GPa is because of the dissociation of the A_1 exciton into free electron–hole pairs [86]. Although the A_1 exciton is unstable in the high pressure range ($\geq$6 GPa), the measured R^S values with the increased pressure ($P \geq 10$ GPa) are shown in the inset of Fig. 7.38b. Livneh et al. attributed this observed high pressure regime of R^S to the pressure-induced change of electronic band structure, and by that the enhancement of the resonant sensitive A_{1g} mode couplings. This point needs to be cleared further from a theoretical point of view.

To address the stability of the B exciton, Livneh and Sterer [86] compared the observed intensity ratio I^{AS}/I^S and the calculated ratio P^{AS}/P^S (normalized by $n(\omega_{ph})/[n(\omega_{ph}) + 1]$, where $n(\omega_{ph}) = 1/[\exp(\omega_{ph}) - 1]$ is the Bose–Einstein factor) of the A_{1g} and E^1_{2g} modes for anti-Stokes and Stokes Raman scattering as a function of 1/T and proposed that as long as the ratio I^{AS}/I^S of the A_{1g} mode follows the resonant feature, B_1 exciton is stable, and hence it is unstable for $T \geq 520$ K.

It has been observed that the b-band shifts to lower frequency with the increase in the laser energy and reaches to the frequency of A_{1g} mode [60]. This dispersive mode was assigned as the two-phonon process having dominant contribution from the dispersive QA mode. In order to address the observed $\omega_b^{S} - \omega_b^{AS} \sim 4.7\,\mathrm{cm}^{-1}$ feature, the temperature and pressure-dependent ω_b^{S} and ω_b^{AS} are plotted with the energy difference $\Delta E = E_L - E_{A1}$ in Fig. 7.37a, b, respectively. The temperature-dependent ω_b^{S} through ΔE data is fitted (see Fig. 7.37a) according to the equation $\hbar\omega_b(q_z, T) = \Delta E(T) - \hbar^2\left(|q_z| \pm \omega_L \frac{n_0}{c}\right)^2/2M_{II}^{A1} + \hbar\omega_{TO}(T)$. Here, n_0 represents the ordinary refractive index. M_{II}^{A1} represents the effective mass of the A_1 exciton along the c-axis. The $\pm$ sign refers to the α and β processes (see Fig. 7.20). The q_z and k_z represent wavevector of the QA mode and the exciton pointing along the c-axis, respectively. The value of $\hbar\omega_b(0, 300) = 52.4\,\mathrm{cm}^{-1}$ is used as a single fitted parameter. The extracted value of the frequency for the QA mode $\sim 56\ \mathrm{cm}^{-1}$ is consistent with the earlier observations from neutron scattering [63]. The calculated temperature coefficient of the QA mode is given by $\left(\frac{\partial\omega_{QA}}{\partial T}\right)_{P=0} \approx -2 \times 10^{-3}$ in cm^{-1}/K for $\Delta E \leq 110$ meV [86], where $\omega_{QA} = \omega_b - \omega_c$. The observed $\left(\frac{\partial\omega_b}{\partial T}\right)_{P=0} \approx -2.66 \times 10^{-2}$ in cm^{-1}/K is much larger than the $\left(\frac{\partial\omega_{QA}}{\partial T}\right)_{P=0}$. Hence, the observed high temperature coefficient of the b-band compared to all the observed modes is attributed to the tuning of the exciton energy with respect to the temperature [86]. The largest shift between the Stokes and anti-Stokes scattering of the b-band for which $\omega_b^{S} = \omega_b^{AS}$ (see Fig. 7.37a) is given by 50 meV for $\Delta E \leq 110\,\mathrm{meV}$, which is close to the energy of c-band (TO mode $\sim 377\ \mathrm{cm}^{-1}$) [47, 86]. Finally, $\omega_b^{S} - \omega_b^{AS}$ is attributed to the reverse order processes for the Stokes and ant-Stokes scattering: the inner photon-like state scatters to the outer exciton–polariton state by absorbing a TO phonon and then scatters back to inner high-energy photon-like state by absorbing a QA mode, i.e., for anti-Stokes process it is TO + QA order, whereas for Stokes scattering it is of QA + TO order. So the redshift between ω_b^{S} and ω_b^{AS} is due to the order of the involvement of the dominant quasi-acoustic (QA) phonon in resonant two-phonon scattering process.

7.9 Photoluminescence of MoS_2: Control of Valley Polarization by Optical Helicity and Observation of Trions

7.9.1 Layer Dependence

The phenomena of radiative recombination of optically excited electron (e) and hole (h) pairs of a bandgap semiconductor are known as photoluminescence (PL). PL study has been used extensively to probe excitons and the bandgap of a semiconductor. As discussed, the bulk MoS_2 undergoes a transition from an indirect bandgap to a direct one in the monolayer limit [27]. PL experiments are useful to probe the transition with the number of layers. Figure 7.39a shows the comparison of the PL spectra for the monolayer and bilayer MoS_2. The striking difference with the addition of another layer is observed for the PL spectra, which is characterized by the photoluminescence quantum yield (QY). The PL spectra with the layer $N = 1$–6 is shown in Fig. 7.39b. The observed peaks are marked as '*A*', '*B*', and '*I*.' The extracted QY for the bright PL of the monolayer is $\sim 4 \times 10^{-3}$, whereas for $N = 2$–6, it is of $\sim 10^{-5}$–10^{-6} [27]. The inset of Fig. 7.39a shows the variations of the QY with the layer numbers. The observed peak A centered at ~1.90 eV corresponds to the direct exciton transition at the *K*-point of the BZ ($K_4 \rightarrow K_5$ optical transition), while the B peak centered at ~2.05 eV corresponds to the direct exciton transition $K_1 \rightarrow K_5$. The B peak is also observed for the monolayer MoS_2 deposited on SiO_2/Si substrate [17, 39]. The A peak having width ~50 meV (for $N = 1$) redshifts very little and broadens with the thickness of the sample. The peak marked as *I* starts to appear with the bilayer and shifts to lower energy with the *N*. The broad peak *I* is attributed to phonon-assisted PL peak of the indirect bandgap MoS_2. Since the states near conduction band edge at the *K*-point of the BZ have d_{z^2} character and Mo atoms reside in-between two planes of S atoms, these states are less affected for the interlayer couplings; whereas at the Γ point, both valence and conduction bands are linear combination of d-states (Mo) and p-states (S) and hence are more affected with the layer stacking. Therefore, indirect bandgap is increased due to the confinement effects, direct bandgap at the *K*-point remains almost constant. The peak energy of the A peak for monolayer and the peak energy of the I peak for all the other layers are plotted together in Fig. 7.39c, which shows that the bandgap decreases and approaches to the bulk value with the layers.

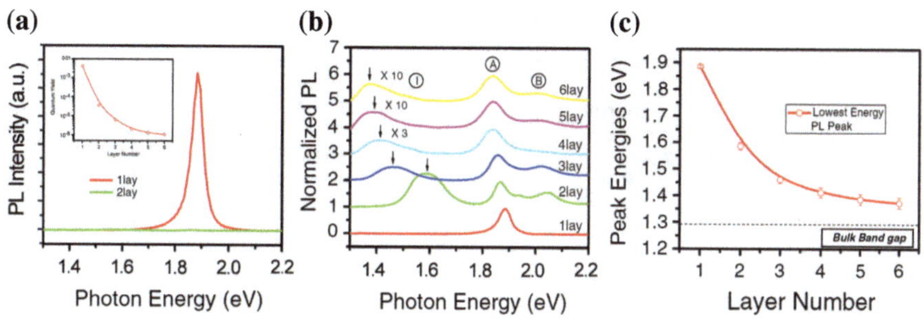

Fig. 7.39 a Room temperature PL spectra for suspended monolayer and bilayer samples. The inset shows the changes in QY with different layers ($N = 1$–6). **b** The normalized PL spectra (with respect to the A-peak intensity) for all the layers of MoS_2. Here, $\times$ 3 and $\times$ 10 indicate the magnification of the I-peak in the corresponding spectra. **c** The A-peak energy for $N = 1$ and the I-peak energy for $N = 2$–6 plotted together showing the bandgap variations with the number of layers. The dashed line represents the indirect bandgap of the bulk-MoS_2. Taken from Ref. [27]

The direct bandgap at K-point does not change with the layers, but the direct excitonic transition at the K-point gives different QY for the monolayer and the thick layer/bulk. In order to address this difference, the quantum efficiency for the MoS_2 is given by $\eta \approx \tau_{rad}/(\tau_{rad} + \tau_{defect} + \tau_{relax})$ [39], where τ_{rad} represents the radiative recombination rate, τ_{defect} the rate of defect induced scattering, and τ_{relax} the rate of intraband relaxation of charge carriers. The phonon-assisted τ_{relax} is very large for the indirect bandgap semiconductor. Because of the presence of the direct bandgap of the monolayer MoS_2, the decay rate via phonons decreases enormously, i.e., $\tau_{relax} \approx 0$. And τ_{rad} is nearly constant for both the bulk and monolayer. Hence, the enhanced QY limited by τ_{defect} for monolayer is due to the quenched channel of τ_{relax} with respect to the bulk value.

7.9.2 Temperature Dependence

At the lowest temperature (~4 K), the observed PL spectra of monolayer MoS_2 for the two different laser powers (200 μW and 40 mW) are shown in Fig. 7.40a. The lowest broad peak is marked as 'L' and the higher narrow peak is marked as 'H'. The L peak is separated from H by ~90 meV (ΔE_s). With the increasing power of the laser line by more than two orders, the shape of the spectra does not change. Figure 7.40b shows the variation of the total integrated intensity of the

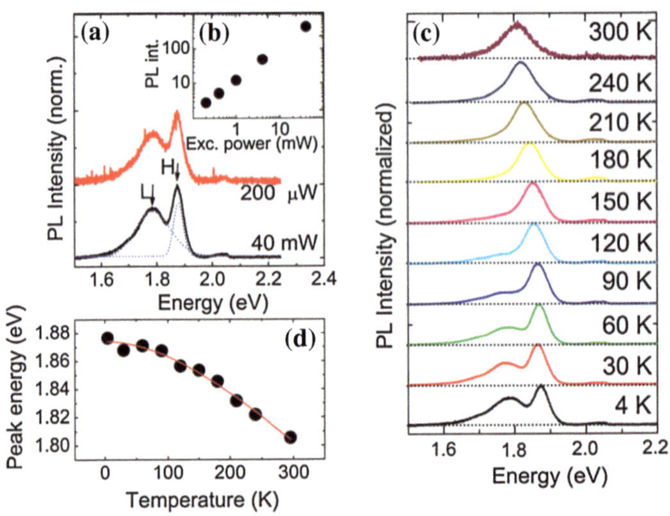

Fig. 7.40 **a** The normalized PL spectra measured at 4 K on monolayer MoS_2 deposited on SiO_2/Si substrate for two different laser (532 nm) powers. The two peaks are fitted with Gaussian functions. **b** The total PL intensity (integrated area of the two peaks) as a function of laser power. **c** The normalized PL spectra for different temperatures. **d** The energy of the H-peak versus temperature. *Solid line* represents the fitting of the Varshni formula. Taken from Ref. [90]

two peaks with the laser power, which shows linear relationship. It does not show the onset of the PL emission due to the presence of the non-radiative recombination channel of charge carriers to the defects at the low power. The constant relative intensity of the two peaks with the laser power indicates that the two radiative recombination channels do not interfere with each other in increasing the population of the charge carriers [90]. PL spectra at various temperatures (*T*) are shown in Fig. 7.40c. The two peaks redshift and broaden with the increased T. The intensity of the L peak decreases and almost vanishes above 120 K. The ΔE_s remains constant before the vanishing of L peak. The redshift of H peak is plotted with T as shown in Fig. 7.40d. The observed data are fitted with the Varshni formula, $E_g(T) = E_g(0) - [\alpha T^2/(T + \beta)]$, where $E_g(0) = 1.874$ eV, $\alpha = 5.9 \times 10^{-4}$eV/K and $\beta = 430$ K. The H peak is attributed (tentatively) to the free excitonic direct transition at the *K*-point. The L peak is assigned due to the bound excitons with the defects. Due to the attraction of the neutral impurities/defects the binding energy of the bound-exciton decreases [68]. Because of the distribution of various defects with different energies in the sample, the observed L peak is broadened. The non-radiative recombination process increases with *T* and it is responsible for the disassociation of the bound-exciton above 120 K rather than the thermal activation processes (the binding energy is larger than the 26 meV).

Figure 7.41a shows the time-resolved PL spectra at different temperatures. It is clear that the PL decays in 5 ps from 4 to 150 K. This fast decay is in accordance with the observation of relative intensity dependence on laser power at low T, i.e., there is no non-radiative relaxation process of the optically generated e–h pairs down to the minimum energy. We can see from Fig. 7.41a that above 150 K, there is an appearance of another long-lived component (LLC). The extracted decay rate and the amplitude of the LLC are plotted in Fig. 7.41b, c, respectively. The decay

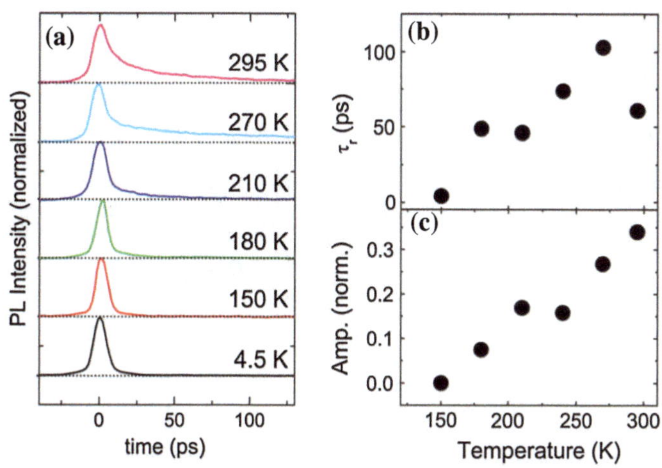

Fig. 7.41 **a** Time-resolved PL for different temperatures using 402 nm laser line. **b** The extracted decay rate and **c** the amplitude of the LLC. Taken from Ref. [90]

rate of LLC increases from 50 ps at 150 K to 100 ps at 270 K and then decreases to 70 ps at 300 K. The amplitude of LLC also increases with *T*, shown in Fig. 7.41c. The appearance of LLC and its increased decay rate are attributed to the onset of non-radiative phonon-assisted scattering channel of exciton-polaritons [90].

Plechinger et al. [91] made two kinds of samples of monolayer of MoS_2 to record the PL spectra. The first kind (A and E) is by mechanical exfoliation method on SiO_2/Si substrate and the second type of sample (B and F) is covered with the oxide layer (HfO_2 or Al_2O_3) of 15 nm thickness. The observed PL spectra at 4 K, shown in Fig. 7.42a, present two kind of features: (1) A and E samples not covered with the oxide layer show two prominent peaks and (2) B and F covered with the oxide layers show only one prominent peak. Since the top oxide layers and the high temperatures are used to make them remove most of the impurities from the surface, the low energy broad peak is attributed to the bound-excitons (to the surface impurities). The vertical dashed line in Fig. 7.42a indicates the redshift of F sample covered with HfO_2 ($\varepsilon_r \approx 20$) and blueshift of B sample covered with Al_2O_3 ($\varepsilon_r \approx 10$) compared to the without oxide layer samples (A and E). The screening of the long range Coulomb interaction does not play the role of shifting of the B and F samples, otherwise for F sample the blueshift could have been more than the B sample. Figure 7.42c shows the temperature variations (4–240 K) of the high energy peak for the two kinds of samples. The largest shift ($\approx$ 55 meV) occurs for the A and E type samples, whereas for the B sample it is $\approx$ 34 meV

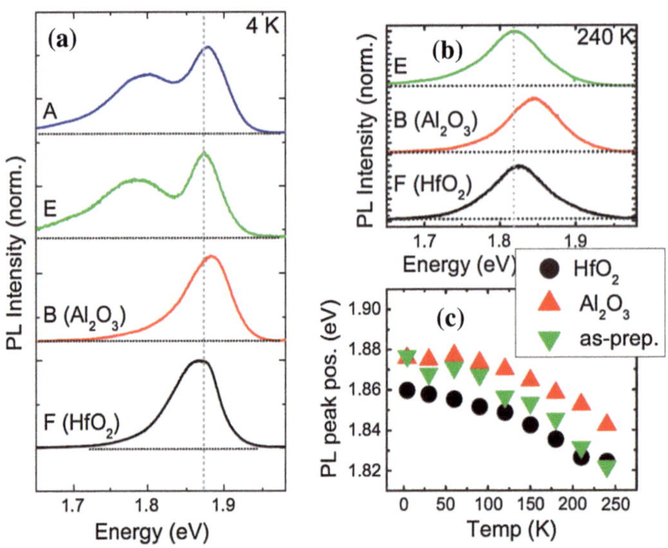

Fig. 7.42 PL spectra of **a** four samples (measured at 4 K) and **b** three sample (measured at 240 K). The *A* and *E* samples (in **a**) are not covered with oxide layers, whereas *B* and *F* are covered. **c** The high-energy peaks for three samples (shown in **b**) as a function of temperature. Taken from Ref. [91]

and for F type it is $\approx$ 36 meV. Around 240 K, the oxide covered samples show the blueshifted peak compared to the without covered samples, shown in Fig. 7.42b. Since the atomic layer deposition process of making oxide layers involves temperature variations from 520 K to below room temperature and the thermal expansion coefficient of MoS_2 is larger than the oxide layers, the sandwiched monolayer MoS_2 remains strained and hence the larger peak shifts are observed for the as-prepared samples than the oxide coated [91].

7.9.3 Effect of Circular Polarization of the Incident Light on Photoluminescence: Valley Selectivity

Even layers of MoS_2 have inversion symmetry, whereas odd layers and hence monolayer MoS_2 $\left(D_{3h}^1\right)$ has no inversion symmetry. Because of no inversion symmetry, spin-orbit coupling splits the valence band of monolayer MoS_2 by $\sim$160 meV at the K and K' points [92, 93] (see Fig. 7.43b). At the band edge of K-points, the symmetry adapted hybridized d-states on Mo are given by $|\phi_c\rangle = |d_{z^2}\rangle$ (1 = 0) and $|\phi_\tau^v\rangle = \frac{1}{\sqrt{2}}\left(|d_{x^2-y^2}\rangle + i\tau|d_{xy}\rangle\right)$ (1 = ±2). The hybridized p-states on S atom are $|\phi_\tau^v\rangle = \frac{1}{\sqrt{2}}\left(|p_x\rangle + i\tau|p_y\rangle\right)$ (1 = ±1); where c and v represent the conduction and valence band, respectively. Here, $\tau = \pm 1$ indicates the valley index for K and K' points connected by time reversal symmetry (TRS). The hybridized states interact with each other to form symmetry adapted linearly combined (SALC) states for the valence and conduction bands of monolayer MoS_2. Now, the effect of the threefold rotational symmetry (C_3) on the states are given as $C_3|v(K,K')\rangle = |v(K,K')\rangle$ and $C_3|c(K,K')\rangle = e^{\mp i2\pi/3}|c(K,K')\rangle$ [32]. Therefore, the SALC states near the top of the conduction band dominated by d_{z^2} states on Mo carry an overall azimuthal quantum number $m_\pm = \pm 1$ at K and K', whereas for the valence band SALC states it is given by $m_\pm = 0$. The optical selection rule for the direct transition at the K and K$'$ becomes $\Delta m_\pm = \pm 1$. Since the carriers have well-defined angular momentum associated with the K and K' states, we need to have laser excitations with particular helicity (i.e., circularly polarized light) to excite them. This phenomena of absorbing left-handed and right-handed circularly polarized lights by the two valleys is known as circular dichroism (CD), see Fig. 7.43c. The TRS requires that $E_\downarrow(k) = E_\uparrow(-k)$. Hence, we have the inherently coupled valley and the spins of the carriers because of the spin-filtered valence band splitting together with the TRS, as shown in Fig. 7.43b. Figure 7.43g shows the PL spectrum (not polarization resolved) observed at $T = 14$ K using 2.33 eV laser excitation. The prominent feature around 1.9 eV is due to the A exciton complexes (AEC) consisting of two peaks: at higher energy it is due to the neutral exciton, and the lower component redshifted by 40 meV is attributed to the charged exciton (discussed later in detail). The PL peak $\sim$2.1 eV is due to the B exciton and $\sim$1.8 eV arises because of excitons bound to the defects, as discussed earlier.

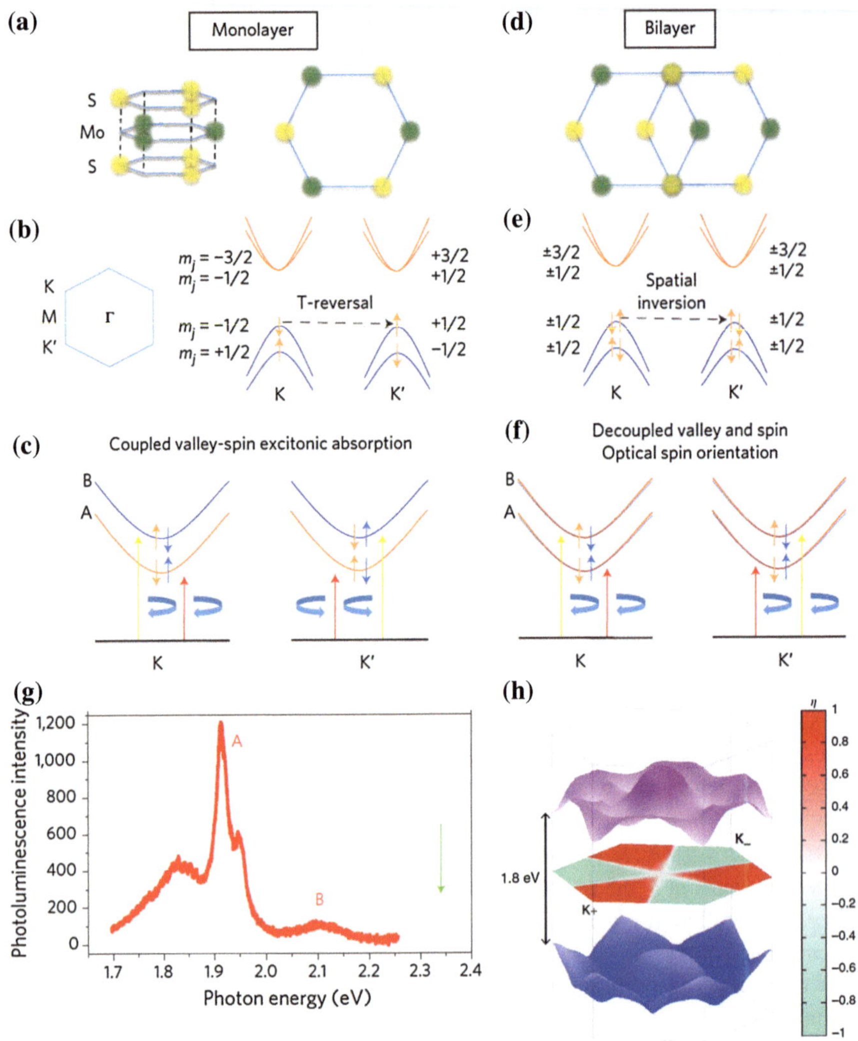

Fig. 7.43 **a** The hexagonal lattice structure for single-layer MoS_2 and **d** the Bernal stacking for bilayer MoS_2. The valence and the conduction bands (marked by the corresponding total angular momentum) across the energy gap at the K and K′ points for (**b**) single layer and **e** for bilayer. Optical selection rules with the applied left-handed and right-handed circularly polarized lights (**c**) for single-layer MoS_2 at the K and K′ points, respectively; whereas for left-handed circularly polarized lights (**f**) in case of bilayer. **g** The unpolarized PL spectrum observed at 14 K using 2.33 eV laser line for single-layer MoS_2. Taken from Ref. [29]. **h** The top and the bottom figures represent the conduction (*pink*) and the valence (*blue*) bands; and the middle figure represents the calculated circular polarization $\eta(k, \omega_{cv})$ at different points of the Brillouin zone. The right scale bar represents the corresponding colors of the calculated $\eta(k, \omega_{cv})$. Taken from Ref. [32]

The degree of circular polarization for the PL emission is given as $\eta(\mathbf{k}, \omega_{cv}) = \frac{|P_{+}^{cv}(\mathbf{k})|^2 - |P_{-}^{cv}(\mathbf{k})|^2}{|P_{+}^{cv}(\mathbf{k})|^2 + |P_{-}^{cv}(\mathbf{k})|^2}$. $P_{\pm}^{cv}(k)$ representing the coupling strength with the circularly polarized lights ($\sigma_{\pm}$) given by, $P_{\pm}^{cv}(\mathbf{k}) = \frac{1}{\sqrt{2}}\left\{P_{x}^{cv}(\mathbf{k}) \pm P_{y}^{cv}(\mathbf{k})\right\}$, where $P_{\alpha}^{cv}(\mathbf{k}) = \left\langle \Psi_{ck} \middle| \mathbf{P}_{a}^{cv} \middle| \Psi_{vk} \right\rangle$. The calculated $\eta(k, \omega_{cv})$ is shown in Fig. 7.43h indicating the exact value of ± 1 at the K and K′ points [32]. The polarization resolved (σ_{+} and σ_{-}) PL spectrum observed at $T = 14$ K using the left-handed circularly polarized (σ_{-}) lights (1.96 eV) is shown in Fig. 7.44a. The observed spectrum consists of only two peaks: The AEC peak is σ_{-} polarized and the peak due to bound-exciton is almost unpolarized. Mak et al. [29] used a parameter to quantify the experimentally observed PL emission polarization, which is given by $\eta^{\exp} = \frac{I(\sigma_{-}) - I(\sigma_{+})}{I(\sigma_{-}) + I(\sigma_{+})}$; where $I(\sigma_{\pm})$ denotes the polarization resolved measured PL intensity. The observed value is given by $\eta^{\exp} = 1.00 \pm 0.05$ and drops to $\sim$ 0.05 below the photon energy 1.8 eV, as shown in Fig. 7.44b. With the excitation of right-handed circularly polarized light ($\sigma_{\pm}$), Mak et al. observed the $\eta^{\exp} = 1.00 \pm 0.05$ (not shown). In general, the quantity $\eta^{\exp}$ depends on the two timescales: (i) exciton lifetime given by $\tau^{-1} = \tau_{r}^{-1} + \tau_{nr}^{-1}$, where r and nr represent the radiative and non-radiative recombination processes, respectively, and (ii) valley lifetime or hole-spin lifetime, τ_{v} [32]. In terms of these two timescales, the helicity for the AEC is

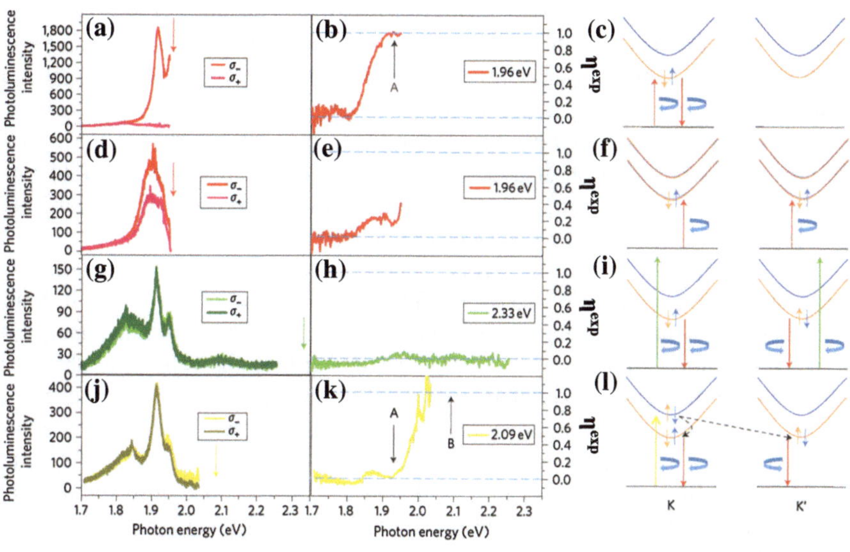

Fig. 7.44 The left column represents the PL spectra for **a, g** and **j** single-layer and **d** for bilayer MoS_2 (at 14 K). Here the arrows indicate the laser energy used in the experiments. The left-handed circularly polarized light (σ_{-}) is used for all the excitations. The middle column represents the corresponding observed helicity in PL emissions. The right column represents the schematics of the corresponding absorption and emission of lights. Taken from Ref. [29]

given by $\eta_A^{\exp} = \frac{1}{1+2\tau_A/\tau_{Av}}$ for A-neutral exciton and $\eta_{A^-}^{\exp} = \frac{\eta_A^{\exp}}{1+2\tau_{A^-}/\tau_{A^-v}}$ for A-charged exciton [29]. Mak et al. estimated that the observed helicity $\eta^{\exp} = 1.00 \pm 0.05$ is in accordance with the exciton lifetime $\tau > 50$ ps from the observed high QY on BN substrate and the valley lifetime $\tau_v > 1$ ns in monolayer MoS_2 [29].

To compare with the monolayer MoS_2, the polarization resolved PL measurement has also been done on the Bernal-stacked (see Fig. 7.43d) bilayer MoS_2. The schematic diagram of the valleys at the K and K′ points with the corresponding spin configurations of the carriers, and the diagram for the optical transitions are shown in Fig. 7.43e, f respectively. Figure 7.44d, e shows the observed PL spectra and the measured helicity for the bilayer, respectively. The observed low helicity $\eta^{\exp}(\text{bilayer}) = 0.25 \pm 0.05$ for the A exciton is in accordance with the hole-spin lifetime (τ_v) of a few hundreds of femtoseconds (from the observed 20 times less QY) [29]. The hole-spin lifetime for the bilayer, therefore, is more than three orders of magnitude less compared to the monolayer MoS_2, say by τ_v^{diff}. The estimated τ_v^{diff} is attributed to the inversion symmetry of the bilayer MoS_2. The σ_- polarization couples charge carriers only from one valley K (see Fig. 7.44c), and excites an exciton with an electron having spin down and a hole of spin up at the K-point. Now the intra-valley scattering involves the spin flip of hole which is forbidden by the spin-filtered energy gap of ~160 meV in absence of magnetic impurity scatterings. The inter-valley scattering (from K to K') involves large momentum transfer with the requirement of the spin-flip as well, and hence this channel is also forbidden. Because of the absence of the inter as well as intra-valley scattering, the high valley lifetime is observed for monolayer MoS_2 [29]. Bilayer MoS_2 has inversion symmetry which ensures that the $E_\uparrow(k) = E_\uparrow(-k)$. The both TRS and the inversion symmetry implies that at each k-point the spin degeneracy remains irrespective of valley index, i.e., $E_\downarrow(k) = E_\uparrow(k)$. Hence, for the bilayer the spin and the valley index are not coupled and with the σ_- excitation populations of both the valleys (K and K') having electron of down-spin and hole of up-spin (see Fig. 7.44f) are achieved, which allows intravalley hole-spin relaxation process via Elliot-Yafet processes [29] leading to the low observed helicity.

The zero helicity is observed from monolayer MoS_2 with the σ_- polarization under the laser excitation of 2.33 eV (532 nm), as shown in Fig. 7.44g, h. Since the probing has been done with the higher energy than the resonance energy of A and B excitons, it populates the charge carriers at both the valleys, as shown in Fig. 7.44i. In this case, no particular spin-valley selection occurrs and hence the observed zero spin-valley polarization. Figure 7.44j, k shows the PL spectrum and the helicity of monolayer MoS_2 with the laser of 2.09 eV energy (594 nm), respectively. Because of resonance with the B excitons, the observed helicity is ~1, whereas the helicity for the A-exciton is zero. In this case, the B-exciton relaxes through non-radiative channel to the A-exciton states (phonon-assisted process) and populates again both the valleys, as shown in Fig. 7.44l. The polarization resolved PL spectra for the monolayer MoS_2 deposited on BN and SiO_2/Si substrates are shown in Fig. 7.45a, c and the corresponding measured helicity in Fig. 7.45b, d respectively. The intensity for BN substrate is ~10 times

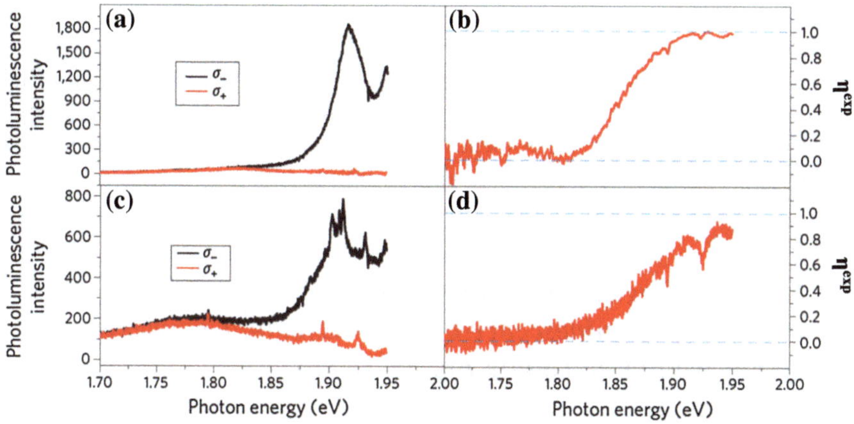

Fig. 7.45 PL spectra observed for single-layer MoS $_2$ **a** on h-BN substrate and **c** on SiO_2/Si substrate, and the corresponding observed helicity in (**b**) and (**d**), respectively. Taken from Ref. [29]

larger than that on SiO_2/Si, whereas the helicity is almost the same for both the substrates. The robustness of the observed valley selective CD indicates how good the charge carriers preserve the information about the valley index and the spin index with them in their lifetimes, i.e., before recombination.

The temperature-dependent helicity of the PL emission spectra for monolayer MoS_2 is shown in Fig. 7.46. The circular polarization remains almost constant in the range of 0–90 K. With the increasing temperature, the observed helicity decreases. As the $\eta^{\exp} \propto \tau_v$, with the increase in temperature the phonon populations and hence the scattering via them also increases by providing the required large momentum transfer from K to K' point in BZ. The inter-valley scattering rate

Fig. 7.46 Temperature-dependent variations in the circular polarizations observed in PL spectra for single-layer MoS_2. The experimental data have been fitted by considering the inter-valley scattering due to the acoustic phonons. Taken from Ref. [94]

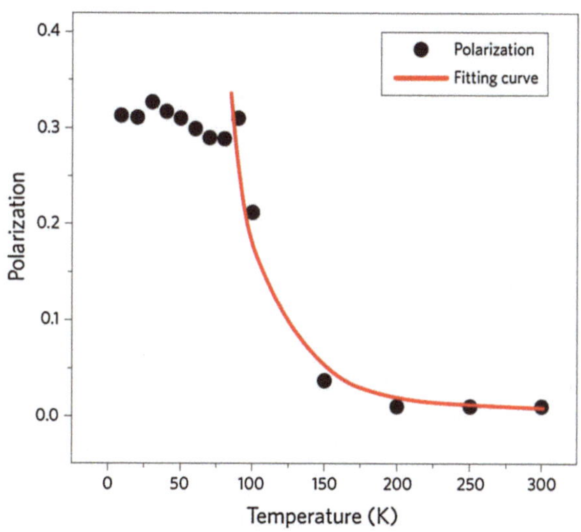

is $\tau_v^{-1} \propto e^{-E_{ph}/K_B T}$, where E_{ph} is the phonon energy at the *K*-point. The extracted value of the phonon frequency by fitting the observed PL helicity is given as $E_{ph} \approx 240\ \text{cm}^{-1}$ in accordance with the acoustic phonon near the *K*-point of the monolayer MoS_2 [63]. Although the fitting is in agreement with the observed helicity above 90 K, more experimental as well as theoretical studies are needed to explore the full understanding of the underlying physics.

7.9.4 Gate-Voltage-Dependent Absorption and Photoluminescence of the A Exciton Complexes: Observation of Tightly Bound Trions

A few absorption spectra for the monolayer MoS_2 at different back gate voltages in the range of −100 to +80 V using 532 nm laser line are shown in Fig. 7.47a. As

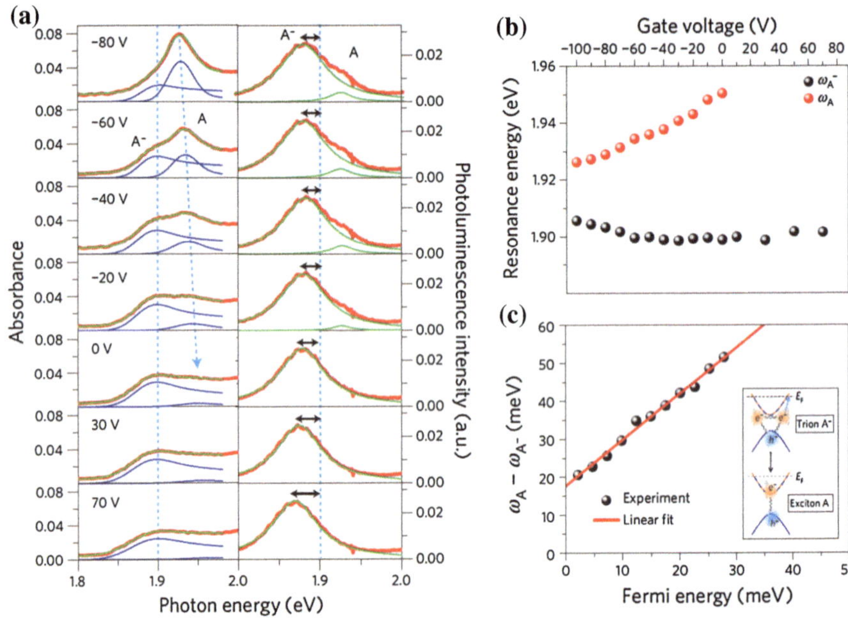

Fig. 7.47 **a** Back-gate voltage-dependent absorption spectra (*left panel*) and the PL spectra (*right panel*) measured at 10 K. The power-law has been used to fit (*blue solid lines*) the data for the absorption spectra and Lorentzian fits has been done for PL spectra. Here the dashed blue lines are a guide to the eye. The arrows in PL spectra indicate the shift of the trion with the gating. **b** The extracted (from absorption spectra) peak positions of trion (ω_{A^-}) and the neutral exciton (ω_A) as a function of gate voltages (*upper scale*) and the corresponding Fermi energy (bottom scale); and the difference $\omega_A - \omega_{A^-}$ in (**c**). The E_F -dependence linear fit has been done to the data. The inset in (**c**) shows the schematics of the breaking of the trion into a neutral exciton and an electron at the Fermi level. Here the offset gate voltage = −107 ± 6 V for $E_F = 0$. Taken from Ref. [30]

discussed, the monolayer MoS_2 device shows n-type transfer characteristics at ambient conditions because of the Fermi level pining. Therefore, the device operated at -100 V has almost the undoped channel. Here we discuss the absorption and PL spectra for the AEC only. The absorption spectrum at -80 V is dominated by an intense neutral A-exciton and a little shoulder due to the negatively charged A-exciton. This charged exciton consisting of one hole of up-spin and two electrons of opposite spins is known as trion (A^-) [30]. As the negative gate voltage decreases, the A^- evolves and around +70 V, the spectrum is dominated completely by A^- exciton. The neutral A-exciton peak gradually diminishes with the decrease in the negative gate voltages, and after a certain voltage (0 V) it is hard to track the peak. The peak positions of these two peaks are plotted versus back gate voltages (Fermi energy in lower axis) in Fig. 7.47b, and their difference $(\omega_A - \omega_{A^-})$ versus Fermi energy (gate voltages in upper axis) is plotted in Fig. 7.47c. The overall reduction in the intensity for the absorption spectra is due to the Pauli blocking associated with the increasing electrons in the conduction band [30]. In the right panel of Fig. 7.47a, the PL spectra at different gate voltages is shown using 532 nm laser line. The neutral A-exciton behaves similarly as in absorption spectra, i.e., after a positive gate voltage it is turned off. The feature of the blueshifts of the neutral A-exciton and the almost constant feature of the A^--exciton with gate voltages are attributed to the combined effect of the many-body interactions and the Pauli blocking [30]. The neutral A-exciton can be thought of as an ionized trion such that the splitting between the peaks is given as $\omega_A - \omega_{A^-} = E_{A^-} + E_F$, where E_{A^-} is the binding energy of the trion A^-. At $E_F = 0, E_{A^-}$ is the separation between the two peaks and the extracted value by fitting the linear relation (see Fig. 7.47c) is $\sim 18.0 \pm 1.5$ meV. The separation $\omega_A - \omega_{A^-}$ behaves linearly with the Fermi energy (E_F), and is in accordance with the earlier observations of trions for the quantum wells [95].

Figure 7.48a, b shows the room temperature PL spectra (with broadened features) and the corresponding gate voltage-dependent measured PL intensity, respectively. The PL intensity for the neutral A-exciton decreases with gating by almost two orders of magnitude, whereas for the trion it remains constant in the range of -100 to +80 V. The decreased PL intensity of the neutral A-exciton is attributed to the spectral weight reduction and its transfer to the trion with the doping as for the absorption spectra at 10 K (see Fig. 7.47a). This high tunability of the PL intensity, which is not observed at 10 K, is attributed to the thermally activated carrier populations from the trion state.

The gating voltage-dependent optical response might be associated with the metal-insulator transition (MIT) as observed for the two-dimensional electron gas (2DEGs) [96]. At ~ -100 V, the channel is undoped ($T = 10$ K) and hence the optically excited excitons dominate the absorption spectra. Now as we increase the doping, for positive gate voltages the conduction band is populated with the free carriers and the neutral A-exciton combines with them to form trion. Hence, at higher positive gate voltages, the absorption spectra is dominated completely by the trions (see Fig. 7.47a). As we deplete the channel more and more, the decrease in the number of carriers leads to the much less screening of the long range

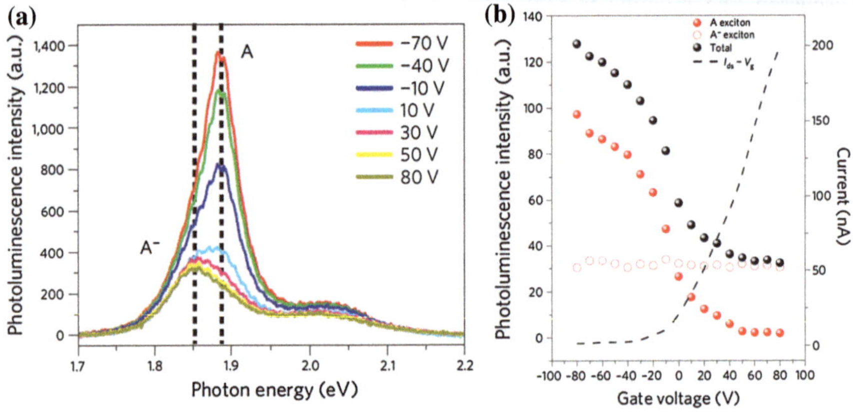

Fig. 7.48 **a** The gating voltage-dependent PL spectra measured at 300 K. **b** The integrated intensity of both the neutral exciton and the trion and their total intensity (*left scale*). Right scale represents the transfer characteristics. Taken from Ref. [30]

Coulomb interactions due to the presence of extrinsic disorders (ionized) in the sample. This implies an increase in potential fluctuations in the sample which in turn localizes the electrons. This feedback mechanism results in nonlinear screening and hence, can reduce the MIT region to a smaller gate voltage. Recently, it has been shown that the monolayer MoS_2 indeed goes to MIT at the top gate voltage of ~ 2.2 V (the carrier concentration $\sim 1 \times 10^{13}$ cm^{-2}) below 80 K [31], as shown in Fig. 7.49. If 2DEGs were uniformly distributed instead of

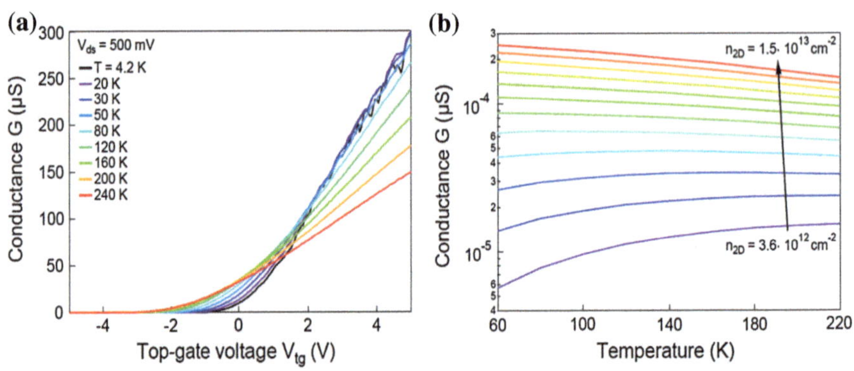

Fig. 7.49 **a** The top gate (V_{tg})-dependent conductance (G) of single-layer MoS_2 for various temperatures. The G shows the thermally activated behavior for low V_{tg} and it decreases with decreasing temperature. Single-layer MoS_2 shows a transition from an insulating regime to a metallic regime above V_{tg} of ~ 1–2 for the decreased temperatures such that the G increases with decreasing temperature and this crossover is clear for the lower temperature curves. **b** The temperature-dependent G of the devices for different values of the electron density (n_{2D}). Taken from Ref. [31]

localized, we should not observe the improvement in the spectral weight of the neutral A-exciton peak in absorption spectra. Since localized electrons do not play an effective role on screening, the more localization means the more prominent neutral A-exciton peak. Now, if the charge carriers would form patches of inhomogeneous carrier density, the associated tuning of the PL intensity should not show any onset. In contrast, experimental results for the PL intensity show an on-state (threshold) with the gate voltage (see Fig. 7.48b), which implies the possibility of the localization of electrons (Wigner crystallization) instead of forming small patches [96]. If the localized electrons were effective in the screening, then we should not observe the enhancement of the A-exciton in absorption spectra; and hence we can say that this MIT may be due to the Mott-Hubbard type instead of Anderson-type. More experimental works are required to explore the physics in the insulating region and the corresponding theoretical supports to explain the observed MIT in the monolayer MoS_2.

7.10 Conclusions

In conclusion, single and a few layer MoS_2 devices show reasonably high carrier mobility ($\sim$700 cm^2/V s) and on–off ratio $\sim 10^8$. The increment of the high temperature (100–300 K) mobility has been engineered by quenching the homopolar and other phonon scattering processes with the help of dielectric coating. The observed robustness of the valley selective circular dichroism suggests that we should explore experimentally valley Hall effect and magnetism in single-layer MoS_2 without applying magnetic field. The observation of negatively charged excitons or trions and the associated metal-insulator-transition at low temperature can be an indication of Wigner-crystal formation in 2-D MoS_2 material, which needs to be further explored. Along with that, the measured dimensionless parameter (ratio of Coulomb potential energy to kinetic energy) $r_s \approx 60$ makes the MoS_2 an ideal system for further exploration of many-body phenomena. In addition, other dichalcogenide materials like WS_2, $MoSe_2$ and WSe_2 [97] are exciting to explore the novel physics in the coming years.

AKS thanks the Department of Science and Technology, India, for funding under the Nanomission Project. AB thanks CSIR for a research fellowship.

References

1. Geim, A.K., Novoselov, K.S.: The rise of graphene. Nature Mat. **6**, 183 (2007)
2. Castro Neto, A.H., Guinea, F., Peres, N.M.R., Novoselov, K.S., Geim, A.K.: The electronic properties of graphene. Rev. Mod. Physics **81**, 109 (2009)
3. Rao, C.N.R., Sood, A.K., Subrahmanyam, K.S., Govindaraj, A.: Graphene: The new two-dimensional nanomaterial. Angew. Chem. Int. Ed. **48**, 7752–7777 (2009)

4. Rao, C.N.R., Sood, A.K., Voggu, R., Subrahmanyam, K.S.: Some novel attributes of graphene. J. Phys. Chem. Lett. **1**, 572–580 (2010)
5. Sood. A.K., Chakraborty, B.: Understanding graphene via Raman scattering. Wiley–VCH Verlag GmbH & Co. KGaA, **chapter–2**, 49–90 (2012)
6. Das, A., Chakraborty, B., Sood, A.K.: Probing single and bilayer graphene field effect transistors By Raman spectroscopy. Mod. Phys. Lett. B **25**, 511–535 (2011)
7. Novoselov, K.S., Jiang, D., Schedin, F., Booth, T.J., Khotkevich, V.V., Morozov, S.V., Geim, A.K.: Two-dimensional atomic crystals. Proc. Natl. Acad. Sci. U S A. **102**, 10451–10453 (2005)
8. Rogers, J.A., Lagally, M.G., Nuzzo, R.G.: Synthesis, assembly and applications of semiconductor nanomembranes. Nature **447**, 45–53 (2011)
9. Bolotin, K.I., Sikes, K.J., Jiang, Z., Klima, M., Fudenberg, G., Hone, J., Kim, P., Stormer, H.L.: Ultrahigh electron mobility in suspended graphene. Solid State Commun. **146**, 351–355 (2008)
10. Radisavljevic, B., Radenovic, A., Brivio, J., Giacometti, V., Kis, A.: Single-layer MoS_2 transistors. Nat. Nanotechnol. **6**, 147–150 (2011)
11. Das, S., Chen, H.Y., Penumatcha, A.V., Appenzeller, J.: High performance multilayer MoS_2 transistors with scandium contacts. Nano Lett. **13**, 100–105 (2013)
12. Zhang, W., et al.: Ultrahigh-gain phototransistors based on graphene-MoS_2 heterostructures. **arXiv:1302.1230** (2013)
13. Liu, K–.K., et al.: Growth of large-area and highly crystalline MoS_2 thin layers on insulating substrates. Nano Lett. **12**, 1538–1544 (2012)
14. Zhou, K.G., Mao, N.N., Wang, H.X., Peng, Y., Zhang, H.L.: A mixed-solvent strategy for efficient exfoliation of inorganic graphene analogues. Angew. Chem. Int. Ed. **50**, 10839–10840 (2011)
15. Li, Y., Wang, H., Xie, L., Liang, Y., Hong, G., Dai, H.: MoS_2 nanoparticles grown on graphene: an advanced catalyst for the hydrogen evolution reaction. J. Am. Chem. Soc. **133**, 7296–7299 (2011)
16. Coleman, N.J., et al.: Two-dimensional nanosheets produced by liquid exfoliation of layered materials. Science **331**, 568–571 (2011)
17. Eda, G., Yamaguchi, H., Voiry, D., Fujita, T., Chen, M., Chhowalla, M.: Photoluminescence from chemically exfoliated MoS_2. Nano Lett. **11**, 5111–5116 (2011)
18. Rao, C.N.R., Nag, A.: Inorganic analogues of graphene. Eur. J. Inorg. Chem. **27**, 4244–4250 (2010)
19. Ramakrishna Matte, H.S.S., Gomathi, A., Manna, A.K., Dattatray, J.L., Ranjan, D., Pati, S.K., Rao, C.N.R.: MoS_2 and WS_2 analogues of graphene. Angew. Chem. Int. Ed. **49**, 4059–4062 (2010)
20. Li, Q., Newberg, J.T., Walter, J.C., Hemminger, J.C., Penner, R.M.: Polycrystalline molybdenum disulfide (2H-MoS_2) nano- and microribbens by electrochemicl/chemical synthesis. Nano Lett. **4**, 277–281 (2004)
21. Balendhran, S., Ou, J.Z., Bhaskaran, M., Sriram, S., Ippolito, S., Vasic, Z., Kats, E., Bhargava, S., Zhuiykov, S., Kalantar-zadeh, K.: Atomically thin layers of MoS_2 via a two step thermal evaporation–exfoliation method. Nanoscale **4**, 461–466 (2012)
22. Peng, Y., Meng, Z., Zhong, C., Lu, J., Yu, W., Jia, Y., Qian, Y.: Hydrothermal synthesis and characterization of single-molecular-layer MoS_2 and MoSe2. Chem. Lett. **8**, 772–773 (2001)
23. Lauritsen, J.V., Kibsgaard, J., Helveg, S., Topsoe, H., Clausen, B.S., Lagsgaard, E., Besenbacher, F.: Size-dependent structure of MoS_2 nanocrystals. Nat. Nanotechnol. **2**, 53–58 (2007)
24. Li, Y., Wang, H., Xie, L., Liang, Y., Hong, G., Dai, H.: MoS_2 nanoparticles grown on graphene: an advanced catalyst for the hydrogen evolution reaction. J. Am. Chem. Soc. **133**, 7296–7299 (2011)
25. Rao, C.N.R., Ramakrishna Matte, H.S.S., Subrahmanyama, K.S., Maitra, U.: Unusual magnetic properties of graphene and related materials. Chem. Sci. **3**, 45–52 (2012)

26. Mathew, S., et al.: Magnetism in MoS_2 induced by proton irradiation. Appl. Phys. Lett. **101**, 102103–102105 (2012)
27. Mak, K.F., Lee, C., Hone, J., Shan, J., Heinz, T.F.: Atomically Thin MoS_2: A new direct-gap semiconductor. Phys. Rev. Lett. **105**, 136805 (2010)
28. Yao, W., Xiao, D., Niu, Q.: Valley-dependent optoelectronics from inversion symmetry breaking. Phys. Rev. B **77**, 235406–235407 (2008)
29. Mak, K.F., He, K., Shan, J., Heinz, T.F.: Control of valley polarization in monolayer MoS_2 by optical helicity. Nature Nanotech. **7**, 494–498 (2012)
30. Mak, K.F., He, K., Lee, C., Lee, G.H., Hone, J., Heinz, T.F., Shan, J.: Tightly bound trions in monolayer MoS_2. Nature Mat. **12**, 207–211 (2013)
31. Radisavljevic, B., Kis, A.: Mobility engineering and metal-insulator transition in monolayer MoS_2. **arXiv:1301.4947** (2013)
32. Cao, T., et al.: Valley-selective circular dichroism of monolayer molybdenum disulphide. Nat. Commun. **3**, 887 (2012)
33. Taniguchi, K., Matsumoto, A., Shimotani, H., Takag, H.: Electric-field-induced superconductivity at 9.4 K in a layered transition metal disulphide MoS_2. Appl. Phys. Lett. **101**, 042603 (2012)
34. Roldan, R., Cappelluti, E., Guineal, F.: Interactions and superconductivity in heavily doped MoS_2. **arXiv:1301.4836** (2013)
35. Ramana, C.V., Becker, U., Shutthanandan, V., Julien, C.M.: Oxidation and metal-insertion in molybdenite surfaces: evaluation of charge-transfer mechanisms and dynamics. Geochem. Trans. **9**, 8 (2008)
36. Bromley, R.A., Murray, R.B., Yoffe, A.D.: The band structures of some transition metal dichalcogenides: III. Group VI A: trigonal prism materials. J. Phys. C: Solid State Phys. **5**, 759–778 (1972)
37. Mattheiss, L.F.: Energy bands for 2H-NbSe and 2H-MoS_2. Phys. Rev. Lett. **30**, 784–787 (1973)
38. Lebegue, S., Eriksson, O.: Electronic structure of two-dimensional crystals from ab initio theory. Phys. Rev. B **79**, 115409 (2009)
39. Splendiani, A., et al.: Emerging photoluminescence in monolayer MoS_2. Nano Lett. **10**, 1271–1275 (2010)
40. Li, T., Galli, G.: Electronic properties of MoS_2 nanoparticles. J. Phys. Chem. C **111**, 16192–16196 (2007)
41. Molina-Sanchez, A., Wirtz, L.: Phonons in single-layer and few-layer MoS_2 and WS_2. Phys. Rev. B **84**, 155413 (2011)
42. Ataca, C., Topsakal, M., Akturk, E., Ciraci, S.: A comparative study of lattice dynamics of three- and two-dimensional MoS_2. J. Phys. Chem. C **115**, 16354–16361 (2011)
43. Verble, J.L., Wieting, T.J.: Lattice mode degeneracy in MoS_2 and other layer compounds. Phys. Rev. Lett. **25**, 362–365 (1970)
44. Wieting, T.J.: Long-wavelength lattice vibrations of MoS_2 and GaSe. Solid State Commun. **12**, 931–935 (1973)
45. Zhang, X., Han, W.P., Wu, J.B., Milana, S., Lu, Y., Li, Q.Q., Ferrari, A.C., Tan, P.H.: Raman spectroscopy of shear and layer breathing modes in multilayer MoS_2. Phys. Rev. B **87**, 115413 (2013)
46. Lee, C., et al.: Anomalous lattice vibrations of single and few-layer MoS_2. ACS Nano **4**, 2695–2700 (2010)
47. Chakraborty, B., Ramakrishna Matte, H.S.S., Sood, A.K., Rao, C.N.R.: Layer-dependent resonant Raman scattering of a few layer MoS_2. J. Raman Spectrosc. **44**, 92–96 (2013)
48. Malarda, L.M., Pimentaa, M.A., Dresselhaus, G., Dresselhaus, M.S.: Raman spectroscopy in graphene. Phys. Rep. **473**, 51–87 (2009)
49. Wang, Y.Y., Ni, Z.H., Shen, Z.X., Wang, H.M., Wu, Y.H.: Interference enhancement of Raman signal of graphene. Appl. Phys. Lett. **92**, 043121 (2008)
50. Zeng, H., Zhu, B., Liu, K., Fan, J., Cui, X., Zhang, Q.M.: Low-frequency Raman modes and electronic excitations in atomically thin MoS_2 films. Phys. Rev. B **86**:241301(R) (2012)

51. Tan, P.H., et al.: The shear mode of multilayer graphene. Nat. Mater. **11**, 294–300 (2012)
52. Benedek, G., Ellis, J., Reichmuth, A., Ruggerone, P., Schief, H., Toennies, J.P.: Organ-pipe modes of sodium epitaxial multilayers on Cu(001) observed by inelastic helium-atom scattering. Phys. Rev. Lett. **69**, 2951 (1992)
53. Luo, N.S., Ruggerone, P., Toennies, J.P.: Theory of surface vibrations in epitaxial thin films. Phys. Rev. B **54**, 5051–5063 (1996)
54. Plechinger, G., Heydrich, S., Eroms, J., Weiss, D., Schuller, C., Korn, T.: Raman spectroscopy of the interlayer shear mode in few-layer MoS_2 flakes. Appl. Phys. Lett. **101**, 101906 (2012)
55. Zhao, Y., et al.: Interlayer breathing and shear modes in few-trilayer MoS_2 and WSe2. Nano Lett. **13**, 1007–1015 (2013)
56. Chen, J.M., Wang, C.S.: Second order Raman spectrum of MoS_2. Solid State Commun. **14**, 857–860 (1974)
57. Stacy, A.M., Hodul, D.T.: Raman spectra of IVb and VIb transition metal disulfides using laser energies near the absorption edges. J. Phys. Chem. Solids **46**, 405–409 (1985)
58. Sourisseau, C., Cruege, F., Fouassier, M.: Second-order Raman effects, inelastic neutron scattering and lattice dynamics in 2H-WS_2. Chem. Phys. **150**, 281–293 (1991)
59. Frey, G.L., Tenne, T., Matthews, M.J., Dresselhaus, M., Dresselhaus,G.: Raman and resonance Raman investigation of MoS_2 nanoparticles. Phys. Rev. B **60**:2883–2892 (1999)
60. Sekine, T., Uchinokura, K., Nakashizu, T., Matsuura, M., Yoshizaki, R.: Dispersive Raman mode of layered compound 2H-MoS_2 under the resonant condition. J. Phys. Soc. Jpn. **53**, 811–818 (1984)
61. Ghosh, P.N., Maiti, C.R.: Interlayer force and Davydov splitting in 2H-MoS_2. Phys. Rev. B **28**, 2237–2239 (1983)
62. Livneh, T., Sterer, E.: Resonant Raman scattering at exciton states tuned by pressure and temperature in 2H-MoS_2. Phys. Rev. B **81**, 195209 (2010)
63. Wakabayashi, N., Smith, H.G., Nicklow, R.M.: Lattice dynamics of hexagonal MoS_2 studied by neutron scattering. Phys. Rev. B **12**, 659–663 (1975)
64. Coehoorn, R., Haas, C., de Groot, R.A.: Electronic structure of $MoSe_2$, MoS_2, and WSe_2. II. The nature of the optical band gaps. Phys. Rev. B **35**, 6203–6206 (1987)
65. Schwierz, F.: Graphene transistors. Nat. Nanotech. **5**, 487–496 (2010)
66. Fivaz, R., Mooser, E.: Electron-phonon interaction in semiconducting layer structures. Phys. Rev. **136A**, 833–836 (1964)
67. Fivaz, R., Mooser, E.: Mobility of charge carriers in semiconducting layer structures. Phys. Rev. **163**, 743–755 (1967)
68. Yu, Y.P., Cardona, M.: Fundamental of semiconductors, 3rd edn. Springer, Heidelberg (2005)
69. Kaasbjerg, K., Thygesen, K.S., Jacobsen, K.W.: Phonon-limited mobility in n-type single-layer MoS_2 from first principles. Phys. Rev. B **85**, 115317 (2012)
70. Novoselov, K.S., Jiang, D., Schedin, F., Booth, T.J., Khotkevich, V.V., Morozov, S.V., Geim, A.K.: Two-dimensional atomic crystals. Proc. Natl. Acad. Sci. U.S.A. **102**, 10451–10453 (2005)
71. Ayari, A., Cobas, E., Ogundadegbe, O., Fuhrer, M.S.: Realization and electrical characterization of ultrathin crystals of layered transition-metal dichalcogenides. J. Appl. Phys. **101**, 014507 (2007)
72. Jena, D., Konar, A.: Enhancement of carrier mobility in semiconductor nanostructures by dielectric engineering. Phys. Rev. Lett. **98**, 136805 (2007)
73. Ming-Wei, L., et al.: Mobility enhancement and highly efficient gating of monolayer MoS_2 transistors with polymer electrolyte. J. Phys. D Appl. Phys. **45**, 345102 (2012)
74. Bao, W., Cai, X., Kim, D., Sridhara, K., Fuhrer, M.S.: high mobility ambipolar MoS_2 field-effect transistors: substrate and dielectric effects. Appl. Phys. Lett. **102**, 042104 (2013)
75. Kim, S., et al.: High-mobility and low-power thin-film transistors based on multilayer MoS_2 crystals. Nat. Commun. **3**, 1011–1017 (2012)

76. Han, S.W., et al.: Band-gap transition induced by interlayer van der Waals interaction in MoS_2. Phys. Rev. B **84**, 045409 (2011)
77. Sze, S.M., Ng, K.K.: Physics of semiconductor devices, 3rd edn. Wiley and Sons, New York (2009)
78. Chakraborty, B., Bera, A., Muthu, D.V.S., Bhowmick, S., Waghmare, U.V., Sood, A.K.: Symmetry-dependent phonon renormalization in monolayer MoS_2 transistor. Phys. Rev. B **85**, 161403(R) (2012)
79. Pisana, S., Lazzeri, M., Casiraghi, C., Novoselov, K.S., Geim, A.K., Ferrari, A.C., Mauri, F.: Breakdown of the adiabatic Born–Oppenheimer approximation in graphene. Nat. Mater. **6**, 198–201 (2007)
80. Yan, J., Zhang, Y., Kim, P., Pinczuk, A.: Electric field effect tuning of electron-phonon coupling in graphene. Phys. Rev. Lett. **98**, 166802–166804 (2007)
81. Das, A., Pisana, S., Chakraborty, B., Piscanec, S., Saha, S.R., Waghmare, U.V., Yiang, R., Krishnamurthy, H.R., Geim, A.K., Ferrari, A.C., Sood, A.K.: Monitoring dopants by Raman scattering in an electrochemically top-gated graphene transistor Nat. Nanotechnol. **3**, 210–215 (2008)
82. Yan, Y., Henriksen, E.A., Kim, P., Pinczuk, A.: Observation of anomalous phonon softening in bilayer graphene. Phys. Rev. Lett. **101**, 136804 (2008)
83. Das, A., Chakraborty, B., Piscanec, S., Pisana, S., Sood, A.K., Ferrari, A.C.: Phonon renormalization in doped bilayer graphene. Phys. Rev. B **79**, 155417 (2009)
84. Malard, L.M., Elias, D.C., Alves, E.S., Pimenta, M.A.: Observation of distinct electron-phonon couplings in gated bilayer graphene. Phys. Rev. Lett. **101**, 257401–257404 (2008)
85. Attaccalite, C., Wirtz, L., Lazzeri, M., Mauri, F., Rubio, A.: Doped graphene as tunable electron–phonon coupling material. Nano Lett. **10**, 1172–1176 (2010)
86. Livneh, T., Sterer, E.: Resonant Raman scattering at exciton states tuned by pressure and temperature in 2H-MoS_2. Phys. Rev. B **81**, 195209 (2010)
87. Aksoy, R., Ma, Y., Selvi, E., Chyu, M.C., Ertas, A., White, A.: X-ray diffraction study of molybdenum disulfide to 38.8 GPa. J. Phys. Chem. Solids **67**, 1914 (2006)
88. Ho, C.H., Wu, C.S., Huang, Y.S., Liao, P.C., Tiong, K.K.: Temperature dependence of energies and broadening parameters of the band-edge excitons of $Mo_{1-x}W_xS_2$ single crystals. J. Phys.: Condens. Matter **10**, 9317–9328 (1998)
89. Connell, G.A.N., Wilson, J.A., Yoffe, A.D.: Effects of pressure and temperature on exciton absorption and band structure of layer crystals: molybdenum disulphide. J. Phys. Chem. Solids **30**, 287–296 (1969)
90. Korn, T., Heydrich, S., Hirmer, M., Schmutzler, J., Schuller, C.: Low-temperature photocarrier dynamics in monolayer MoS_2. Appl. Phys. Lett. **99**, 102109 (2011)
91. Plechinger, G., Schrettenbrunner, F.-X., Eroms, J., Weiss, D., Schüller C, C., Korn T, T.: Low-temperature photoluminescence of oxide-covered single-layer MoS_2. Phys. Status Solidi (RRL) **6**, 126–128 (2012)
92. Xiao, D., Liu, G.-B., Feng, W., Xu, X., Yao, W.: Coupled spin and valley physics in monolayers of MoS_2 and other group-VI dichalcogenides. Phys. Rev. Lett. **108**, 196802–196805 (2012)
93. Ramasubramaniam, A.: Large excitonic effects in monolayers of molybdenum and tungsten dichalcogenides. Phys. Rev. B **86**, 115409 (2012)
94. Zeng, H., Dai, J., Yao, W., Xiao, D., Cui, X.: Valley polarization in MoS_2 monolayers by optical pumping. Nature Nanotech. **7**, 490–493 (2012)
95. Huard, V., Cox, R.T., Saminadayar, K., Arnoult, A., Tatarenko, S.: Bound states in optical absorption of semiconductor quantum wells containing a two-dimensional electron gas. Phys. Rev. Lett. **84**, 187–190 (2000)
96. Finkelstein, G., Shtrikman, H., Bar-Joseph, I.: Optical spectroscopy of a two-dimensional electron gas near the metal-insulator transition. Phys. Rev. Lett. **74**, 976–979 (1995)
97. Tonndorf, P., et al.: Photoluminescence emission and Raman response of monolayer MoS_2, MoSe2, and WSe2. Opt. Express **21**, 4908–4916 (2013)

Chapter 8
Optical Characterization, Low-Temperature Photoluminescence, and Photocarrier Dynamics in MoS_2

T. Korn, G. Plechinger, S. Heydrich, F.-X. Schrettenbrunner, J. Eroms, D. Weiss and C. Schüller

Abstract In recent years, the dichalcogenide MoS_2 has gained attention as an interesting material system for basic research and possible optoelectronic applications. Here, we report on optical spectroscopy of few- and single-layer MoS_2 flakes. We use Raman spectroscopy to characterize our samples. The energy of the characteristic phonon modes in MoS_2 depends on the number of layers, so that the thickness of individual flakes can be mapped in scanning Raman experiments. While bulk MoS_2 is an indirect-gap semiconductor, single-layer MoS_2 has a direct band gap and emits strong photoluminescence. We investigate the photoluminescence in single-layer MoS_2 for different experimental conditions. Additionally, we study the photocarrier dynamics in time-resolved photoluminescence experiments.

8.1 Introduction

In recent years, graphene, a two-dimensional crystal structure prepared from bulk graphite was extensively studied by the solid-state physics community, and a number of unusual electronic and mechanical properties were discovered. Just a few years after their initial experiments on graphene, A. Geim and K. Novoselov were awarded the Nobel prize in physics in 2010 [1].

Just like graphite, MoS_2, and related dichalcogenides such as WS_2, $MoSe_2$, and WSe_2, are layered crystal structures, so that fabrication of few- and single-layer flakes from bulk crystals is possible by mechanical exfoliation techniques [2]. Unlike graphene, MoS_2 has a finite band gap, which is indirect for the bulk crystal

T. Korn (✉) · G. Plechinger · S. Heydrich · F.-X. Schrettenbrunner · J. Eroms · D. Weiss · C. Schüller
Institut für Experimentelle und Angewandte Physik, Universität Regensburg, Regensburg 93040, Germany
e-mail: tobias.korn@physik.uni-regensburg.de

Z. M. Wang (ed.), *MoS_2*, Lecture Notes in Nanoscale Science and Technology 21, DOI: 10.1007/978-3-319-02850-7_8, structure. In 2010, two publications reported observation of strong photoluminescence (PL) from single-layer MoS_2 [3, 4], and first-principles calculations using density functional theory (DFT) showed a transition from an indirect to a direct band gap in single-layer MoS_2 due to confinement perpendicular to the layer plane [3–5]. Additional interest in the material system was created by transport experiments, published in 2011, demonstrating high room-temperature mobility and large on/off ratio in single-layer MoS_2 field effect transistors (FET) covered with high-k dielectrics [6]. These initial experiments sparked a large research effort by experimentalists and theorists, which has already yielded remarkable progress. While recent advances related to optical spectroscopy will be mentioned in conjunction with our own results, a brief overview of progress in some other areas is given below.

8.1.1 MoS_2 Preparation, Synthesis, and Growth Methods

Bulk MoS_2 crystals are readily available and allow for easy preparation of single- and few-layer flakes by means of mechanical exfoliation. Recently, laser-induced thinning of few-layer flakes to single-layer thickness was demonstrated [7], and XeF_2 etching also yielded single-layer MoS_2 when applied to few-layer flakes [8]. Large-scale fabrication and commercial application of MoS_2-based devices requires alternative methods, however. Here, chemical exfoliation techniques were demonstrated that, in combination with mild annealing, allow for preparation of large-area films which show strong photoluminescence PL at room temperature [9], and the performance of transistors fabricated from this material rivals that of devices prepared from exfoliated flakes [10]. Physical and chemical vapor deposition of MoS_2 on different substrates, such as copper [11], graphene [12], and SiO_2 [13, 14] was also shown to yield high-quality films. Vapor–solid growth was used to produce large-area flakes that show a large valley polarization in low-temperature PL measurements [15].

8.1.2 MoS_2-Based Devices

The initial experiment on single-layer MoS_2 FETs [6] sparked a large number of follow-up studies. In dual-gate transistors fabricated from few-layer MoS_2 using Al_2O_3 as top gate dielectric, a large on/off ratio of more than 10^8, comparable to that observed in single-layer transistors, could be observed [16]. While MoS_2 transistors using solid gate dielectric typically show n-type unipolar operation, ambipolar transistors were demonstrated using an ionic liquid in an electric double layer transistor structure [17]. By combining several transistors, functional electronic devices, including a small-signal amplifier [18] and larger integrated circuits such as logic gates and oscillators [19] were fabricated from exfoliated single- and bilayer flakes. The combination of MoS_2 and graphene in a heterostructure was used to create a nonvolatile memory cell [20].

8.2 Crystal and Band Structure of MoS_2

Figure 8.1a and b show the single-layer MoS_2. A single-layer consists of a plane of hexagonally arranged Mo atoms, sandwiched between top and bottom planes of hexagonally arranged S atoms. Here, a Mo atom is situated at the center of an equilateral prism, with S atoms at the bottom and top corners of the prism. In the 2H configuration, which is the most common stacking order for multilayer MoS_2, layers are stacked in such a way that the corner of a prism is stacked above the center of the prism in the adjacent layer. It is clearly visible that single-layer MoS_2 does not have an inversion center. By contrast, even-numbered stacks of 2H–MoS_2 show inversion symmetry.

While bulk MoS_2 has an indirect band gap, with the valence band maximum at the Γ point and the conduction band minimum at a point along the $\Gamma - K$ direction, single-layer MoS_2 has a direct gap at the K point. Both conduction band and valence band states close to the band gap consist predominantly of Mo d orbitals [5], the weak curvature of both bands around the K point indicates large effective masses. A simplified schematic of the Brillouin zone of single-layer MoS_2 (Cf. [21]) is shown in Fig. 8.2a. Here, the conduction band and the topmost spin-split valence band are indicated around the inequivalent K^+ and K^- points at the corners of the hexagonal Brillouin zone. At K^+ and K^-, the spin splitting of the valence band is opposite. This leads to valley-selective optical absorption, as shown in Fig. 8.2b: at the K^+ point, interband transitions from the valence to the conduction band are only possible using σ^+ excitation, while at K^-, only σ^- radiation is absorbed. This coupling of spin and valley degrees of freedom can lead

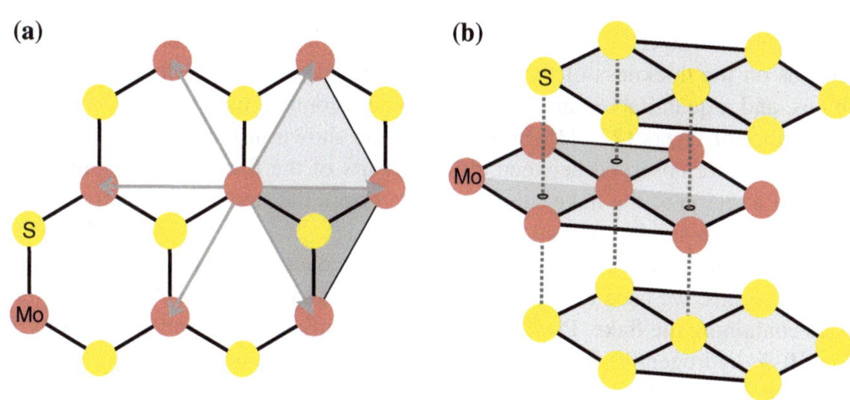

Fig. 8.1 **a** Top view of the MoS_2 crystal structure for a single-layer. **b** 3D view of the MoS_2 crystal structure for a single-layer

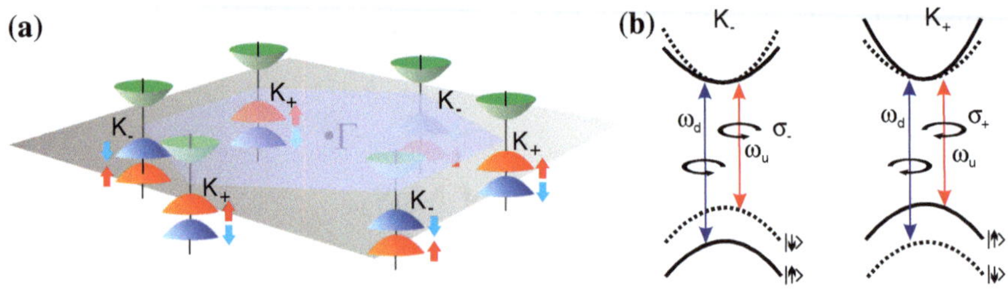

Fig. 8.2 **a** Schematic view of the Brillouin zone of single-layer MoS_2. **b** Simplified band structure and optical transitions for single-layer MoS_2. Cf. [21]

to optically induced valley and spin Hall effects [21]. Due to the large valence-band spin splitting and the coupling of spin and valley indices, these effects should be robust and may be useful for future device applications.

8.3 Experimental Setups, Sample Preparation, and Initial Characterization

The MoS_2 flakes used in our experiments were prepared with the mechanical exfoliation method well-established for graphene, from natural MoS_2 bulk crystals (SPI supplies). A transparent adhesive tape (Scotch "magic tape") was used to cleave and re-recleave flakes collected from a bulk crystal, then pressed firmly onto a substrate and peeled off. A *p*-doped silicon wafer with a 300 nm SiO_2 layer on top and lithographically defined metal markers was used as substrate. After initial characterization with an optical microscope, some samples were investigated with an atomic force microscope (AFM). Figure 8.3a shows an optical microscope image of a typical MoS_2 flake. Attached to a large, opaque bulk-like crystal, smaller, semi-transparent parts of the flake are visible, which correspond to single- and few-layer thickness of the MoS_2. The visibility of thin flakes strongly depends on the thickness of the transparent SiO_2 layer and the illumination conditions, and a quantitative analysis of the optical contrast may be used to identify the number of layers [22, 23]. In the AFM scan shown in Fig. 8.3b, well-defined height steps are observed between different parts of the flake, and the number of layers of a region of the flake can be extracted from the step height, where 0.7 nm correspond to a single-layer step.

Some of the flakes identified in this way were covered by Al_2O_3 or HfO_2 oxide layers. For this, we used atomic layer deposition (ALD) to cover the whole wafer piece containing the flake. Prior to the ALD, the samples were annealed at 450 °C in a 10 % hydrogen/90 % nitrogen atmosphere, then immediately transferred to the ALD chamber. Both dielectrics were grown at 250° C with alternating pulses of trimethylaluminum (TMA) and water for the Al_2O_3 layer, and tetrakis(dimethylamino)hafnium (TDMAH) and water for the HfO_2 layer, respectively.

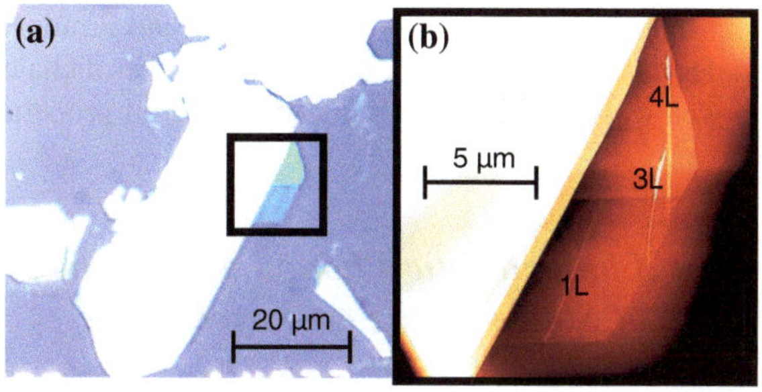

Fig. 8.3 **a** Optical microscope image of a MoS_2 flake on a Si/SiO_2 substrate. The *black square marks* the area of the AFM scan shown in (**b**). **b** AFM image of the MoS_2 flake shown in (**a**). The thickness of different parts of the flake is indicated

All of the Raman spectroscopy experiments on the MoS_2 flakes were performed at room temperature. For this, we utilized a microscope setup, in which a 532 nm continuous wave (cw) laser was coupled into a 100X microscope objective, which also collected the scattered light in backscattering geometry. The scattered light was recorded using a triple grating spectrometer equipped with a liquid nitrogen-cooled charge-coupled device (CCD) sensor. The sample was mounted on a piezo-stepper table and scanned under the microscope. The spatial resolution of this setup is about 500 nm. For some Raman measurements, a cross-polarized detection geometry was used. For this, the laser was linearly polarized, and the back-scattered light was analyzed using a second polarizer aligned perpendicular to the polarization of the laser.

For low-temperature photoluminescence (PL) measurements, the sample was mounted in a He-flow cryostat. A microscope setup with a 40X objective, into which either a 532 nm or 473 nm cw laser was coupled, was used to collect the PL. The spatial resolution of this setup is about 1 μm. The PL was recorded using a single-grating spectrometer equipped with a CCD sensor. Various low-pass filters were used in front of the spectrometer slit to suppress stray light from the laser sources.

For near-resonant excitation of PL, a cw He:Ne laser with 632.8 nm wavelength was used. The laser was circularly polarized using an achromatic quarter wave plate. To measure the circular polarization degree, the PL from the sample was first passed through another achromatic quarter wave plate, then analyzed using a linear polarizer placed in front of the spectrometer slit.

Time-resolved PL (TRPL) measurements were performed using the same microscope setup. The second harmonic (wavelength 402 nm) from a picosecond Ti: Sapphire laser was used to excite the sample, the PL was collected with a streak camera system. The time resolution of this setup is about 5 ps.

8.4 Raman Spectroscopy

Raman spectroscopy is a highly useful technique to study two-dimensional crystal structures. It provides fast, nondestructive characterization with submicron spatial resolution. In graphene, Raman spectroscopy has been utilized to determine, e.g., the number of layers [24], the carrier concentration [25–27], and the number of defects [28]. For the study of MoS_2 and other dichalcogenides, it is also becoming a well-established tool to study the material properties [29–33].

8.4.1 Characteristic Phonon Modes and Raman Spectra of Single- and Few-Layer MoS_2

Figure 8.4 shows three characteristic Raman-active modes of MoS_2. The A_{1g} mode corresponds to an out-of-plane optical vibration of the sulfur atoms, while the E^1_{2g} mode is an in-plane optical vibration of molybdenum and sulfur atoms. The shearmode is a rigid-layer oscillation of adjacent layers. In 2010, Lee et al. [29] showed that the Raman frequencies of A_{1g} and E^1_{2g} modes depend on the number of layers of a flake. The A_{1g} mode stiffens with an increasing number of layers, as additional layers increase the effective restoring forces acting on the atoms oscillating perpendicular to the layer plane. By contrast, the E^1_{2g} mode anomalously softens. This mode softening was shown to stem from enhanced dielectric screening provided by additional layers [34]. The different frequency shifts of the A_{1g} and E^1_{2g} allow the frequency *difference* between the two modes to be used as a fingerprint for the number of layers.

Figure 8.5a shows Raman measurements of MoS_2 flakes whose thickness was determined by AFM. It is clearly visible that the E^1_{2g} mode shifts to lower frequencies with increasing number of layers, while the A_{1g} mode stiffens. For a monolayer, the frequency difference is about 18 cm^{-1}, while bulk-like flakes show larger values of about 26 cm^{-1}. The change of the frequency difference is largest from monolayer to bilayer flakes, for thicker flakes, it quickly approaches the bulk value. Frequency-difference Raman measurements therefore, allow for a reliable identification of flakes with thickness between one and four layers. The amplitudes of the A_{1g} and E^1_{2g} modes show a nonmonotonic dependence on the number of

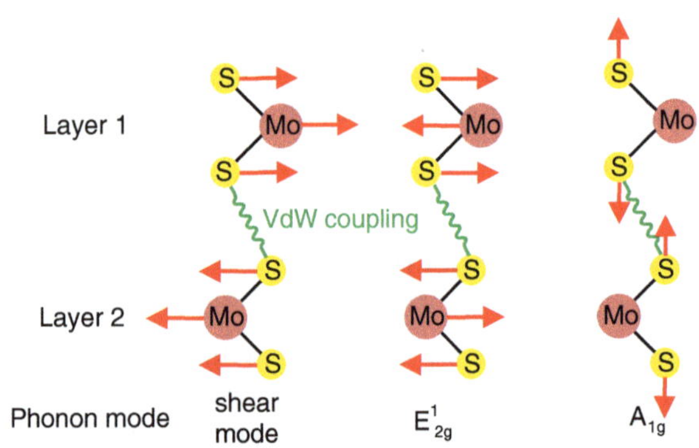

Fig. 8.4 Characteristic phonon modes in MoS_2

layers, which is in part due to interference effects caused by the SiO_2 substrate [35]. Similar frequency shifts of the A_{1g} and E^1_{2g} modes were also reported for WS_2, while WSe_2 does not show pronounced shifts [36], even though both dichalcogenides have the same crystal structure as MoS_2.

While most of the phonon modes in two-dimensional crystals depend on the specific material system, two types of oscillation are generic for layered structures. These are rigid-layer oscillations, in which adjacent layers oscillate, with the oscillatory amplitude either perpendicular (compression/breathing modes) or parallel (shear modes) to the layer planes. Shear modes have been observed in a number of bulk-layered crystals, including graphite [37], MoS_2 [38], and hexagonal boron nitride [39]. Recently, they were detected in few-layer graphene [40]. Due to the weak van der Waals coupling between adjacent layers in layered crystals, the shear modes are at very low Raman shifts, with values between 22 cm^{-1} (bulk GaS) [41] and 52 cm^{-1} (bulk hexagonal BN) [39]. Using a linear-chain model, Tan et al. calculated the frequency of the shear mode [40] as a function of the number of layers and found that the frequency for a bilayer decreases from the bulk value by a factor of $\sqrt{2}$.

We investigated the shear modes in few-layer MoS_2 [42] using a triple grating Raman spectrometer to suppress the elastically scattered laser light. Additionally, we used cross-polarized detection geometry to suppress the background of inelastically scattered light from the *p*-doped silicon substrates [43], which is polarized in parallel to the laser. Figure 8.5b shows low-energy Raman spectra for

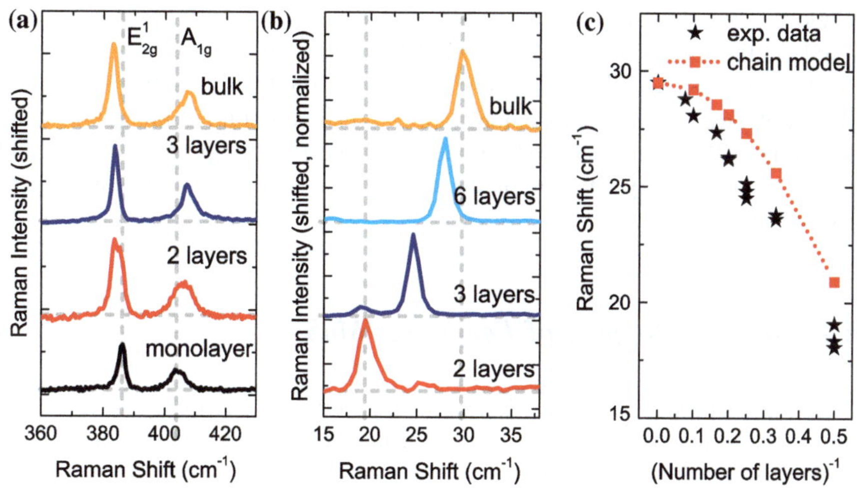

Fig. 8.5 **a** Raman spectra of the E^1_{2g} and A_{1g} modes measured on MoS_2 flakes of known thickness. **b** Raman spectra of the shear mode in few-layer MoS_2 flakes. The *vertical lines* in both panels serve as guide to the eye. **c** Spectral position of the shear mode as a function of (number of layers)$^{-1}$. Multiple datapoints indicate measurements on different flakes. The *dotted line* indicates the mode position calculated using a linear-chain model

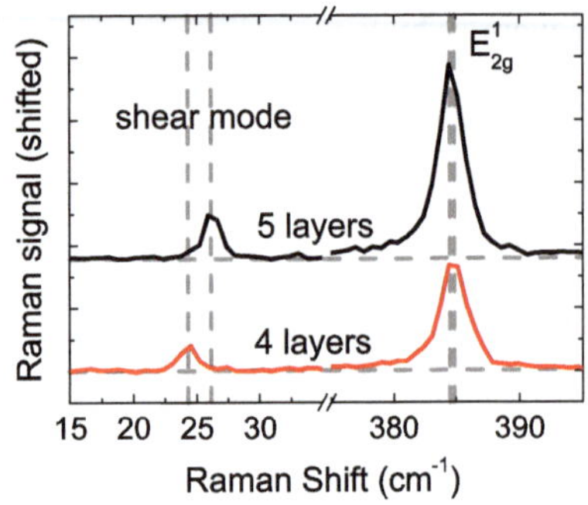

Fig. 8.6 Raman spectra of the E^1_{2g} and shear modes measured on four- and five layer MoS_2 flakes. The *vertical lines* serve as guide to the eye, the scales for the Raman shift in both spectral regions are identical

several flakes of known thickness. The shear mode shifts from less than 20 cm^{-1} for a bilayer to about 30 cm^{-1} for bulk-like flakes, as indicated in Fig. 8.5c. Here, multiple datapoints represent measurements on different flakes. This large relative shift makes Raman measurements of the shear mode an ideal tool to identify MoS_2 flakes of up to six layers. Additionally, the absence of a shear mode peak clearly identifies a monolayer of MoS_2. We observe a larger relative frequency shift than calculated for graphene using the linear-chain model. In Fig. 8.5c, the values calculated by this model are compared to the experimental data with the shear mode frequency for bulk MoS_2 as the only free parameter. The observed discrepancy may stem from the fact that all of our measurements are carried out under ambient conditions, so that the MoS_2 flakes are covered with adsorbates. The large frequency shift of the shear mode also for thicker flakes is demonstrated in Fig. 8.6. Here, Raman spectra of four- and five layer flakes are directly compared. While the blueshift of the shear mode from four to five layers is immediately visible in the spectra, the redshift of the E^1_{2g} mode is far less pronounced.

More recently, several additional studies on the rigid-layer oscillations in few-layer MoS_2 were carried out, demonstrating observation of the compression mode and additional, lower energy shear modes in the Raman spectra [44–46], and giving a group-theory analysis of the different shear modes in odd- and even-layered MoS_2.

8.4.2 Raman Mapping of Few-Layer MoS_2

Due to the dependencies of the Raman modes on the number of layers, few-layer MoS_2 can be mapped with high contrast using Raman microscopy. For this, the flake is scanned under the focus of the excitation laser. The Raman spectra recorded for each position are analyzed using an automated routine, and a selected value is encoded in a false color scale. Figure 8.7a is a false color map of the MoS_2 flake shown using optical and AFM microscopy in Fig. 8.3. The intensity of the A_{1g} mode is plotted as a function of position. The outline of the flake, as well as regions of different Raman intensity, is clearly visible. As discussed above, the

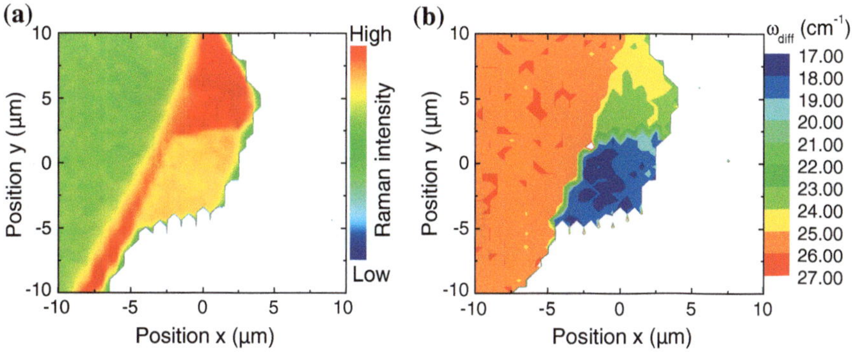

Fig. 8.7 **a** False color map of the A_{1g} mode intensity. **b** False color map of the difference between the A_{1g} and E^1_{2g} mode frequencies. The scan region is identical to (**a**). The Raman scans were performed on the flake shown in Fig. 8.3

Raman intensity depends nonmonotonously on the number of layers, giving a larger signal for the regions containing three and four layers at the top of the flake than for the bulk-like region on the left side of the flake. A map of the same flake, depicting the frequency *difference* between the A_{1g} and E^1_{2g} modes, is shown in Fig. 8.7b. This map gives an even higher contrast between regions of different thickness, and the transition between three and four layers at the top of the flake becomes apparent. The absolute values of the frequency difference can be associated with a certain number of layers, so that maps prepared in this way allow for a *quantitative* identification of individual flakes.

The shear mode in MoS_2 can also be used for Raman microscopy [42]. Although its integrated intensity is about eight times weaker than that of the E^1_{2g} mode, sufficient Raman signal for mapping can be collected with integration times of about 3 s for each position. The large relative shift of the shear mode with the number of layers allows for high-contrast identification of thicker flakes, as depicted in Fig. 8.8. Here, the false color map of the shear mode position in Fig. 8.8b shows a large contrast between the large regions with four and five layers and even larger contrast with the bulk-like region on the top-right corner of the flake.

8.5 Excitonic Transitions in Single-Layer MoS_2

The direct band gap of single-layer MoS_2 and the corresponding strong photoluminescence emission make this material system highly interesting for the study of excitons in the limit of a two-dimensional crystal. A large number of experimental techniques, which were previously utilized to study, e.g., quantum wells, quantum

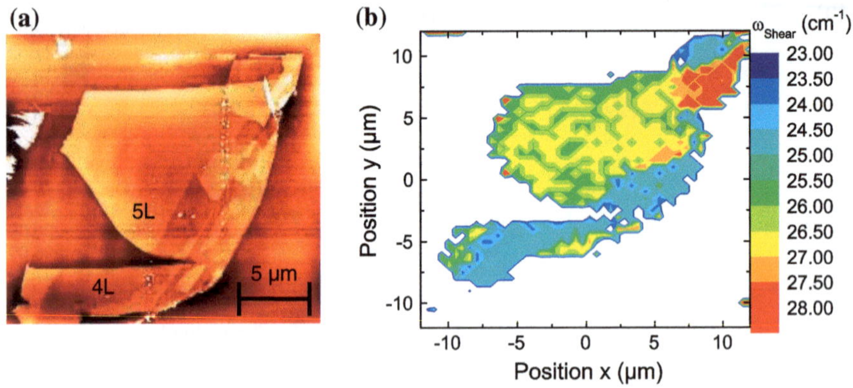

Fig. 8.8 **a** AFM image of a few-layer MoS_2 flake. The thickness of different parts of the flake is indicated. **b** False color map of the shear mode frequency

dots, or other nanostructures based on direct-gap semiconductor material systems such as GaAs/AlGaAs, can be readily applied to MoS_2.

8.5.1 Photoluminescence Measurements

We investigated the photoluminescence of single-layer MoS_2 as a function of various experimental parameters. First, we discuss the temperature dependence. Figure 8.9a shows PL spectra of single-layer MoS_2 as a function of temperature. At liquid-helium temperature, we observe three peaks in the PL spectra, a broad lower energy peak, a more narrow peak at medium energy, and an additional, weaker peak at higher energy. As the sample temperature is increased, all peaks redshift and broaden, the lowest-energy peak is suppressed and not visible for temperatures above 150 K. The medium-energy peak position at room temperature corresponds to the value reported previously [3, 4] for the A exciton, the transition from the upper level of the spin-split valence band to the conduction band. The high-energy peak corresponds to the B exciton, the transition from the lower spin level of the valence band to the conduction band. The lower energy peak S arises from excitons bound to surface adsorbates, as will be shown below. Its quenching with temperature cannot be described by thermal activation, therefore, we infer that it is due to an increased probability for nonradiative recombination with temperature. The spectral shift of the A exciton peak is well described by the Varshni [47] equation, $E_g(T) = E_g(0) - \frac{\alpha T^2}{T+\beta}$, which describes the band gap reduction with temperature due to thermal expansion of the crystal lattice for many semiconductors. We observe that the shift of the A exciton peak matches this equation, with $E_g(0) = 1.874$ eV, $\alpha = 5.9 \cdot 10^{-4}$ eV/K and $\beta = 430$ K [48], as Fig. 8.9b shows. While the B exciton also shows a redshift with temperature, it is

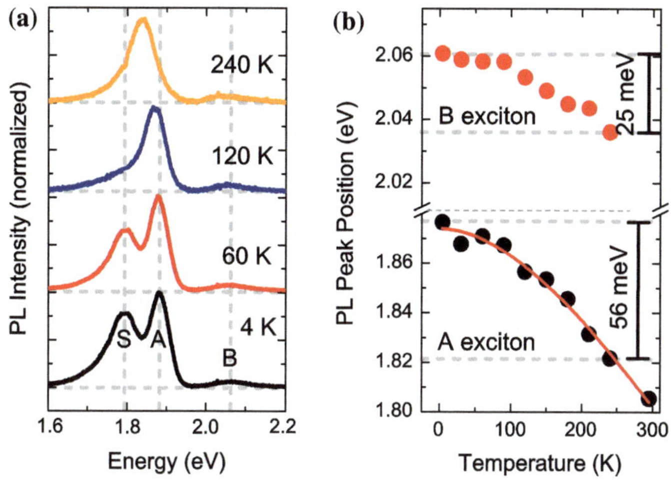

Fig. 8.9 **a** PL spectra of a single-layer MoS_2 flake as a function of temperature. **b** Spectral positions of the A and B exciton peaks as a function of temperature. The *solid line* is a fit to the Varshni formula

far less pronounced. In the temperature range from 4 to 240 K, in which the A exciton energy shifts by 56 meV, the B exciton only shifts by 25 meV, as indicated by the vertical bars in Fig. 8.9b. This different behavior suggests that not only the band gap, but also the valence-band spin splitting decreases with increasing temperature due to thermal expansion of the crystal lattice.

Next, we investigate the temperature-dependent PL in oxide-covered MoS_2 flakes [49]. Figure 8.10a shows PL spectra of HfO_2- and Al_2O_3-covered single-layer MoS_2 flakes compared to an as-prepared MoS_2 flake. The measurements were performed at liquid-helium temperature. While the as-prepared flake PL shows two PL peaks corresponding to the S and A excitons, the PL of both oxide-covered flakes only has a single peak corresponding to the A exciton. This observation confirms that the S exciton is related to surface-bound excitons, which form in MoS_2 due to surface adsorbates. The oxide-covered flakes are annealed in Hydrogen hydrogen/nitrogen atmosphere at 450° C prior to the ALD process, so that the most surface adsorbates are removed and the cleaned MoS_2 flake is subsequently protected by the top oxide layer. Additionally, we observe that the A exciton peak in the oxide-covered flakes is shifted with respect to the position in the as-prepared flakes. We observe a blueshift for the Al_2O_3-covered flake and a redshift for the HfO_2-covered flake. These spectral shifts cannot be explained by a change of the dielectric environment: both oxides have dielectric constants larger than 1, so that enhanced dielectric screening should *decrease* the exciton binding energy, resulting in a blueshift for both oxide-covered flakes, which should be larger for HfO_2 ($\varepsilon_r \approx 20$) than for Al_2O_3 ($\varepsilon_r \approx 10$). The different shifts for the different flakes can be explained, however, by strain induced due to the different thermal expansion coefficients of the MoS_2 flake and the top and bottom oxide layers. Clear evidence for the presence of strain can be gained from temperature-dependent PL measurements. Figure 8.10b shows the A exciton peak position as a function of temperature for the as-prepared and oxide-covered flakes. While the as-prepared flake shows a blueshift of the A exciton of 55 meV as the flake is cooled from 240 K to liquid-helium temperature, both oxide-covered flakes show significantly smaller shifts of about 35 meV. This clearly indicates that MoS_2 flakes sandwiched between bottom SiO_2 and top ALD oxide layers cannot contract freely as the temperature is reduced, resulting in *tensile* strain. The influence of dielectric screening on the excitonic binding energy is masked by this strain. It also follows that the MoS_2 flakes strongly adhere to the oxide layers if they are sandwiched between top and bottom oxide. Recently, the effects of biaxial strain on the band structure of MoS_2 have been theoretically studied by several groups [50–53], and a metal-insulator transition has been predicted. Controlled application of uniaxial strain was also demonstrated in a recent experiment [54].

We now discuss the excitation power-dependence of the excitonic transitions in single-layer MoS_2. Figure 8.11a shows PL spectra of an as-prepared MoS_2 flake measured at liquid-helium temperature for two different excitation powers. We clearly see that the shape of the PL spectrum, containing the S and A exciton peaks, does not change with the excitation power even as it is increased by a factor of 200. From this, we infer that the A and S excitons form independently of each other, and that there is no relaxation from A to S excitons during the photocarrier lifetime. Additionally, we have studied the integrated PL intensity as a function of the excitation power (Fig. 8.11b). In the power range investigated by us, we observe a linear dependence of the PL intensity on the excitation power, indicating

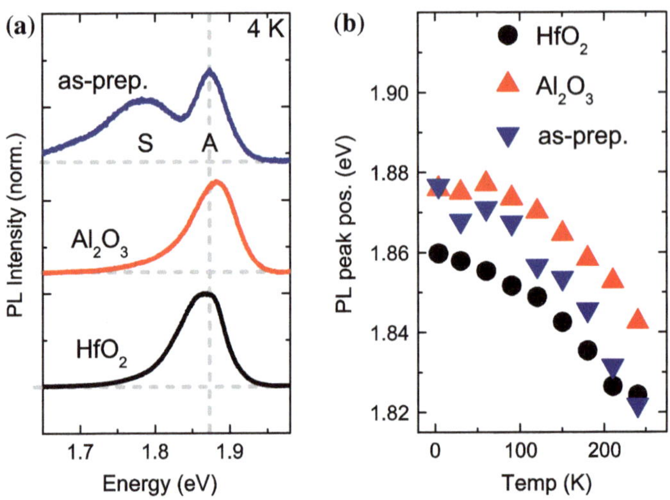

Fig. 8.10 **a** PL spectra of single-layer MoS_2 flakes with and without top ALD-deposited oxide layers on top measured at liquid-helium temperature. **b** Spectral position of the A exciton peak for the flakes shown in (**a**) as a function of temperature

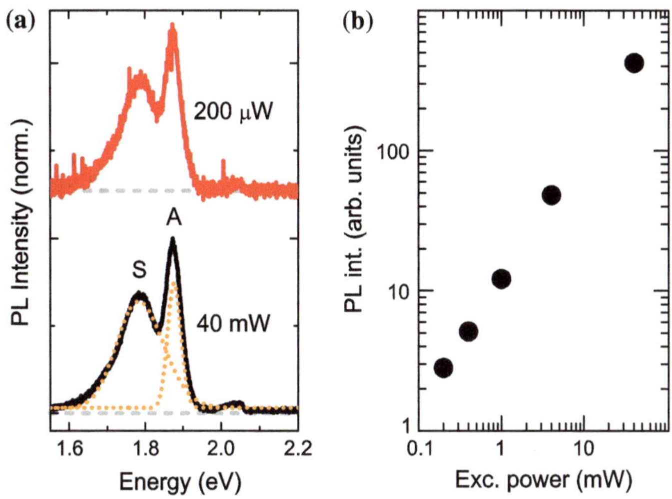

Fig. 8.11 **a** PL spectra of a single-layer MoS_2 flake for different excitation powers measured at liquid-helium temperature. The *dotted lines* indicate fits to the A and S exciton peaks. **b** Integrated PL intensity of the A exciton as a function of excitation power

that there is no saturation of the PL due to bleaching of the absorbtion of the MoS_2 flake. Also, there is no threshold for the onset of PL emission observable, which would indicate the presence of saturable defects that allow for nonradiative recombination.

By contrast, we observe a strong power dependence of the B exciton intensity in single-layer MoS_2. Figure 8.12a shows PL spectra of single-layer MoS_2 measured at room temperature for different excitation powers. For weak excitation, the high-energy B exciton is only weakly visible as a shoulder of the A exciton, which dominates the PL emission. As the laser excitation power is increased, the B exciton peak becomes more pronounced, and for the largest excitation powers used in the experiment its integrated intensity even surpasses that of the A exciton (Fig. 8.12b). We may explain these observations as follows: due to the nonresonant excitation well above the band gap, electron-hole pairs can form, both, A and B excitons. For low excitation densities, however, most holes can relax from the lower valence band into the upper valence band before photocarrier recombination takes place, so that the PL shows mainly A excitons. For higher excitation densities, the increased hole density in the upper valence band limits the available states for relaxation from the lower valence band, so that B excitons are observed in the PL as well. The additional broadening of the A exciton peak and the redshift of both peaks for high excitation powers indicates laser-induced heating of the flake.

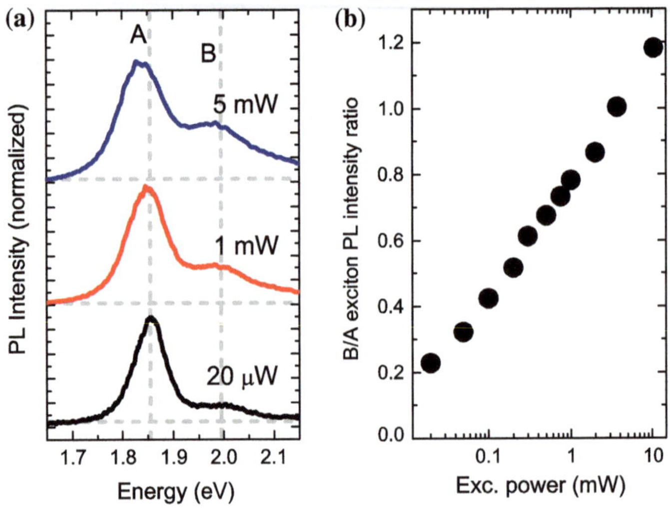

Fig. 8.12 **a** PL spectra of a single-layer MoS_2 flake for different excitation powers measured at room temperature. **b** Ratio of the integrated PL intensities of A and B excitons as a function of excitation power

8.5.2 Valley Polarization

Recently, the optical creation of a valley polarization using circularly polarized excitation in dichalcogenides and its observation by photoluminescence have been studied by several groups [15, 55–59]. Optical orientation with circularly polarized light is also a well-established tool to study spin physics in direct-gap semiconductors, and the circular polarization degree of the PL is used as a measurement of *spin* polarization in these systems. For MoS_2 monolayers, however, it was demonstrated that the origin of the PL polarization is a valley polarization. To show this, Sallen et al. applied a large in-plane magnetic field to their MoS_2 samples during near-resonant PL measurements [58]. They observed no decrease of the PL polarization degree with the applied magnetic field. For a spin polarization, a depolarization due to the Hanle effect would be observed, as the magnetic field induces spin precession, which randomizes the spin orientation during the photocarrier lifetime.

Here, we present our results on optically generated valley polarization. Figure 8.13a shows PL spectra of single-layer MoS_2 measured at liquid-helium temperature using circularly polarized, near-resonant excitation. It is clearly visible that co-circular excitation and PL detection lead to a larger PL signal than contra-circular excitation and detection. The circular polarization of the PL is a clear indication of a valley polarization, i.e., the optical generation of electrons and holes only at the K^+ or at the K^- points in the Brillouin zone, due to the selection rules [21]. This valley-selective photocarrier population must be partially

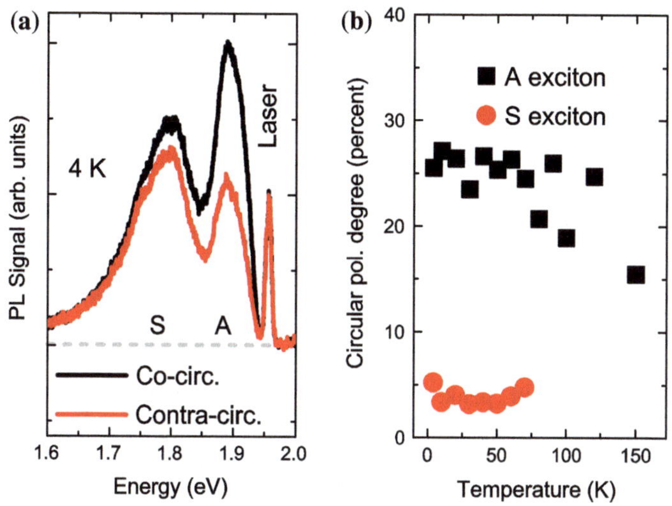

Fig. 8.13 **a** Polarization-resolved PL spectra of a single-layer MoS_2 flake collected under near-resonant excitation at liquid-helium temperature. **b** Circular polarization degree of the A and S excitons as a function of sample temperature

preserved during the photocarrier lifetime in order to be observable in the circular polarization of the PL. We calculate the circular polarization degree, depicted in Fig. 8.13a, by spectrally integrating the A and S exciton PL emission intensities, I_A and I_S. The circular polarization degree P is then given by $P = \frac{I(\sigma^+) - I(\sigma^-)}{I(\sigma^+) + I(\sigma^-)}$. We observe values of $P = 25\,\%$ for the A exciton and significantly smaller values of about $P = 5\,\%$ for the S exciton. A similar discrepancy between A and S exciton polarization was observed by Sallen et al. [58], but the maximum values of the circular polarization of the A exciton reported by different groups vary, with the highest values reported approaching 100 % at low temperatures [15]. With increasing temperature, the PL polarization degree we measure for the A exciton slightly decreases. Recently, the dependence of the A exciton polarization degree the excitation energy was investigated by Kioseoglou et al. using a tunable laser source [59]. The authors observed a strong decrease of P with increasing excitation energy, which they relate to phonon-mediated intervalley scattering, which becomes more pronounced with increasing excess energy of the photocarriers. This mechanism may also explain the large variance of reported values for the A exciton polarization at low temperatures, as the excess energy for a fixed excitation wavelength depends on the A exciton transition energy, which fluctuates between individual samples due to surface adsorbates and strain [49]. The temperature-dependent reduction of the A exciton polarization may stem, in part, from increasing photocarrier excess energy, as the band gap decreases with temperature. Additionally, the photocarrier lifetime increases with temperature [48], as we will discuss below, so that even for a constant intervalley scattering rate, the valley polarization will decrease during the photocarrier lifetime.

8.5.3 Photocarrier Dynamics

We investigated the photocarrier dynamics of single-layer MoS_2 in time-resolved PL (TRPL) measurements. Highly nonresonant excitation with a frequency-doubled Ti: Sapphire laser was used for all experiments. Figure 8.14a shows time-resolved PL traces measured on an as-prepared MoS_2 flake for different temperatures. The PL traces were generated by spectrally averaging a 20 nm wide region around the A exciton PL peak. At low temperatures, the PL emission shows rise and decay times which are close to the temporal resolution of our experimental setup of about 5 ps. The fast risetime of the PL, even under nonresonant excitation with an excess energy of more than 1 eV indicates a very rapid relaxation of the optically generated electron-hole pairs to the band extrema. As the sample temperature is increased to 150 K, the PL decay becomes bi-exponential, with a longer lived component. This longer lived decay component becomes more pronounced with increasing temperature, as indicated by its amplitude extracted from a bi-exponential decay fit to the data (Fig. 8.14c). The decay constant of this component raises from 50 ps at 180 K to 100 ps at 270 K, then decreases again slightly as the temperature is increased further. This longer lived PL component can be attributed to exciton-phonon scattering, which becomes more probable at elevated temperatures due to an increasing phonon population. Exciton-phonon-scattering events may give an exciton a large center-of-mass momentum, so that radiative

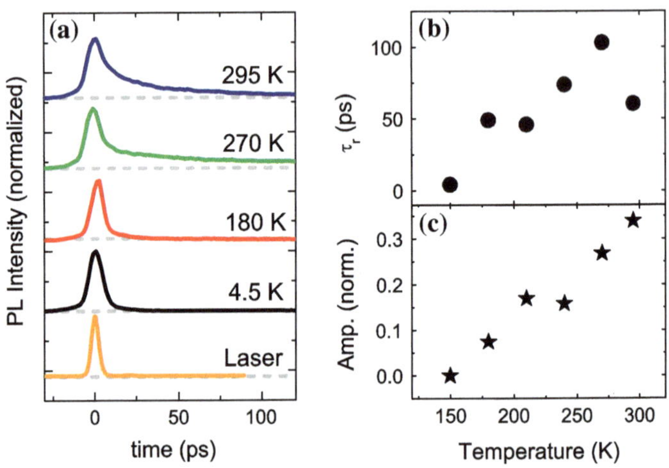

Fig. 8.14 a Time-resolved traces of the A exciton PL for different temperatures. **b** Decay constant and **c** amplitude of the long-lived PL component

recombination is forbidden due to momentum conservation [60]. After such a scattering event, the exciton has to reduce its momentum again via additional scattering processes before it may recombine radiatively. Temperature-dependent TRPL measurements on oxide-covered flakes (not shown) demonstrate a similar temporal behavior of the photocarrier dynamics, indicating that the A exciton recombination is not significantly influenced, i.e., by surface adsorbates, strain, or dielectric screening. More recently, time-resolved pump-probe spectroscopy was performed on single- and few-layer MoS_2 to study the photocarrier dynamics [61, 62]. PL measurements on gated single-layer MoS_2 revealed the coexistence of neutral and negatively charged A excitons (trions), and showed temperature-dependent population effects due to the changing photocarrier dynamics [63].

8.6 Conclusion and Outlook

In conclusion, we investigated single- and few-layer MoS_2 by means of optical spectroscopy techniques. We used Raman microscopy to study the phonon modes of MoS_2 and their dependence on the number of layers. In addition to previously reported higher energy modes, we also observed the low-energy interlayer shear mode, which shows a strong blueshift with increasing number of layers. We demonstrate that Raman microscopy using either the high- or the low-energy Raman modes can be used to map MoS_2 flakes with high contrast. In low-temperature photoluminescence measurements, we observe surface-bound and free excitons. Temperature-dependent photoluminescence reveals a suppression of bound-exciton emission, and different redshifts of the A and B excitons, indicating changes in the valence band splitting with temperature. In oxide-covered MoS_2 flakes, surface-bound exciton emission is absent even at low temperatures. However, the oxide-covered flakes are strained due to different thermal expansion coefficients of MoS_2 and the surrounding layers. Power-dependent PL measurements revealed that there is no population transfer from A excitons to surface-bound S excitons during the photocarrier lifetime. By contrast, relaxation from the B into the A exciton state readily occurs, so that the B exciton PL emission for low excitation is weak. It increases with the excitation density due to valence-band state filling. In PL experiments using near-resonant, circularly polarized excitation, we observed a pronounced circular polarization of the A exciton PL emission and a small circular polarization degree for S exciton PL. Time-resolved PL measurements show that the photocarrier recombination, even under highly nonresonant excitation conditions, occurs on the few-ps timescale at low temperatures. With increasing temperature, the PL decay becomes bi-exponential, with a longer lived component on the 50 ps timescale, due to increased exciton-phonon scattering.

This chapter only reviews our recent activity regarding the optical spectroscopy of MoS_2, in the context of research done by other groups. As discussed in the introduction of this chapter, the study of two-dimensional crystal structures, especially transition-metal dichalcogenides, is an active and rapidly growing field of research, which is driven, on one hand, by application-oriented investigation of transport and optoelectronic devices, and on the other hand, by basic research efforts to study material properties and novel effects related to the peculiar band structure. Remarkable progress has been made in a short period of time. However, many basic issues that are well known for common semiconductor materials, such as techniques for controlled *p*- and *n*-type doping, or the influence of various defects on the electronic structure, still need to be investigated in detail. The first devices utilizing dichalcogenides in conjunction with, e.g., graphene, indicate that a new area of layer-by-layer materials design using two-dimensional crystals is just a beginning.

Acknowledgments The authors gratefully acknowledge financial support by the DFG via SFB689, SPP 1285, GRK 1570, and KO3612/1-1 as well as fruitful discussion with L. Wirtz, A. Molina-Sánchez, J. Maultsch, and N. Scheuschner.

References

1. Geim, A.K.: Angew. Chem. Int. Ed. **50**(31), 6966 (2011)
2. Novoselov, K.S., Jiang, D., Schedin, F., Booth, T.J., Khotkevich, V.V., Morozov, S.V., Geim, A.K.: G **102**(30), 10451 (2005)
3. Mak, K.F., Lee, C., Hone, J., Shan, J., Heinz, T.F.: Phys. Rev. Lett. **105**(13), 136805 (2010)
4. Splendiani, A., Sun, L., Zhang, Y., Li, T., Kim, J., Chim, C.Y., Galli, G., Wang, F.: Nano Lett. **10**(4), 1271 (2010)
5. Lebègue, S., Eriksson, O.: Phys. Rev. B **79**, 115409 (2009)
6. Radisavljevic, B., Radenovic, A., Brivio, J., Giacometti, V., Kis, A.: Nat. Nanotech. **6**, 147 (2011)
7. Castellanos-Gomez, A., Barkelid, M., Goossens, A.M., Calado, V.E., van der Zant, H.S.J., Steele, G.A.: Nano Lett. **12**, 3187 (2012)
8. Huang, Y., Wu, J., Xu, X., Ho, Y., Ni, G., Zou, Q., Koon, G., Zhao, W., Castro Neto, A., Eda, G., Shen, C., Zyilmaz, B.: Nano Res. **6**(3), 200 (2013)
9. Eda, G., Yamaguchi, H., Voiry, D., Fujita, T., Chen, M., Chhowalla, M.: Nano Lett. **11**(12), 5111 (2011)
10. Hwang, W.S., Remskar, M., Yan, R., Kosel, T., Park, J.K., Cho, B.J., Haensch, W., Xing, H.G., Seabaugh, A., Jena, D.: Appl. Phys. Lett. **102**(4), 043116 (2013)
11. Kim, D., Sun, D., Lu, W., Cheng, Z., Zhu, Y., Le, D., Rahman, T.S., Bartels, L.: Langmuir **27**(18), 11650 (2011)
12. Shi, Y., Zhou, W., Lu, A.Y., Fang, W., Lee, Y.H., Hsu, A.L., Kim, S.M., Kim, K.K., Yang, H.Y., Li, L.J., Idrobo, J.C., Kong, J.: Nano Lett. **12**(6), 2784 (2012)
13. Zhan, Y., Liu, Z., Najmaei, S., Ajayan, P.M., Lou, J.: Small **8**(7), 966 (2012)
14. Lee, Y.H., Zhang, X.Q., Zhang, W., Chang, M.T., Lin, C.T., Chang, K.D., Yu, Y.C., Wang, J.T.W., Chang, C.S., Li, L.J., Lin, T.W.: Adv. Mater. **24**(17), 2320 (2012)
15. Wu, S., Huang, C., Aivazian, G., Ross, J.S., Cobden, D.H., Xu, X.: ACS Nano **7**(3), 2768 (2013)
16. Liu, H., Ye, P.: IEEE Elect. Device Lett. **33**(4), 546 (2012)
17. Zhang, Y., Ye, J., Matsuhashi, Y., Iwasa, Y.: Nano Lett. **12**(3), 1136 (2012)
18. Radisavljevic, B., Whitwick, M.B., Kis, A.: Appl. Phys. Lett. **101**(4), 043103 (2012)
19. Wang, H., Yu, L., Lee, Y.H., Shi, Y., Hsu, A., Chin, M.L., Li, L.J., Dubey, M., Kong, J., Palacios, T.: Nano Lett. **12**(9), 4674 (2012)

20. Bertolazzi, S., Krasnozhon, D., Kis, A.: ACS Nano **7**(4), 3246 (2013)
21. Xiao, D., Liu, G.B., Feng, W., Xu, X., Yao, W.: Phys. Rev. Lett. **108**, 196802 (2012)
22. Castellanos-Gomez, A., Agraït, N., Rubio-Bollinger, G.: Appl. Phys. Lett. **96**(21), 213116 (2010)
23. Benameur, M.M., Radisavljevic, B., HTron, J.S., Sahoo, S., Berger, H., Kis, A.: Nanotechnology **22**(12), 125706 (2011)
24. Ferrari, A.C., Meyer, J.C., Scardaci, V., Casiraghi, C., Lazzeri, M., Mauri, F., Piscanec, S., Jiang, D., Novoselov, K.S., Roth, S., Geim, A.K.: Phys. Rev. Lett. **97**(18), 187401 (2006)
25. Yan, J., Zhang, Y., Kim, P., Pinczuk, A.: Phys. Rev. Lett. **98**, 166802 (2007)
26. Casiraghi, C., Pisana, S., Novoselov, K.S., Geim, A.K., Ferrari, A.C.: Appl. Phys. Lett. **91**(23), 233108 (2007)
27. Heydrich, S., Hirmer, M., Preis, C., Korn, T., Eroms, J., Weiss, D., Schüller, C.: Appl. Phys. Lett. **97**(4), 043113 (2010)
28. Ferrari, A.C.: Solid State Commun. **143**, 47 (2007)
29. Lee, C., Yan, H., Brus, L.E., Heinz, T.F., Hone, J., Ryu, S.: ACS Nano **4**(5), 2695 (2010)
30. Najmaei, S., Liu, Z., Ajayan, P.M., Lou, J.: Appl. Phys. Lett. **100**(1), 013106 (2012)
31. Scheuschner, N., Ochedowski, O., Schleberger, M., Maultzsch, J.: Phy. Status solidi B **249**(12), 2644 (2012)
32. Rice, C., Young, R.J., Zan, R., Bangert, U., Wolverson, D., Georgiou, T., Jalil, R., Novoselov, K.S.: Phys. Rev. B **87**, 081307 (2013)
33. Tonndorf, P., Schmidt, R., Böttger, P., Zhang, X., Börner, J., Liebig, A., Albrecht, M., Kloc, C., Gordan, O., Zahn, D.R.T., de Vasconcellos, S.M., Bratschitsch, R.: Opt. Express **21**(4), 4908 (2013)
34. Molina-Sánchez, A., Wirtz, L.: Phys. Rev. B **84**, 155413 (2011)
35. Li, S.L., Miyazaki, H., Song, H., Kuramochi, H., Nakaharai, S., Tsukagoshi, K.: ACS Nano **6**(8), 7381 (2012)
36. Zeng, H., Liu, G.B., Dai, J., Yan, Y., Zhu, B., He, R., Xie, L., Xu, S., Chen, X., Yao, W., Cui, X.: Sci. Rep. **3**, 1608 (2013)
37. Nemanich, G.L.R.J., Solin, S.A.: In: Proceedings of the International Conference on Lattice Dynamics. In: M. Balkanski (ed.), Flammarion (1975)
38. Verble, J., Wietling, T., Reed, P.: Solid State Commun. **11**(8), 941 (1972)
39. Kuzuba, T., Era, K., Ishii, T., Sato, T.: Solid State Commun. **25**(11), 863 (1978)
40. Tan, P.H., Han, W.P., Zhao, W.J., Wu, Z.H., Chang, K., Wang, H., Wang, Y.F., Bonini, N., Marzari, N., Pugno, N., Savini, G., Lombardo, A., Ferrari, A.C.: Nat. Mater. **11**, 294 (2012)
41. Hayek, M., Brafman, O., Lieth, R.M.A.: Phys. Rev. B **8**, 2772 (1973)
42. Plechinger, G., Heydrich, S., Eroms, J., Weiss, D., Schüller, C., Korn, T.: Appl. Phys. Lett. **101**(10), 101906 (2012)
43. Chandrasekhar, M., Rössler, U., Cardona, M.: Phys. Rev. B **22**, 761 (1980)
44. Zeng, H., Zhu, B., Liu, K., Fan, J., Cui, X., Zhang, Q.M.: Phys. Rev. B **86**, 241301 (2012)
45. Zhao, Y., Luo, X., Li, H., Zhang, J., Araujo, P.T., Gan, C.K., Wu, J., Zhang, H., Quek, S.Y., Dresselhaus, M.S., Xiong, Q.: Nano Lett. **13**(3), 1007 (2013)
46. Zhang, X., Han, W.P., Wu, J.B., Milana, S., Lu, Y., Li, Q.Q., Ferrari, A.C., Tan, P.H.: Phys. Rev. B **87**, 115413 (2013)
47. Varshni, Y.: Physica **34**, 149 (1967)
48. Korn, T., Heydrich, S., Hirmer, M., Schmutzler, J., Schüller, C.: Appl. Phys. Lett. **99**(10), 102109 (2011)
49. Plechinger, G., Schrettenbrunner, F.X., Eroms, J., Weiss, D., Schnller, C., Korn, T.: Phy. Stat. Solidi (RRL) û Rap. Res. Lett. **6**(3), 126 (2012)
50. Scalise, E., Houssa, M., Pourtois, G., Afanas'ev, V., Stesmans, A.: Nano Research **5**, 43 (2012)
51. Yun, W.S., Han, S.W., Hong, S.C., Kim, I.G., Lee, J.D.: Phys. Rev. B **85**, 033305 (2012)
52. Johari, P., Shenoy, V.B.: ACS Nano **6**(6), 5449 (2012)
53. Scalise, E., Houssa, M., Pourtois, G., Afanas'ev, V., Stesmans, A.: Physica E (2012). (In Press)

54. Rice, C., Young, R.J., Zan, R., Bangert, U., Wolverson, D., Georgiou, T., Jalil, R., Novoselov, K.S.: Phys. Rev. B **87**, 081307 (2013)
55. Cao, T., Wang, G., Han, W., Ye, H., Zhu, C., Shi, J., Niu, Q., Tan, P., Wang, E., Liu, B., Feng, J.: Nat. Comm. **3**, 887 (2012)
56. Zeng, H., Dai, J., Yao, W., Xiao, D., Cui, X.: Nat. Nanotech. **7**, 490 (2012)
57. [57] K. Kin Fai Mak, K. He, J. Shan, T. Heinz, Nature Nanotech. **7**, 494 (2012)
58. Sallen, G., Bouet, L., Marie, X., Wang, G., Zhu, C.R., Han, W.P., Lu, Y., Tan, P.H., Amand, T., Liu, B.L., Urbaszek, B.: Phys. Rev. B **86**, 081301 (2012)
59. Kioseoglou, G., Hanbicki, A.T., Currie, M., Friedman, A.L., Gunlycke, D., Jonker, B.T.: Appl. Phys. Lett. **101**(22), 221907 (2012)
60. Sanvitto, D., Hogg, R.A., Shields, A.J., Whittaker, D.M., Simmons, M.Y., Ritchie, D.A., Pepper, M.: Phys. Rev. B **62**(20), R13294 (2000)
61. Wang, R., Ruzicka, B.A., Kumar, N., Bellus, M.Z., Chiu, H.Y., Zhao, H.: Phys. Rev. B **86**, 045406 (2012)
62. Shi, H., Yan, R., Bertolazzi, S., Brivio, J., Gao, B., Kis, A., Jena, D., Xing, H.G., Huang, L.: ACS Nano **7**(2), 1072 (2013)
63. Mak, K.F., He, K., Lee, C., Lee, G.H., Hone, J., Heinz, T.F., Shan, J.: Nat. Mater. **12**, 207 (2013)

Chapter 9
The Application of Nanostructure MoS_2 Materials in Energy Storage and Conversion

Xue Zhang, Jin Liang and Shujiang Ding

Abstract A series of environmental problems have emerged owing to the excess consumption of fossil fuels. Development of clean alternative energy has turned into an urgent issue facing to all the nations. Nanostructured MoS_2, with particular chemical and physical properties, has been studied extensively and intensively over the past years. A comprehensive overview of the progress achieved within the application of MoS_2 in energy storage and conversion will be given, which is composed of lithium ion batteries, Mg ion batteries, dye-sensitized solar cells and photocatalytic hydrogen evolution.

9.1 Introduction

MoS_2 is a semiconducting, photovoltaic, and photocatalytic material with an indirect band gap of ~ 1.2 eV in the bulk form [1]. As a typical transition metal dichalcogenides (TMDs), MoS_2 possess an analogous structure to graphite. In the crystal structure of MoS_2, each Mo (IV) sits in the center of a triangular prism and is bound to six S atoms; meanwhile, each S atom is connected to three Mo centers [2]. The Mo atoms are sandwiched between two layers of closely packed S atoms, forming S-Mo-S layers that are held together by weak van der Waals forces and this particular structure allows foreign ions or molecules to be introduced between the layers through intercalation, such as Li^+ and Mg^{2+} [3]. Various nanostructured MoS_2, such as nanoflower [4], nanoflake [5], hierarchical microsphere [6], hollow nanoparticle [7], have been synthesized by a wide range of techniques, such as gas-phase reaction of MoO_3 and H_2S [8–10], thermal decomposition of ammonium thiomolybdate [11, 12], solid-state reaction of $MoCl_5$ and Na_2S [13], sonochemistry

X. Zhang · J. Liang · S. Ding (✉)
MOE Key Laboratory for Nonequilibrium Synthesis and Modulation of Condensed Matter and Department of Applied Chemistry, School of Science, Xi'an Jiaotong University, Xi'an 710049, China
e-mail: dingsj@mail.xjtu.edu.cn

Z. M. Wang (ed.), *MoS_2*, Lecture Notes in Nanoscale Science and Technology 21, DOI: 10.1007/978-3-319-02850-7_9, © Springer International Publishing Switzerland 2014

[14], pulsed laser deposition [15], microwave irradiation [16], liquid exfoliation in aqueous surfactant solutions [17], ball milling [18], chemical vapor deposition [19], electrochemical lithium intercalation [20], laser thinning [21], and hydrothermal reactions at low temperature [22]. Due to the important mechanical, electrical, and optical properties of MoS_2, it has been demonstrated to have enormous applications in the fields of electrochemical devices, hydrogen storage, catalysis, capacitors, solid lubricant, and intercalation host [23–28].

In this chapter, we will focus on the application of MoS_2 in the area of energy storage and conversion. It covers the following four main aspects. First, the application of MoS_2 as anode material in lithium ion batteries with high power density and long cycle life, which is a promising candidate for electric vehicles and hybrid electric vehicles to solve global warming and energy crisis. Second, Mg ion batteries, with high theoretical specific capacity of 2,205 mA h g^{-1}, are considered as a promising choice for energy storage as well. To date, there are several reports about the application of MoS_2 as the cathode material in Mg ion batteries and they will be displayed in this part. Third, due to the low cost, simple fabrication process, and high power conversion efficiency, dye-sensitized solar cells (DSCs) have attracted considerable attention. The materials as the counter electrode in DSCs obtained extensive attentions will be summarized in this section. In the final part, the application of MoS_2 as catalyst in photocatalytic hydrogen evolution will be illustrated.

9.2 The Application of Nanostructured MoS_2 Materials in Energy Storage

9.2.1 Lithium Ion Batteries

Global warming, increasing toxicity in environment, and escalating energy shortage, derived from the consumption of fossil fuels, are the urgent issues facing all the nations. Many alternative energy technologies have been developed in an attempt to alleviate the critical problems above and electrochemical energy storage is considered as one of the most promising solutions [29–33]. As one of the most promising candidates in electrochemical energy storage area, a possible solution to provide a clean storage option for portable instruments and the grid, the rechargeable lithium ion batteries (LIBs), have been investigated worldwide these years for their high energy density, high voltage, and long cycle life compared to other types of batteries [31, 34–36]. They have become the dominant power in portable consumer devices, such as cell phones, laptop, and so on. However, their application in electric vehicles and hybrid electric vehicles which demands even higher power energy sources that can operate under much higher current condition (tens of Amperes) than the operating current of mobile phones (about 100 mA) is still a challenge because of the electrode materials suffering from stability, safety, and capacity issues at high rate [37]. The intensively used commercial anode materials in LIBs are usually graphitic materials due to their low cost, flat potential profile, and good structural stability during cycling [35, 38–40]. Nevertheless, graphitic materials suffer from a relatively low theoretical specific capacity of 372 mA h g^{-1} and cannot meet the requirement of high capacity storage. As a result, a series of alternative anode materials with higher specific capacity as a substitution for graphite-based electrodes have been proposed up to now, such as metals, metal oxides, and metal sulfides-based negative electrodes [41, 42]. Layer-structured transition metal dichalcogenides (TMDs), MX_2 (M = Ti, Nb, Mo, Ta; X = S, Se, Te), with unique chemical and physical properties, have attracted much attention as a type of novel anode material, owing to their low cost, high safety, and facile synthesis process [24, 43–46].

As the representative member in the TMDs family, molybdenum disulfide (MoS_2) has been extensively investigated in many fields, including electrochemical devices, hydrogen storage, catalysis, capacitors, solid lubricant, nanotribology, and intercalation host [23–27, 47–49]. It has an analogous structure to graphite, [32] which facilitates reversible Li^+ intercalation/extraction in their lattices and enables MoS_2 to be a kind of good electrode material for lithium ion batteries as well [32]. Although MoS_2 has emerged as a promising material for application as an anode in lithium ion batteries, the cycling stability, and high rate property of the bulk MoS_2 are still unsatisfactory for its practical application, causing by the intrinsic large volume expansion of the metal sulfides during alloying/dealloying [24, 29, 50]. However, extensive research has proved that nanostructured materials can achieve higher specific capacity and better cycling performance compared with their bulky counterparts, due to the decreased diffusion lengths, enhanced kinetics, and large ionic contact area [32, 33, 51–53]. It is believed that the electrochemical performance of MoS_2 is significantly influenced by the morphology, microstructure, and particle size of the materials [44]. So far, several strategies have been reported to enhance the cycling performance of MoS_2-based materials as the anode in lithium ion batteries: (1) enlarging the interlayer distance between MoS_2 layers [45]; (2) synthesizing nanostructured MoS_2 with different morphologies [4, 40]; (3) preparing nanocomposites consisting of MoS_2 and conductive carbonaceous materials such as amorphous carbon [54], graphene [55], nanocarbon tubes [46], and polyanline, [32] which are already known to be mechanically stable buffers and can prevent the large volume expansion of various metal or metal oxide nanoparticles [29],

Herein, we will induce you the research progresses about the application of nanostructured MoS_2 as the anode material mainly in lithium ion batteries in recent years.

9.2.1.1 Nanostructured MoS_2 as Anode in LIB

Pure MoS_2, with various nanostructures, such as nanoflakes, nanospheres, nanosheets, has been investigated as the anode in lithium ion batteries.

In 2009, Wang et al. [4] reported the synthesis of 3D flowerlike MoS_2 sample built from MoS_2 nanoflakes through a facile ionic liquid (IL) assisted hydrothermal method. In this research work, the ionic liquid was used as the additive, for it is not only a kind of environmentally friendly solvent acting as a reaction medium for reactants; but also as a morphology directing agent for the final products. The obtained product was further annealed at 800 °C in the atmosphere N_2/H_2 to increase the crystallinity of MoS_2. The electrochemical performance of the 3D flowerlike MoS_2 prepared with IL, the MoS_2 nanosheets without IL, the annealed MoS_2 sample with IL and the annealed MoS_2 without IL as the anode in lithium ion battery was evaluated. During the first cycle, they possessed the discharge capacity of 869, 724, 807, and 652 mA h g^{-1} with a Coulombic efficiency of 65.9, 58.65, 60.2, and 59.3 %, respectively. At the 50th cycle, both MoS_2 samples exhibit the close discharge capacity (about 633 mA h g^{-1} for MoS_2 prepared with IL and 605 mA h g^{-1} for the MoS_2 prepared without IL, respectively). At the 100th cycle, the former and the latter retains 531 and 395 mA h g^{-1} capacity, with 54.9 and 47.3 % of the initial maximal capacity, respectively. At the 150th cycle, the capacity of the MoS_2 samples prepared with and without IL are 439 mA h g^{-1} (45.4 % of the initial maximal capacity) and 283 mA h g^{-1} (33.9 % of the initial maximal capacity), respectively. And, both MoS_2 samples exhibit worse cycling durability after their annealing. It's said that the enhanced electrochemical lithiation/delithiation performances can be attributed to more diffusion paths and larger electrode–electrolyte contact area for lithium ions that the amorphous MoS_2 with 3D flowerlike nanostructure provided in comparison to plane MoS_2 nanosheets prepared without IL. What's more, the 3D flowerlike architecture can better accommodate the structural strains caused by the lithium insertion–extraction.

Liu et al. [5] reported a simple synthesis method (rheological phase reaction) to synthesize MoS_2 nanoflakes. The MoS_2 electrode delivered an initial lithium insertion capacity of about 1174.7 mA h g^{-1} at the discharge current density of 60 mA g^{-1} and retained a reversible capacity of 851.5 mA h g^{-1} after 20 cycles. It is suggested that there were four possible reasons for the high lithium intercalation capacity of the MoS_2 nanoflakes electrodes: (1) lithium ions intercalate into nanoflake clusters; (2) lithium ions intercalate into defect sites in nanoflakes (partly formed nanotubes); (3) lithium ions intercalate into intratubal sites (the hollow core) through the open end; (4) lithium ions intercalate into the MoS_2 layer sites to form Li_xMoS_2.

Piao et al. [40] have reported a facile hydrothermal preparation method for MoS_2 nanospheres consisting of disordered layers assisted by L-cysteine. This method is environmentally friendly, not requiring toxic reagents and solvents. In this synthesis process, L-cysteine, a cheap and common amino acid, plays fundamental roles of both the sulfur source and the directing molecule in the formation of the MoS_2 nanospheres. During the synthesis reaction, L-cysteine reacts with metal ions to form a complex, and a polypeptide that serves as the template is formed by peptide bonding between each of the functional groups of the L-cysteine molecule. When the MoS_2 nanospheres were evaluated as the anode for lithium ion batteries, they showed excellent cycle stability with a high specific capacity compared to the MoS_2 nanosheets obtained without the addition of L-cysteine. The enhanced electrochemical properties of the MoS_2 nanospheres might be attributed to the increased interlayer distance providing more space and fast conductive channels for the lithium ions, the shortened paths for fast lithium ion diffusion resulting from the disordered layers, structural stability provided by the spherical structure during the charge–discharge process. The first charge capacities of the nanospheres and nanosheets were 751.4 and 872.0 mA h g^{-1} at a current density of 100 mA g^{-1}, respectively. Further, after the 30th cycles, the MoS_2 nanospheres exhibited a high charge capacity of 706 mA h g^{-1} with excellent retention. Even at a high current density of 1,000 mA g^{-1}, the specific capacity of the MoS_2 nanospheres remained at 658.1 mA h g^{-1} after 30 cycles, which was 93 % of that at a current density of 100 mA g^{-1}.

Lou et al. [6] have reported a facile synthesis of uniform hierarchically structured MoS_2 nanosheets microspheres with the addition of polystyrene microspheres (PSs) to facilitate the assembly of MoS_2-NSs to form uniform microspheres (see Fig. 9.1). The assembly of nanosheets could provide high surface area for storing more lithium ions and the large void space between

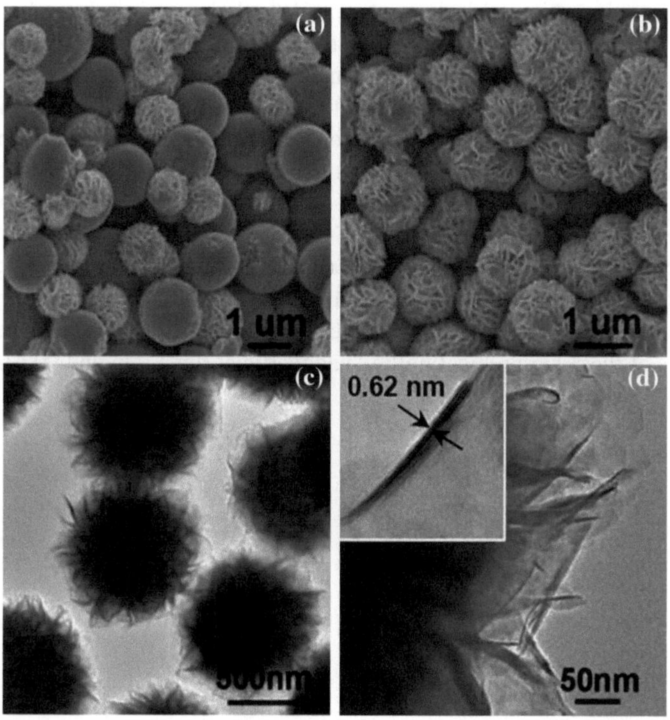

Fig. 9.1 **a** Scanning electron microscopy image (SEM) of the as-synthesized material; **b** SEM; and **c** transmission electron microscopy (TEM) images of the MoS_2-NS microspheres; **d** a high-resolution TEM image of several MoS_2-NSs; the inset shows a HRTEM image of a single MoS_2 nanosheet [6]

nanosheets could buffer the volume variation accompanying the charge–discharge process, thus leading to improved cycling stability. In comparison to MoS_2 flakes attained without the addition of PSs, the MoS_2-NS microspheres manifest improved lithium storage properties with better cycling performance and higher capacity (see Fig. 9.2). The initial discharge and charge capacities of MoS_2-NS microspheres are found to be 1,160 and 791 mA h g^{-1}. Such a high initial lithium storage capacity can be associated with the unique hierarchical structure of the stacked MoS_2 nanosheets. The irreversible capacity loss of around 31.8 % may be mainly attributed to the irreversible processes, such as formation of solid-electrolyte interface (SEI) layer and trapping of some lithium in the lattice of MoS_2. After 50 discharge–charge cycles at a current density of 100 mA g^{-1}, the MoS_2-NS microspheres can still deliver a capacity of 672 mA h g^{-1}, while the MoS_2 flakes only deliver 390 mA h g^{-1}. After 70 cycles, the capacities of MoS_2-NS microspheres and MoS_2 flakes further decrease to 585 and 163 mA h g^{-1}, respectively. At different current densities of 200, 500, and 1,000 mA g^{-1}, the capacities of MoS_2-NS microspheres are 726, 581, and 353 mA h g^{-1}, respectively, as compared to 311, 207, and 134 mA h g^{-1} only for pure MoS_2 flakes. The enhanced lithium storage properties of the MoS_2-NS microspheres can be attributed to the unique hierarchical structure: (1) the large surface area offered by the ultrathin NSs endows the MoS_2-NS microspheres with high specific capacity as a result of increased reactive sites and interface between the active material and electrolyte; (2) the presence of sufficient void space between the organized MoS_2 nanosheets not only allows fast lithium ion diffusion but also effectively buffers the mechanical stress and volume variation accompanying the lithium inserting–de-inserting process; (3) the flexible nature of such ultrathin MoS_2-NSs enhances the robustness of the electrode structure.

Yang et al. [7] have prepared hierarchical hollow MoS_2 nanoparticles composed of nanosheets with an increased interlayer distance via a simple solvothermal reaction at a low temperature. The initial lithiation and delithiation capacities of these hierarchical hollow nanoparticles are 1,236 and 907 mA h g^{-1},

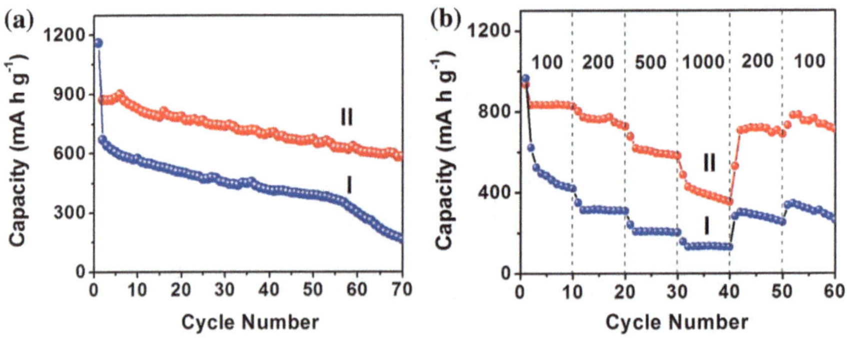

Fig. 9.2 **a** Comparative cycling performance of MoS_2 flakes *I* and MoS_2-NS microspheres *II* at a current density of 100 mA g^{-1}. **b** Cycling performance of MoS_2 flakes *I* and MoS_2-NS microspheres *II* at different current densities indicated (mA g^{-1}) [6]

at a current density of 100 mA g^{-1}, corresponding to a Coulombic efficiency of 74 %. After 80 cycles, the discharge capacity of the hierarchical hollow MoS_2 nanoparticles could be still maintained at 902 mA h g^{-1}, much higher than that of solid MoS_2 nanoparticles at 410 mA h g^{-1}. The high specific capacity and superior cycling stability of the hierarchical hollow nanoparticles could be attributed to their large specific surface area, hollow structural feature, and increased layer distance of S-Mo-S. The large surface area indicates the increased reactive sites and interfaces, and then favors the active material to achieve a high specific capacity. The hollow structure could effectively tolerate the volume change caused by the discharge–charge processes, reduce the diffusion distance of lithium ions, and then improve the cycling stability. The increased layer distance between neighboring S-Mo-S layers would facilitate the quick diffusion of lithium ions and improve the reaction reversibility. As a result of these advantages, the hierarchical hollow nanoparticles exhibit a better rate performance than the solid ones as well. At the current densities of 100, 200, 300, 500, and 1,000 mA g^{-1}, the specific capacities of the hierarchical hollow MoS_2 nanoparticles are 1,030, 950, 910, 850, and 780 mA h g^{-1}, respectively. When the current density goes back to 100 mA g^{-1}, the specific capacity of the hierarchical hollow MoS_2 nanoparticles also returns to 1,010 mA h g^{-1}.

In Liu et al. report [45], the layered MoS_2 was exfoliated and restacked with enlarged C-axis spacing for the first time via a simple hydrothermal procedure at 160 °C for 48 h, and tested as the anode material for lithium ion batteries. When compared to raw MoS_2, the restacked MoS_2 shows much better cycling stability. For raw MoS_2, the charge capacity obviously decreases with the cycle number, from over 800 mA h g^{-1} for the first three cycles to 226 mA h g^{-1} at the 50th cycle, however, the restacked MoS_2 with a slight decrease that may be caused by electrochemically driven electrolyte degradation over long cycling after the 40th cycle and above 750 mA h g^{-1} kept at the 50th cycle. What's more, even at a high charge current density of 1,000 mA g^{-1}, the restacked MoS_2 electrodes can still exhibit a reversible specific capacity as high as 710 mA h g^{-1},which can be attributed to the enlarged c parameter that can relax the strain and lower the barrier for lithium intercalation. The specific surface areas of raw and restacked MoS_2 are 4.89 and 9.83 m^2 g^{-1}, respectively, so another possible reason for improved electrochemical performances of restacked MoS_2 is the increase of the surface area, particularly at edges, which would facilitate interfacial reactions. It is demonstrated that the strategy of enlarging the space between layers and increasing the surface area (especially edge area) for electrode materials with layered structures could be a common way to improve the intercalation kinetics in terms of rechargeable batteries.

Qiao et al. [37] have performed a modified vacuum-assisted impregnation route to synthesize highly ordered mesoporous MoS_2 with a narrow pore size distribution, a high pore volume and a high specific surface area. Moreover, the obtained mesoporous MoS_2 product exhibits extended d-spacing of (002) crystal plane (0.66 nm), which can provide sufficient space for ultrafast lithium ion intercalation. The electrochemical performance of the bulk and mesoporous MoS_2

electrodes was investigated. At the current density of 0.1 A g^{-1}, the initial discharge capacities of the bulk and mesoporous MoS_2 electrodes are 1,055 and 1,052 mA h g^{-1}, respectively. However, in subsequent cycles, the discharge capacity of bulk MoS_2 electrode decreased continuously and only delivered a capacity of 346 mA h g^{-1} after 100 cycles. In contrast, the discharge capacity of is as high as 876 mA h g^{-1} after 100 cycles. What's more, the mesoporous MoS_2 also exhibits excellent high rate capacity of 608 mA h g^{-1} at the discharge current of 10 A g^{-1} (15C), which places MoS_2 as a viable next generation high power source for electric vehicles. It can be found that unique mesostructure of mesoporous MoS_2 is beneficial to the improvement of the cycling capacity and rate performance, and the reason can be described as follows: (1) narrow pore size distribution, high specific surface area, and high pore volume of the highly ordered mesopores facilitate electrolyte diffusion, and afford a large electrolyte/electrode interphase for lithium ion intercalation as a result; (2) the expanded (002) interlayer can provide sufficient space for ultrafast lithium ion intercalation; and (3) the intrinsic properties of mesoporous MoS_2 can provide a short pathway for lithium ion diffusion and low activation energy for lithium ion intercalation reaction.

In Mitre sagar et al.' report [31], cellulose (CMC) was used as the binder in substitution for PVDF. It has been reported that CMC binder shows better stability in comparison to PVDF in case of alloying reaction where huge volume change is accounted due to incorporation of large amount of Li in the matrix. It's believed that the combination of MoS_2 and CMC binder could successfully overcome the effect of volume change and improve the electrochemical performance with excellent cycling stability and rate capability. MoS_2 nanomaterial was obtained via a simple cost-effective chemistry, used as lithium battery anode, and demonstrated a superior cyclic stability using CMC as a binder. The MoS_2 nanomaterial showed a stable capacity of 880 mA h g^{-1} for 50 cycles at a current rate of 100 mA g^{-1}, and exhibited stable capacity of 676 mA h g^{-1} at 1,000 mA g^{-1}, which can be considered as one of the best rate capability of MoS_2 at that time.

9.2.1.2 CNT@MoS_2 Composites as Anode in LIB

Due to the various unique properties associated with high anisotropy, one-dimensional and intriguing tube-like structure, large surface area, high electric conductivity, mechanical performance, chemical stability, carbon nanotubes (CNTs), a new crystalline form of carbon, have been intensively investigated as a support material for the fabrication of semiconductor/carbon nanotube nanocomposites for energy conversion and storage devices. Especially, various CNTs-based hybrid materials have been reported with an aim to obtain materials with higher capacity and good cycling performance for LIBs.

Li et al. [46] first reported the application of MoS_2@CNT nanocomposites in lithium ion batteries in 2007. MoS_2 overlayers supported on coaxial carbon nanotubes were synthesized at low temperature via an alternative, simple, and effective solution-phase route, which can be potentially extended to the preparation of other coaxial metal dichalcogenide/carbon nanotubes. When tested for lithium storage capabilities, the CNT@MoS_2 hybrid structure shows a relatively good cyclic capacity retention with a reversible capacity of only up to 400 mA h g^{-1}, probably due to the low mass fraction of MoS_2 in the composite.

Afterward, Lou et al. [24] reported the growth of ultrathin MoS_2 nanosheets (NSs) on the CNT backbone (CNT@MoS_2-NSs) through a simple glucose-assisted hydrothermal method and obtained as high as 75 % content of MoS_2 in the CNT@MoS_2 (see Fig. 9.3). Glucose here plays a crucial role in the formation of the CNT@MoS_2-NS hybrid structures that might mediate the formation of MoS_2-NSs with smaller size and thickness and serve as a binder to help the MoS_2-NSs to grow on the surface of CNTs with uniform coverage along the longitudinal axis. The MoS_2 sample prepared without either glucose or CNTs was composed of micrometer-sized platelet-like structures, which probably consisted of densely stacked sheet-like subunits. What's more, if only CNTs were added, a mixture of CNTs and MoS_2 flakes was obtained in which the CNTs just acted as wires that connected discrete MoS_2 flasks. All the samples above were tested as the anode in lithium ion batteries (see Fig. 9.4). The initial discharge and charge capacities of CNT@MoS_2-NSs were found to be 1,434 and 862 mA h g^{-1} at a constant current rate of 100 mA g^{-1}, respectively, which might be associated with the unique

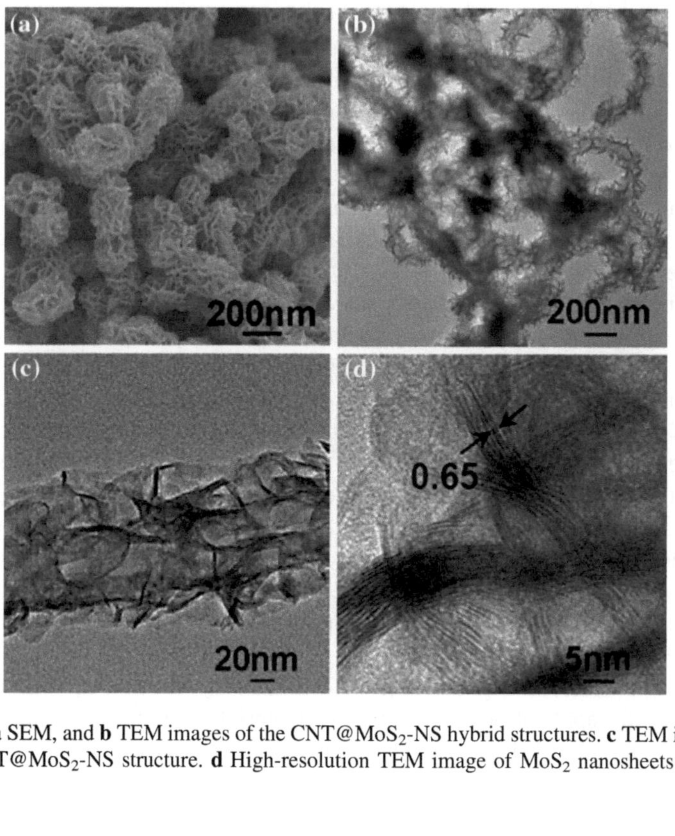

Fig. 9.3 **a** SEM, and **b** TEM images of the CNT@MoS_2-NS hybrid structures. **c** TEM image of a single CNT@MoS_2-NS structure. **d** High-resolution TEM image of MoS_2 nanosheets [24]

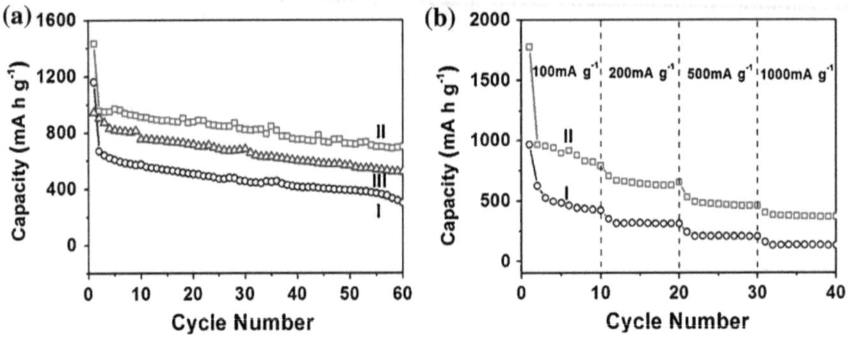

Fig. 9.4 **a** Cycling performance of: *I* MoS_2-NSs, *II* CNT@MoS_2-NSs, and *III* CNT@MoS_2-NSs without glucose at a current rate of 100 mA g^{-1}. **b** Cycling performance of: *I* MoS_2-NSs, and *II* CNT@MoS_2-NSs at different current rates [24]

structure of the thin MoS_2-NSs. After 60 discharge–charge cycles, a reversible capacity as high as 698 mA h g^{-1} can still be retained by the CNT@MoS_2-NSs. In contrast, the pure MoS_2 flakes can only deliver a much lower capacity of 300 mA h g^{-1}, and the mixture of MoS_2 flakes and CNT (synthesized without addition of glucose) delivers a capacity of 521 mA h g^{-1}. At the current density of 200, 500, and 1,000 mA g^{-1}, the CNT@MoS_2-NSs displayed reversible capacities of 653, 459, and 369 mA h g^{-1}, respectively. However, the capacities of pure MoS_2 flakes at these current rates are only 309, 202, and 129 mA h g^{-1}, respectively. The enhanced lithium storage performance of CNT@MoS_2 NS structures compared to pure MoS_2 flakes can probably attribute the excellent integration of CNT@MoS_2-NSs.

In Piao et al.' report [29], biomolecule-assisted synthetic routes have been used because they are green chemistry approaches without toxic reagents and solvents and have obtained unique structures. 1D structured CNTs-MoS_2 hybrid materials consisting of curved MoS_2 nanosheets with 1:2 and 1:4 molar ratio of MoS_2 to CNTs were attained on the basis of a facile hydrothermal method assisted by L-cysteine that serve as a sulfur source as well as a binder to help MoS_2 nanosheets grow on the surface of the CNTs for the first time. At a current density of 100 mA g^{-1}, the CNTs-MoS_2 hybrid materials (1:2 and 1:4) delivered initial capacities of 917.4 and 1096.5 mA h g^{-1} with the Coulombic efficiencies of 78.81 and 67.17 %, respectively. At the highest current density (1,600 mA g^{-1}), discharge capacity of the CNTs-MoS_2 hybrid materials (1:4) remained at 530.6 mA h g^{-1} after 30 cycles. The enhanced electrochemical performance of the CNTs-MoS_2 hybrid materials can be attributed to stabilization role of the CNTs and synergistic effects of the amorphous carbon formed by L-cysteine and the few-layered MoS_2.

Except for these, Feng et al. [3] and Mahanaty et al. [50] carried relative research on the application of MoS_2@CNT materials in lithium ion battery as well. In Feng et al.' report [3], a solvothermal method was developed for the synthesis of

MoS_2/CNT composite at relatively low temperature. The MoS_2/CNT composite delivered an initial discharge capacity of 1256.1 mA h g^{-1} and retained 757.5 mA h g^{-1} after 50 cycles with a current density of 50 mA g^{-1}. Mahanaty et al. [50] have applied dry physical grinding to prepare MoS_2-MWCNT hybrids with various weight ratios of MoS_2 to MWCNT. At the current density of 100 mA g^{-1}, MoS_2-MWCNT (1:1) hybrid they exhibited a reversible discharge capacities of 1,090 mA h g^{-1} after 30 cycles.

9.2.1.3 Amorphous Carbon@MoS_2 Composites as Anode in LIB

In Lee et al.' work [54], a series of graphene-like MoS_2/amophous-C composites have been firstly synthesized through a facile hydrothermal route with the addition of glucose and annealing at 800 °C for 2 h in H_2/N_2 atmosphere. The graphene-like MoS_2 with single-layer were uniformly dispersed in amorphous carbon (see Fig. 9.5). Among all the samples, the graphene-like MoS_2/aC–1.0 exhibits the best charge/discharge performances with a reversible capacity of 912 mA h g^{-1} after 100 cycles, which corresponds to a capacity retention rate of 95 %. It is suggested that such a high capacity should be attributed to two factors: The MoS_2 graphene-like structure and the synergistic effect between the graphene-like MoS_2 and amorphous carbon. The graphene-like MoS_2/a–C composites have larger inter-laminar distances between the graphene-like MoS_2 nanosheets. Because the growth of MoS_2 crystals, especially in the (002) plane, was greatly inhibited by the

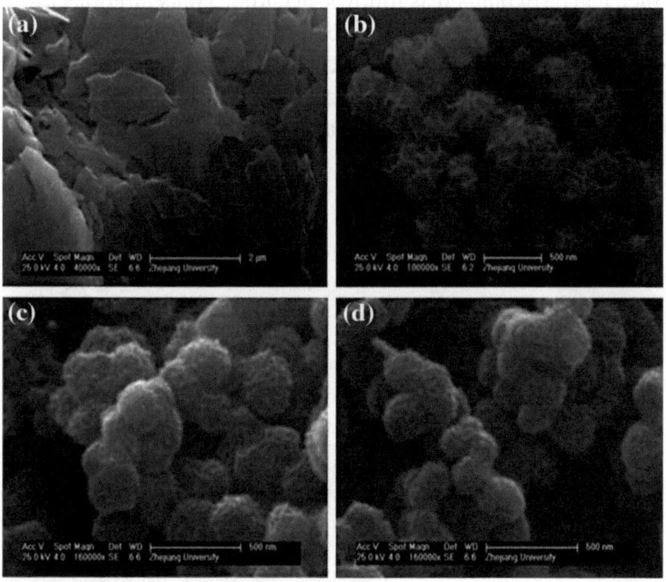

Fig. 9.5 SEM images of **a** annealed MoS_2 and graphene-like MoS_2/a–C composites prepared by hydrothermal method with adding **b** 0.5 g, **c** 1.0 g, and **d** 2.0 g of glucose [54]

amorphous carbon, the graphene-like MoS_2 nanosheets should have more defect sites. Moreover, more nanoclusters, defect sites and hollow core should exist due to the composite between the graphene-like MoS_2 nanosheets and amorphous carbon. The composite of graphene-like MoS_2 with amorphous carbon inhibits the restack of the graphene-like MoS_2 nanosheets and stabilizes the electrode structure. Additionally, the amorphous carbon can stabilize the disordered structure of the graphene-like MoS_2 nanosheets throughout the cycling regime to accommodate more Li^+ ions, and also keep the active materials electrically connected.

Archer et al. [56] have demonstrated the effect of the ratio of the carbon on the cyclability and rate capability of MoS_2 as lithium ion battery anode and ascertain the importance of optimization of inactive carbon in electrode materials. Pure MoS_2 and 11, 22, 32, 41 wt% carbon containing MoS_2 materials were prepared through a simple one-pot hydrothermal synthesis procedure and designated as MS-0, MS-11, MS-22, MS-32, and MS-41, respectively. At a constant current density of 100 mA g^{-1} and in the voltage range of 0.05–3 V, pure MoS_2 particle exhibited a high discharge capacity of 2,362 mA h g^{-1} in the first discharge cycle, however, it shows very poor 1st charge and 2nd discharge capacities of 247 and 53 mA h g^{-1}, respectively. The first discharge capacities of MS-11, MS-22, MS-32, and MS-41 were 2,108, 1,462, 1,130, and 1,078 mA h g^{-1} with Coulombic efficiencies of 79, 62, 63, and 61 % respectively. After 100th cycle, MS-22 showed a discharge capacity of 755 mA h g^{-1} with Coulombic efficiencies of 98 %, in contrast, MS-11, MS-32, and MS-41 showed capacities of 10,517 and 354 mA h g^{-1} (see Fig. 9.6). It's not difficult to find that MS-22 exhibits the best battery performance in terms of showing higher capacity and long-term stability among all the carbon-MoS_2 composites, which can be explained as follows: at carbon concentrations less than 22 %, the MoS_2 particles are not effectively coated with carbon that promotes possible electrochemical reaction of Li_2S and electrolyte during the first discharge process; on the other hand, increasing carbon concentrations more than 22 % results in thicker carbon coating and increase in inactive mass in the electrode.

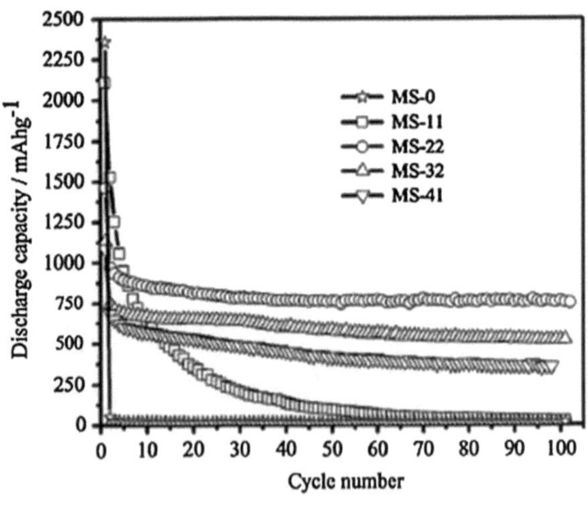

Fig. 9.6 Cycling stability of pure MoS_2 and various MoS_2-carbon composites [56]

Lou et al. [44] have reported the rational design and facile synthesis of novel carbon-coated MoS_2 (C-MoS_2) nanorods with high uniformity. α-MoO_3 nanorods used as the precursor were prepared by a hydrothermal method, then sulfidated by H_2S-gas with the subsequent chemical vapor deposition (CVD) of a thin carbon layer to obtain the uniform C-MoS_2 nanorods. When evaluated as LIB anodes, the as-obtained C-MoS_2 nanorods exhibit enhanced stability and reversibility upon extended cycling compared with the bare MoS_2 nanorods without carbon coating. There is a high reversible capacity of 621 mA h g^{-1} retained after 80 cycles at a current density of 200 mA h g^{-1} for C-MoS_2 nanorods electrode, which corresponds to a capacity loss of only 0.25 % per cycle. Additionally, the rate capability of the C-MoS_2 nanorods is also much better than that of the bare MoS_2 nanorods. At a high current density of 1,000 mA g^{-1}, the C-MoS_2 sample manifests a capacity of 550 mA h g^{-1}, whereas a capacity of only 430 mA h g^{-1} is delivered by the MoS_2 sample. The improved electrochemical performance of the C-MoS_2 sample probably originates from the enhanced structural integrity and improved conductivity brought by the uniform carbon coating. Furthermore, the 1D and nanoscale features grant the C-MoS_2 nanorods exceptional capabilities to accommodate the large volume variation and phase transition, and realize facile lithium uptake and release.

9.2.1.4 Graphene@MoS_2 Composites as Anode in LIB

Graphene, a flat one-atom-thick monolayer of carbon atoms exfoliated from graphite and arranged in a two-dimensional honeycomb network, has attracted enormous research interest energy storage devices, such as lithium ion batteries, owing to its large specific surface area, extraordinary electronic conductivity, excellent mechanical flexibility, high thermal and chemical stability. In particular, when used in lithium ion batteries, graphene could cushion the volume changes of the Li^+ storage host during charge and discharge operations. As a desirable matrix, graphene composites with metal (Sn, Si, etc) or metal oxides (SnO_2, etc) have been widely investigated in the field of lithium ion batteries with high specific capacity and improvement of the cycling performance obtained. Nevertheless, the cycle stability of these composites was not satisfactory and 15–30 % of capacity loss was typical after 30–50 cycles, which may be resulted from the structural and morphological mismatch between graphene and Sn, Si and metal oxides.

These years, there are some reports about the application of MoS_2 and graphene composites as anode material in lithium ion batteries with increasing electron conductivity and electrochemical performance and Chen et al. have done intensively study. In Chen et al. previous report [54], graphene-like MoS_2/a–C composites have been synthesized. However, the rate capability and cyclic stability of the graphene-like MoS_2/a–C composite is poor at a high current density of 1,000 mA g^{-1}, which may be attributed to the low conductivity of MoS_2 as well as the mediocre conductivity of amorphous carbon. In order to certify the above hypothesis, novel kind of nanocomposites comprised of single-layer MoS_2, graphene, and amorphous carbon (SL-MoS_2/G@a–C) were prepared by a facile solution-phased method and then annealing in H_2/N_2 atmosphere at 800 °C (see Fig. 9.7) [57]. The significant improvement in electrochemical performances was investigated on the basis of the composite microstructures and functions of the three components (see Fig. 9.8). The SL-MoS_2/G@a–C composite with 1:1, 1:2, and 1:4 molar ratio of MoS_2 to graphene was obtained, and was marked with MoS_2/G@a–C (1:1), (1:2), and (1:4). At a current density of 100 mA g^{-1}, the SL-MoS_2/G@a–C (1:1) composite exhibited the highest lithiation and delithiation capacity of 1,760 and 1,116 mA h g^{-1} during the first cycle respectively, with the Coulombic efficiency of 63.4 %, which is larger than that of graphene-like SL-MoS_2/a–C. For the SL-MoS_2/G@a–C (1:2) composite, the lithiation and delithiation capacity was 1,510 and 908 mA h g^{-1} with a Coulombic efficiency of 60.2 %, 1,563 and 837 mA h g^{-1} with a Coulombic efficiency of 53.5 % for the SL-MoS_2/G@a–C (1:4) composite, 1,468 and 777 mA h g^{-1} with a Coulombic

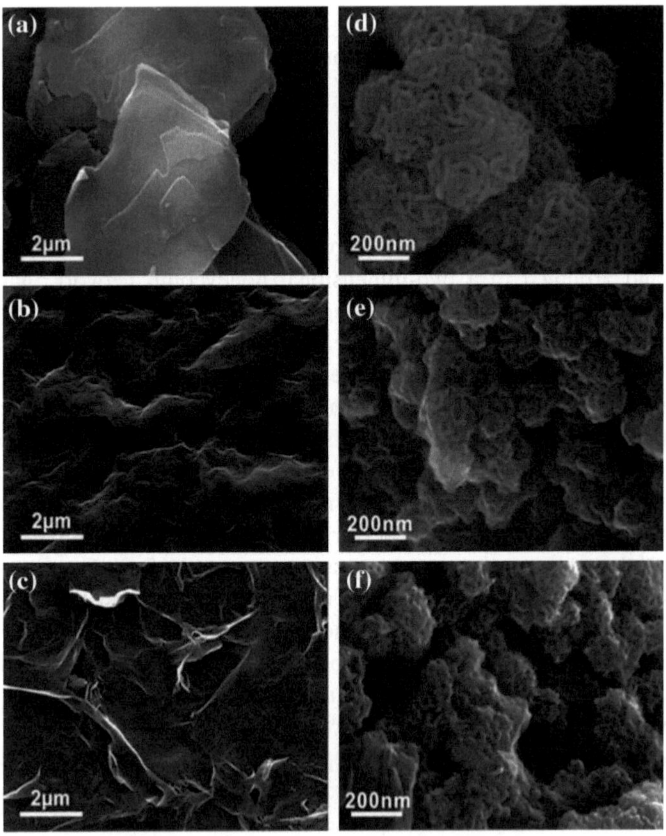

Fig. 9.7 SEM images of **a** graphite powers, **b** GO sheets, **c** graphene sheets and **d** SL-MoS_2/G@a–C (1:1), **e** SL-MoS_2/G@a–C (1:2), and **f** SL-MoS_2/G@a–C (1:4) after the annealing in H_2/N_2 at 800 °C for 2 h [57]

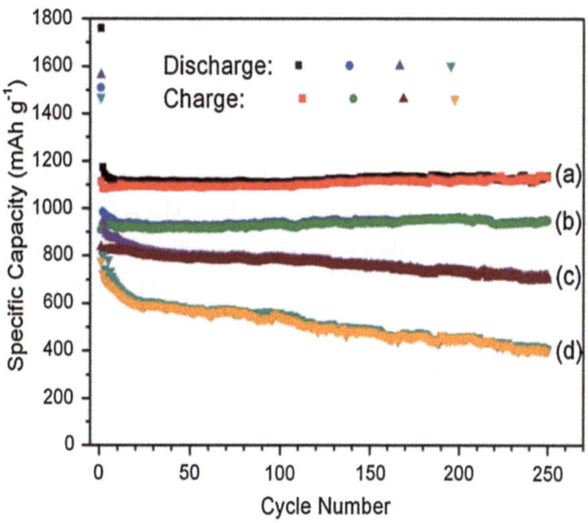

Fig. 9.8 Cycling behavior of three kinds of SL-MoS_2/G@a–C nanocomposites and graphene sheets at a current density of 100 mAh g^{-1}, **a** SL-MoS_2/G@a–C (1:1), **b** SL-MoS_2/G@a–C (1:2), **c** SL-MoS_2/G@a–C (1:4) and **d** graphene sheets [57]

efficiency of 52.9 % for the graphene electrode. When compared to the graphene-like SL-MoS_2/a–C composite without graphene, SL-MoS_2/G@a–C (1:1 and 1:2) composites exhibited an improved Coulombic efficiency at the first cycle, while remaining a high reversible capacity (900–1,100 mA h g^{-1}). The lithiation and delithiation process for SL-MoS_2/a–C and SL-MoS_2/G@a–C composites during charge/discharge was schematized by the authors. The single-layer MoS_2 sheets in both composites prepared by the hydrothermal method are highly dispersed in amorphous carbon and contain many defect sites. The presence of these defect sites indeed facilitate more lithium ions intercalation, but there are always fractions of lithium ions trapped in these defect sites and hardly extracted, which leads to additional irreversible capacity loss. If adding graphene in the composites, the graphene can greatly increase the conductivity of the composite and enhance the electron rapid transfer during the electrode reaction due to the extremely electronic conductivity and large specific surface area of graphene, thus reducing the resistance and electrochemical polarizations. The extraction of those lithium ions retained in the defect sites was facilitated, thereby improving the Coulombic efficiency at the first cycle. The SL-MoS_2/G@a–C nanocomposites also displayed excellent cyclic stability and high-rate capability. During the 250 cycles, the SL-MoS_2/G@a–C (1:1) and SL-MoS_2/G@a–C (1:2) electrodes have hardly any capacity fading, which can be attributed to the robust composite structure, the synergistic interactions of the compositions, and the unique structure of single-layer MoS_2. Even at a high current density of 1,000 mA g^{-1}, SL-MoS_2/G@a–C (1:1) exhibits high specific capacity of 850 mA h g^{-1} and good cycling stability.

At the same year, MoS_2/GNS composites with MoS_2 layers growing on a graphene nanosheet (GNS) have been synthesized via a facile in situ solution-phase reduced method, which made the MoS_2 anode material achieve the highest specific capacity initially [58]. Electrochemical tests indicate that the MoS_2/GNS composites exhibit very high capacity and excellent rate capability and cycling stability as anode materials for lithium ion batteries. The first discharge and charge capacity were 2,200 and 1,300 mA h g^{-1} repectively, which is much higher than the capacity of free MoS_2 obtained through the same method. After 50 cycles, MoS_2/GNS composites exhibited extraordinary cycling behavior and still retained the reversible capacity of 1,290 mA h g^{-1}, however, only a reversible capacity of 605 mA h g^{-1} was retained for free MoS_2 electrode. What's more, the MoS_2/GNS composites also demonstrate good rate performance (see Fig. 9.9). Even at a high current density of 1,000 mA g^{-1}, the specific capacity remains at 1,040 mA h g^{-1}, which is still higher than that of free MoS_2. At various current densities, even though the current changes from 1,000 to 100 mA g^{-1}, the specific capacity of the MoS_2/GNS composites returned to 1,300 mA h g^{-1} at once and does not ultimately change in the following cycles. The enhanced electrochemical performance of the MoS_2/GNS composites could be attributed to the intimate interaction between the graphene substrates and MoS_2 layers grown on them, which was facilitated to MoS_2 electrochemically active since charge carriers could be effectively and rapidly conducted back and forth from MoS_2 layers to graphene layers.

In 2013, Lee et al. [43] have proposed a new concept that a tighter integration between graphene-like MoS_2 and GNS could synergize their interaction to significantly improve the electrochemical performance of the MoS_2–GNS composites as lithium ion battery anodes. A facile and scalable aqueous solution chemistry process have been applied for the synthesis of SL-MoS_2-GNS composites in the presence of a cationic surfactant which are known for their ability to control crystal morphology in the synthesis of various inorganic nanostructured materials, cetyltrimethylammonium bromide (CTAB). Two composites were synthesized, namely SL-MoS_2-GNS02 and SL-MoS_2-GNS05, which were prepared with 0.02 and 0.05 M solutions of CTAB-functionalized GOS. At a current density of 100 mA g^{-1}, annealed MoS_2 and the SL-MoS_2-GNS02 composite delivered a discharge capacity of 1,091 and 1,367 mA h g^{-1} in the first cycle with the Coulombic efficiencies of 75.6 and 66.7 %. After 100 cycles, there was only 446 mA h g^{-1} capacity retained for annealed MoS_2. On the contrary, the SL-MoS_2-GNS composites display significantly improved cycling stability, with the reversible

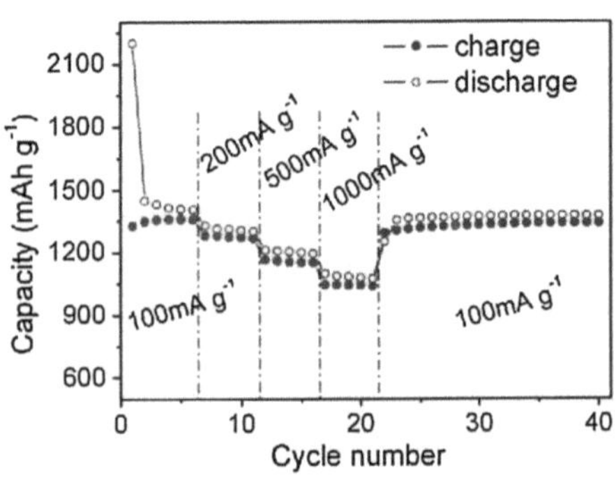

Fig. 9.9 Cycling behavior of MoS_2/GNS at various current densities [58]

capacities of 808 mA h g^{-1} retained after 100 cycles. Except for a slight decrease noticeable only for the first 10 cycles, the reversible capacity of the SL-MoS_2-GNS composites is fairly constant. What's more, the capacity of SL-MoS_2-GNS05 is lower than that of SL-MoS_2-GNS02, which probably due to the higher carbon content of the former. The excellent electrochemical reversibility for Li^+ storage in the SL-MoS_2-GNS composites could be attributed to the robust composite structure and the increase in synergistic interactions between SL-MoS_2 and graphene.

Expect for the achievement above, Guo et al. [35] have reported the cycling capacity and rate performance of MoS_2-graphene hybrid material that synthesized via a simple combination of a lithiation-assisted exfoliation process and a hydrazine monohydrate vapor reduction technique as the anode in lithium ion batteries. It's said that the as-prepared MoS_2-NS–RGO cannot only prevent the agglomeration of MoS_2 nanosheets but also restrict the growth of Mo nanoparticles during lithiation by the bond formed between MoS_2 nanosheets and graphene. At a current density of 0.5 A g^{-1}, the MoS_2 NS–RGO electrode delivered a reversible capacity of 915 mA h g^{-1} after 700 cycles. When the current densities increase from 0.5 to 1, 2, 5, 10, and 20 A g^{-1}, the electrode exhibits high capacity retention, as the specific capacities change from 635 to 581, 539, 464, 403, and 339 mA h g^{-1}, respectively. When the current density returns to 0.5 A g^{-1}, the specific capacity can recover to 605 mA h g^{-1}.

9.2.1.5 Polyaniline@MoS_2 Composites as Anode in LIB

Polyaniline (PANI), a kind of conductive and soft polymer, can effectively improve the conductivity and stability of active materials.

Schmidt et al. [32] firstly reported the constructing of uniform hybrid structures of MoS_2 evenly embedded in PANI matrix. In this research, hierarchical MoS_2/PANI nanowires structure composed of various building blocks on the nanoscale was fabricated in a mild condition (see Fig. 9.10). In such structures, the nanoscale dimension of building blocks can promote the kinetics of Li^+ storage due to the shortened diffusion paths, while the primary architecture at the micrometer scale can effectively avoid aggregation of the active nanomaterials and facilitate the transport of electrons and ions. In this report, a series of MoS_2/PANI nanowires with various MoS_2 contents were prepared and denoted as MoS_2/PANI-I, -II, and -III. At a current density of 100 mA g^{-1} between 0.01 and 3.0 V, the nanowires of MoS_2/PANI-I, -II, and -III deliver much improved charge capacities of 827.5, 1062.7, and 1015.3 mA h g^{-1} compared to the commercial MoS_2 microparticles with a charge capacity of 684.9 mA h g^{-1}. After 20 cycles, the reversible capacity of MoS_2 microparticles drops to around 184.9 mA h g^{-1}. In contrast, MoS_2/PANI-I, -II, and -III retain reversible higher capacities of 474.2, 952.6, and 748.8 mA h g^{-1} after 50 cycles, respectively. Obviously, MoS_2/PANI-II presents the highest specific capacity and the best cyclability among these hybrid nanowires (see Fig. 9.11a). It is proven that the content of PANI plays a crucial role in the electrochemical performance of the MoS_2/PANI nanowires electrode. As a result,

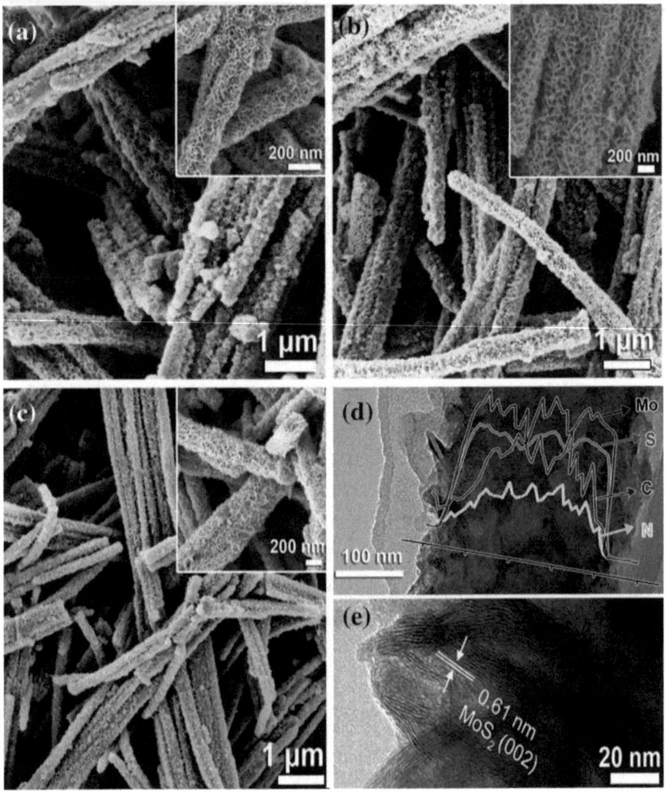

Fig. 9.10 Typical SEM images of the as-obtained hierarchical **a** MoS_2/PANI-I, **b** MoS_2/PANI-II, and **c** MoS_2/PANI-III nanowires. **d** TEM image with elemental (Mo, S, C, and N) distribution. **e** HR–TEM image of a MoS_2/PANI–II nanowire obtained at R_{AHM}/ATM of 0.43 [32]

it's necessary to optimize the weight ratio between MoS_2 and PANI. MoS_2/PANI-II displayed enhanced rate capability in comparison to commercial MoS_2 as well. MoS_2/PANI–II delivered an initial capacity of 1006.4 mA h g^{-1} at a current density of 200 mA g^{-1}, and retained a capacity of around 320 mA h g^{-1} as the

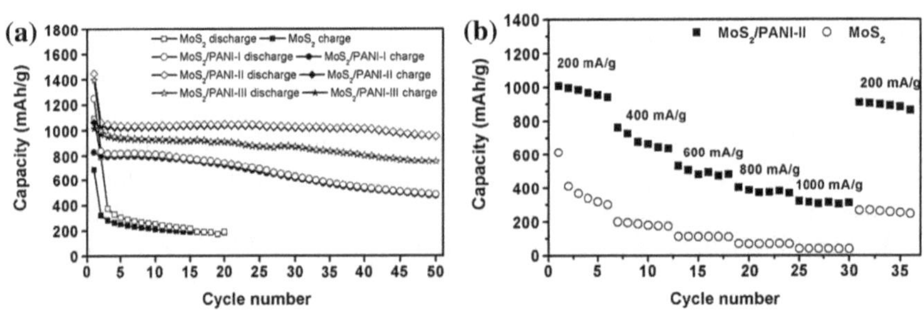

Fig. 9.11 **a** cycling performance of the MoS_2/PANI nanowires and the commercial MoS_2 microparticles tested in the range of 0.01–3.0 V versus Li^+/Li at the current density of 100 mA/g. **b** Rate performances (charge capacity is presented) [32]

current density increases to 1,000 mA g^{-1}. When the current density decreases to 200 mA g^{-1} after cycling under high current densities, it can still regain a reversible capacity near 900 mA h g^{-1} (see Fig. 9.11b). As for the MoS_2 microparticles, the reversible capacity fades to less than 50 mA h g^{-1} as the current density is 1,000 mA g^{-1}, and regains only around 40 % of the initial capacity of 250 mA h g^{-1}, when the current density drops back to 200 mA g^{-1}. It is demonstrated that the high capacity, excellent cycle stability, and rate capability of MoS_2/PANI nanowires could be attributed to their hybrid structure with conductive PANI and the hierarchical features. The enhanced conductivity facilitates the electron and Li^+ transfer in the electrode. Moreover, the MoS_2/PANI nanowires are hierarchically built up with nanosheets, which avoid the aggregation and retain small dimensions and large surface area. Such ultrathin MoS_2 nanosheets shorten the diffusion paths of Li^+, thus improving the dynamic performance of Li^+ storage. The large contact area between the building blocks and the electrolyte offers more active sites for Li^+ insertion/distraction, resulting in high specific capacity. Also, the voids between the nanosheets and the soft PANI chains in the composite accommodate the volume change, which effectively mitigates the stress and protects active materials from pulverization during the discharge/charge process. After all, a new protocol for developing electrode materials based on organic–inorganic nanohybrids was pointed out.

9.2.2 Mg Ions Battery

Except for the utilizing of the lithium ions in secondary batteries, Mg ions can be used for rechargeable batteries as well.

Owing to the characters of high theoretical specific capacity (2,205 mA h g^{-1}), great raw material abundance, environmentally friendly, and good operational safety in comparison to lead–acid and nickel–cadmium systems, rechargeable Mg batteries, have shown the potential for reversible electrochemical energy storage and gained increasing interest especially after work pioneered by Aurbach's group [53, 59, 60]. However, there are two major problems for rechargeable Mg batteries: lower voltage and lower specific energy density.

The working voltage of rechargeable Mg batteries is still lower than that of lithium ion batteries, which can be reduced to two main issues: (1) the kinetically sluggish Mg intercalation/insertion and diffusion in cathode materials which refines the choice of cathode materials as hosts [61–63]; (2) the anode/electrolyte incompatibility due to the high polarizing ability of the divalent Mg^{2+} cation [61, 62]. The specific energy density of rechargeable Mg batteries achieved so far is low, suffering from the generally low working voltage, which is mainly limited by the cathode material [64].

Up to now, a few materials have been investigated as the cathode in rechargeable Mg batteries and exhibit reasonable discharge capacity and cycling endurance, such as Chevrel phase $M_xMo_6T_8$ (M = metal, T = S, Se) [59, 61, 62],

TiS_2 nanotubes [65], and mesoporous $Mg_{1.03}Mn_{0.97}SiO_4$ [66]. MoS_2, with layered structure consisting of covalently bound S-Mo-S tri-layers, have attracted some attention as intercalation host for the weak interlayer interactions formed by van der Waals force allowing foreign atoms or molecules to be introduced between the layers through intercalation [67]. Nevertheless, there are only a few reports about the application of MoS_2 as cathode in Mg batteries.

It is demonstrated that Novak [63] has made a great step to carry out the electrochemical intercalation of Mg^{2+} into MoS_2 using a molten salt electrolyte based on $MgCl_2/AlCl_3$/1-ethyl-3-methylimidazolium chloride (EMIC) at 80 °C [67]. Then, the electrochemical properties of MoS_2 electrode was initially investigated by Li et al. [67]. It is believed that nanosized materials with the novel morphologies have somewhat better performance than bulk materials. MoS_2 nanostructures including hollow-cage fullerene-like particles, fibrous floccus, and spherical nanovesicles were successfully synthesized through designed solution chemical reactions of Na_2MoO_4 and sulfurization reagents such as CS_2, Na_2S, CH_3CSNH_2, CSN_2H_4, KSCN, etc, at low temperature in 200 °C and then annealed at 700 °C. As expected, Mg^{2+} ions can reversibly intercalate/deintercalate into MoS_2 samples during the electrochemical charge/discharge processes. At a current of 0.05 mA (for charge) and 0.02 mA (for discharge) and cutoff voltage of 2.80 V (for charge) and 0.6 V (for discharge), an obvious reversible charge/discharge phenomenon could be observed for the first charge/discharge curve of the cell and a large number of cycles could be repeated which indicated that the electrode structure did not break up during the electrochemical procedure. Control experiment of electrochemical charge/discharge measurement using the commercial MoS_2 powders was carried out as well. The charge/discharge process was usually fast and the curve was quite sharp. The voltage changed so quickly with time that the capacity of the cell was almost zero.

Recently, the combination of highly exfoliated, graphene-like MoS_2 (G-MoS_2) cathode, and ultra small Mg nanoparticle (N-Mg) anode for rechargeable Mg batteries have been firstly reported by Chen et al. [64]. The highly exfoliated, graphene-like MoS_2 was prepared via a solvothermal route from the reaction of MoO_3 and thioacetamide with pyridine as the solvent. The G-MoS_2 is composed of single- and several-layer structures, with the lattice fringe spacings of G-MoS_2 being 5–0.70 nm that are larger than those of bulk MoS_2 (B-MoS_2, 0.63 nm), indicating a spacing enlargement for the layers after exfoliation. Meanwhile, the ultra small Mg nanoparticles with an average diameter of 2.5 nm were prepared by ionic liquid-assisted chemical reduction to be used as the anode material. Such a configuration exhibits an operating voltage of 1.8 V and a well reversible discharge capacity of about 170 mA h g^{-1}. In the report, other three types of cell were fabricated with the four samples (G-MoS_2, B-MoS_2, N-Mg, and B-Mg) combined in an orthogonal manner as well, to separately investigate the effect of different active materials on cell performance. As is shown in Fig. 9.12, after 50 cycles, 95 % of the initial capacity is maintained with the Coulombic efficiency close to 99 % for the G-MoS_2/N-Mg cell. In comparison, the B-MoS_2/B-Mg cell maintains merely 60 % of the initial capacity after 5 cycles. Regardless of the type

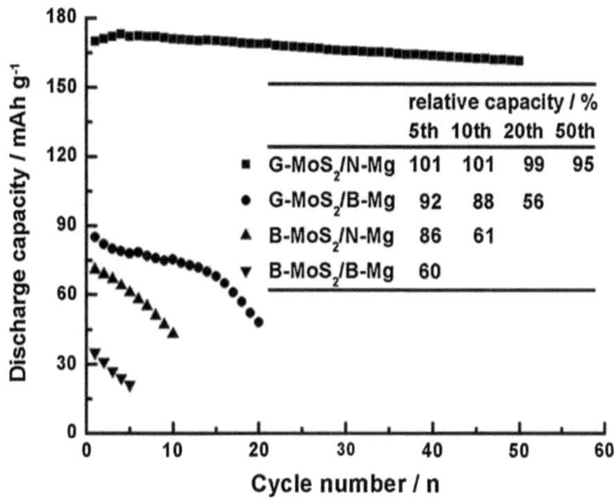

Fig. 9.12 Typical cycling behavior of the cells fabricated with B- or G-MoS_2 cathode and B- or N-Mg anode with a discharge rate of 20 mA g^{-1}. In the inset table, the relative capacity at a certain cycle refers to the ratio of the discharge capacity of the cells at the corresponding cycle to that at the first cycle [64]

of Mg anode used, the cells with B-MoS_2 cathode also show inferior cycling stability, which is likely caused by the trapping of Mg ions in the dense layers of pristine MoS_2. It is supposed that the interlayer separation of restacked MoS_2 was enlarged upon Mg ions insertion, resulting in more available intercalation sites. The findings have therefore opened up new opportunities for improving the electrochemical performance of rechargeable Mg ions batteries.

9.3 The Application of Nanostructured MoS_2 Material in Energy Conversion

9.3.1 Solar Cell

In the development of clean energy, solar energy is a plentiful and environmentally safe energy source [68]. Owing to the easy fabrication, good plasticity, environmental friendliness, low cost, and high power conversion efficiency, dye-sensitized solar cells (DSCs), considered as an alternative to conventional silicon-based solar cells, have attracted much attention since the report by Oregan and Gratzel in 1991 [1, 68–71]. Typically, DSCs are composed of a photoanode (a nanocrystalline semiconductor film on an FTO substrate sensitized by a dye), an electrolyte containing the I^{3-}/I^- redox couple, and a counter electrode (CE) [68]. The counter electrode, mainly catalyzing the I^-/I^{3-} redox couple and completing the electric circuit in the DSCs, should possess both high electrical conductivity and excellent electrocatalytic activity [1, 72]. Nowadays, Pt is still the most widely used CE for DSCs. However, Pt is an expensive metal and can be decomposed to Pt I_4 by I^-/I^{3-} redox couple, thus restricting the commercial production and resulting in the critical issues on the cell performance, long-term stability, and commercialization of DSCs [1, 73, 74]. Therefore, a variety of alternative materials have been proposed, including carbon-based materials [75–77], conducting polymers [78], metals in the form of sulfides [79–81], nitrides [82, 83], and carbides [72, 84, 85]. MoS_2, an indirect band gap semiconductor with an energy gap of about 1.2 eV in the bulk form, has attracted enormous interest as photovoltaic and photocatalytic materials. However, only a few studies have been performed to investigate the use of MoS_2 as the CE in a DSC system [1, 68, 86].

In 2011, Wu et al. [86] synthesized MoS_2 by a hydrothermal method and employed it as CE materials for DSCs. The DSC with the MoS_2 CE exhibited a power conversion efficiency of 7.59 %. Then, there are some relative reports on the application of MoS_2 as CE material for the Pt-free DSCs in the following years.

MoS_2/graphene nanocomposite used as the CE in DSCs was initially reported by Lin et al. [68]. The MoS_2/graphene nanocomposite was synthesized by simply mixing graphene oxide (GO) nanosheets with a solution of ammonium tetrathiomolybdate, and then MoS_2 nanoparticles with a hexagonal structure deposited onto the reduced GO were obtained after thermal annealing in H_2 flow at 650 °C. It is found that the functional groups in GO are critical to the immobilization of MoS_2 nanoparticles onto the graphene surface, and the density of MoS_2 nanoparticles on graphene can be modulated by the concentration of ammonium tetrathiomolybdate. DSCs with sputtered Pt, RGO, MoS_2, and MoS_2/RGO CEs were assembled. The DSC assembled with a MoS_2/RGO CE shows a comparable power conversion efficiency of 6.04 % to the DSC with a Pt CE (6.38 %), due to the large active surface area, the superior electrical conductivity of RGO. It is suggested that the MoS_2/RGO CE shows great potential as a low cost alternative to Pt in DSCs.

Then, in Wu et al. report [1], a catalytic film of MoS_2/graphene composite have been successfully prepared by mixing the commercially available MoS_2 nanoparticles and graphene flakes and employed as CE material for I^{3-} reduction in DSCs. The effect of the MoS_2/graphene CEs' thickness controlled by the adhesive plaster was evaluated for DSCs. The open-circuit voltage (Voc) and a short-circuit current (Jsc) increase with increasing the adhesive plaster layers from 1 to 3. When the number of adhesive plaster layers increases to 4 or 5 layers, it holds a negative effect on the FF and Jsc of DSCs. When the thickness and the graphene content were controlled at about 6um and 1.5 wt% for the MoS_2/graphene CE respectively, the DSC with the MoS_2/graphene exhibited an impressive power conversion efficiency of 5.98 % under solar simulator 1.5 G, which is comparable to that using the Pt CE (6.23 %). It is suggested that the excellence of photoelectric performances of DSC with MoS_2/graphene CE mainly can be attributed to following aspects: first, the lamellar structure of MoS_2 and graphene dispersed each other uniformly and provides a large active surface area on the electrode, which benefits transport for the liquid electrolyte, particularly, it is favorable to increase the I^-/I^{3-} redox reaction rate on the CE side, thus possibly enhancing the Jsc value; second, the excellent conductive and electrocatalytic activity of graphene and MoS_2 supported a low Rct at the interface between electrolyte and the CE for the I^-/I^{3-} redox reaction, which benefits for the increase in FF value. Nevertheless, this CE was not transparent optically.

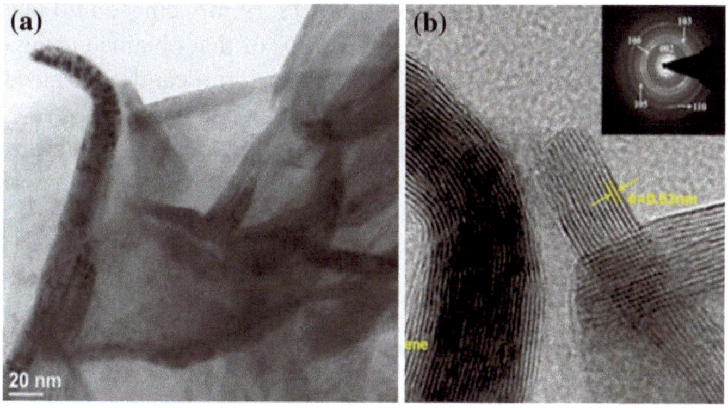

Fig. 9.13 **a** and **b** TEM images of the MoS_2–GNS composite [72]

The application of optically transparent MoS_2-graphene nanosheet (GNS) CEs in DSCs have been presented by Chou et al. for the first time in 2013 (see Fig. 9.13) [72]. It is said that the characteristic of optical transparency is a substantial benefit of DSCs for certain practical applications, like roof panels, windows, or various decorative facilities. The MoS_2–GNS composites were synthesized via a facile one-step thermal hydrolysis/dissociation of thiourea in the presence of graphene oxide (GO) in an aqueous solution consisting of sodium molybdate as a source of molybdenum, and then subjected to electrophoresis onto a fluorinated tin oxide (FTO) glass substrate at room temperature to obtain the MoS_2–GNS composite films with controlled thickness, excellent uniformity, large-scale production without binders. The photovoltaic performance of DSCs with GNS, MoS_2, MoS_2–GNS, and Pt CEs was measured under an illumination of 100 mW cm^{-2} and the corresponding photovoltaic parameters were summarized (see Fig. 9.14). The DSC using the GNS CE exhibits a poor conversion efficiency of 2.68 % due to its poor electrocatalytic activity, while the DSC based on the

Fig. 9.14 Photovoltaic characteristics of the DSCs with the Pt, MoS_2, GNS, and MoS_2–GNS CEs, respectively [72]

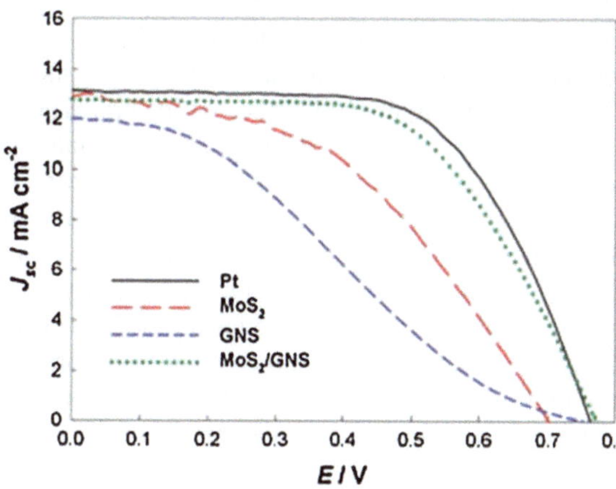

MoS_2–GNS CE has an Voc of 773 mV, Jsc of 12.79 mA cm^{-2}, a fill factor (FF) of 0.59, and a cell efficiency (Z) of 5.81 up to 93 % of that obtained using the Pt CE. When compared with the MoS_2 and GNS CEs, a significantly enhanced FF value for the MoS_2–GNS CE can be observed. It is concluded that the MoS_2–NGS CE is a promising candidate as a highly efficient and low-cost CE material in Pt-free transparent DSCs and the facile low-temperature electrophoretic deposition from suspension can be applied into large-scale flexible substrates.

9.3.2 Photocatalytic Hydrogen Evolution

For burning of fossil fuels has brought harmful influence on the global environment, it is necessary and urgent to exploit new clean energy, such as low-carbon sources of energy [87, 88]. Hydrogen, a clean energy carrier giving only water when burnt, have attracted enormous and increasing attention, for it is not only an environmentally benign fuel for the generation of power, heat, and electricity, but also a versatile reagent for the synthesis of ammonia and for the conversion of carbon dioxides into methanol and hydrocarbons [89–91]. Hydrogen, as a real alternative to hydrocarbons, must be produced in a sustainable fashion. One promising strategy is to use sunlight directly or indirectly (through wind power, for instance) to split water [90]. Solar energy, the most abundant and sustainable energetic resource, can both be captured and converted into electricity by employing a solar cell and be converted to chemical energy stored within chemical bonds of highly energetic molecules via a photoelectrochemical process [88]. Since the photoelectrochemical water splitting on the TiO_2 electrode was pioneering reported by Fujishima and Honda [92], this photocatalytic process has received worldwide attention as a solar energy driven pathway to the clean, low-cost, and environmentally friendly hydrogen production [87]. Simultaneously, large-scale and sustainable hydrogen evolution by photocatalytic water splitting has been vigorously pursued [93]. To proceed the hydrogen evolution reaction (HER), the photocatalytic system usually requires photocatalysts with high catalytic activities under the visible light [87]. The photocatalysts are usually composed of semiconductor materials and co-catalysts. In particular, the photocatalytic HER efficiency significantly relies on the properties of the hydrogen evolution co-catalyst [94]. Generally speaking, an appropriate co-catalyst can effectively promote the separation of photogenerated electron and hole pairs, offer the low activation potential for H_2 or O_2 evolution and suppress the recombination of photogenerated charges by the efficient separation of charges on different sites [95, 96]. Currently, Pt is the main co-catalyst used for HER due to its negligible overpotential and excellent kinetics for the hydrogen evolution reaction. However, owing to the scarcity and high cost, the widespread use of Pt is limited [87]. It is therefore important to develop highly efficient and low-cost nonnoble metal alternatives. Studies indicate that metal sulfides with bandgaps allowing absorption in the visible region of the spectrum can be good candidates to meet the above demands, for example, NiS [97], SrS [98], CuS [99], and WS_2 [100], etc [95]. Recently, studies have shown that molybdenum disulfide (MoS_2) can efficiently catalyze the electrochemical and photochemical HER as well [93]. Here, the application of MoS_2 as co-catalyst in HER will be reviewed.

In the early studies, metal sulfides have demonstrated high activity in H_2 involved reactions in heterogeneous catalysis; however, they are rarely used as co-catalysts in photocatalytic H_2 production. Li et al. [101] have found MoS_2 as an efficient co-catalyst for CdS in the photocatalytic H_2 production reactions. The MoS_2/CdS catalysts with highly dispersed MoS_2 on CdS were prepared by impregnating CdS with an aqueous solution of $(NH_4)_2MoS_4$, followed by a treatment in H_2S flow at high temperatures for 2 h. They found that the rate of H_2 evolution on CdS is significantly enhanced by loading MoS_2 as a co-catalyst on CdS for the H_2 production from a lactic acid solution under visible light. The activity of CdS is increased by up to 36 times when loaded with only 0.2 wt% of MoS_2, and the activity of 0.2 wt% MoS_2/CdS is even higher than that of 0.2 wt% Pt/CdS under the same reaction conditions. Nevertheless, factors influencing the photocatalytic activity of MoS_2/CdS and the reasons for the enhanced activity of MoS_2/CdS were not clear yet [96]. Afterward, Li et al. [96] have done a series of relative experiments to explore the factors mentioned above. It is suggested that the excellent catalytic property of MoS_2, together with the junctions formed between MoS_2 and CdS that leads to the more efficient electron transfer between the two components, is mainly responsible for the greatly enhanced activity of MoS_2/CdS in the photocatalytic H_2 production.

However, the photocatalytic activity of MoS_2 as co-catalyst applied toward the photocatalyst CdS is still unsatisfactory [95]. As reported, an efficient mixing, requiring fresh interfacial contacts between the reactants, may be beneficial to the enhancement of the composite photocatalysts' activities for H_2 evolution. Ball milling (BM) is very effective in realizing a highly efficient mixing of solid–solid reagents under solvent-free condition. Particularly, ball milling can lead to intimate contacts between the starting materials on molecular scale, thus enhance the reactivity of the reagents (mechanical activation) due to an induced mechanical breaking of molecular bonds (mechanochemistry). Besides, this method is more cost-efficient, eco-friendly, and can be used in mass production. For the reasons above, MoS_2/CdS photocatalyst was prepared by Meng et al. [95] via a simple ball milling combined calcination method. The optimized 0.9 mol % MoS_2/CdS photocatalyst can exhibit a H_2 evolution rate of 1,315 μmol h^{-1}, much larger than the H_2 evolution rate reported by Li et al. due to the pre-thermal treatment of CdS, the uniform loading of MoS_2 on the surface of CdS and intimate contact between the two components by ball milling and calcination.

Unfortunately, cadmium is a widespread environmental pollutant that is toxic and harmful to human beings [102]. Titania has proven to be a suitable candidate for photocatalytic water splitting among various oxide semiconductor photocatalysts, for its biological and chemical inertness, cost effectiveness, environmental friendliness, availability, and long-term stability against photo- and chemical corrosion. Typically, bare TiO_2 has poor photocatalytic activity. It is the type and amount of co-catalyst that decide the photocatalytic H_2 production activity of TiO_2. It has been reported that a MoS_2-TiO_2 system shows high photocatalytic performance for pollutant degradation. Then, TiO_2 nanoparticles (NPs) on a layered MoS_2/graphene (MG) hybrid for use in photocatalytic H_2 production was synthesized by Jaroniec et al. [102] for the first time through two-step hydrothermal process. The TiO_2/MG composite photocatalysts showed high photocatalytic H_2 production activity with a rate as high as 165.3 μmol h^{-1} for the sample containing 0.5 % MG hybrid co-catalyst consisting of 95 % MoS_2 and 5 % graphene with corresponding apparent quantum efficiency 9.7 % at 365 nm even without a noble-metal co-catalyst (see Fig. 9.15). It is believed that the positive synergetic effect between the MoS_2 and graphene sheets as the components of co-catalyst on the photocatalytic H_2 production activity can efficiently suppress charge recombination, improve interfacial charge transfer, and provide a greater number of active adsorption sites and photocatalytic reaction centers. Zhang et al. [87] have initially reported the fabrication of few layer MoS_2 nanosheet-coated TiO_2 nanobelt heterostructures with 3D hierarchical configuration consisting of a core of TiO_2 nanobelt and shell of MoS_2 nanosheets by a simple hydrothermal method using the TiO_2 nanobelts with the rough surface as the template (see Fig. 9.16). The TiO_2@MoS_2 heterostructure showed an excellent photocatalytic hydrogen production activity, giving the highest hydrogen production rate of 1.6 mmol h^{-1} g^{-1} when 50 wt% of MoS_2 was loaded on the TiO_2 nanobelts, which can be attributed to the matched energy band of TiO_2@MoS_2 heterostructure favoring the charge transfer and suppressing the photoelectron-hole recombination between MoS_2 and TiO_2.

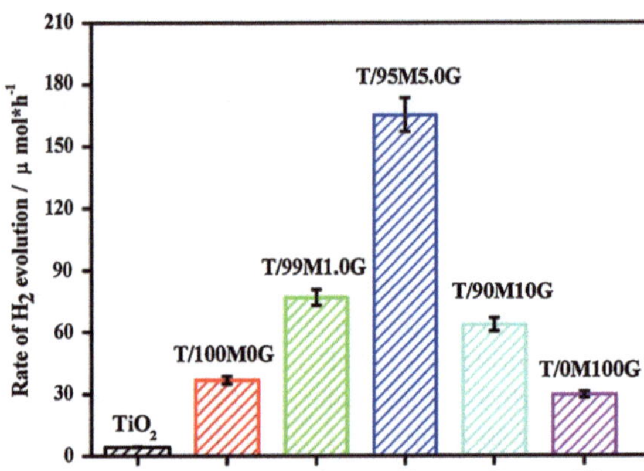

Fig. 9.15 Photocatalytic H2 evolution of TiO_2/MG composites. Photocatalytic H2 production experiments were performed in 25 % (v/v) ethanol/water solutions under UV irradiation using the photocatalyst TiO_2/MG composites with different MoS_2 and graphene contents in the MG hybrid as co-catalyst. The T/95M5.0G composite photocatalyst containing 95 % MoS_2 and 5 % graphene in the co-catalyst showed the highest H2 production rate [102]

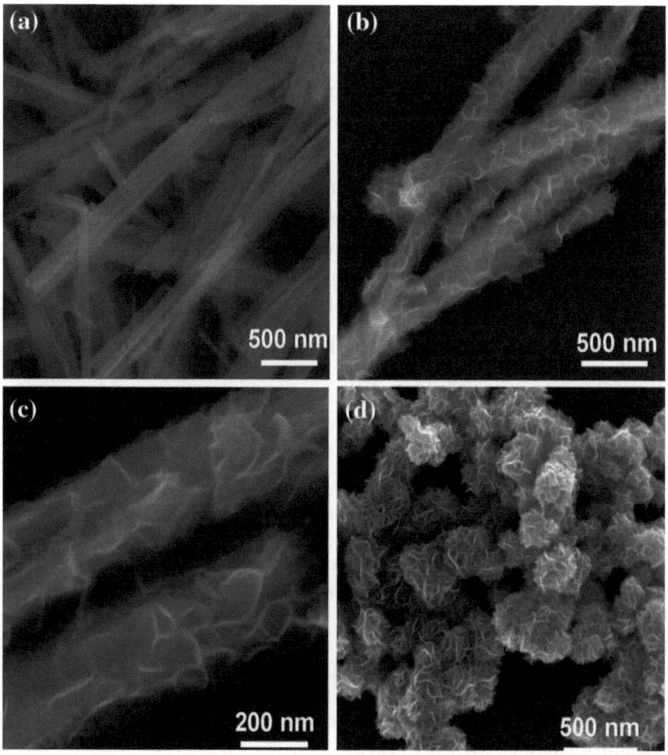

Fig. 9.16 SEM images of **a** pure TiO_2 nanobelts, **b**, **c** TiO_2 @ MoS_2heterostructures (50 wt% of MoS_2) and **d** pure MoS_2microspheres assembled from MoS_2 nanosheets [87]

Except for the CdS and TiO_2 as the photocatalyst, graphene, a novel two-dimensional carbon nanomaterial, have been reported as the promising platform for growing of MoS_2 materials due to its high surface area, high charge mobility, and good stability. Recently, the MoS_2/reduced graphene oxide (RGO) nanohybrid catalyst with limited-layered MoS_2 confinedly grown on RGO sheets has been prepared by Lu et al. [93] via a facilely hydrothermal method for efficiently photocatalytic HER. The two-dimensional RGO sheets not only provide a confined substrate for selective growth of limited-layer MoS_2 co-catalyst with a large number of exposed catalytic sites but also form the interconnected two-dimensional conductive networks for efficiently transferring photogenerated electrons from excited dye to catalytic active sites of MoS_2, thus suppressing the recombination processes and enhancing the photocatalytic efficiency of HER. As a result, a high apparent quantum efficiency of 24.0 % has been obtained at 460 nm over the Eosin Y-sensitized MoS_2/RGO photocatalyst.

9.4 Conclusions

In the past few years, considerable progress has been achieved in the synthesis and application of nanostructured MoS_2 materials in energy storage and conversion, including lithium ion batteries, Mg ion batteries, dye-sensitized solar cells and photocatalytic hydrogen evolution. It can be found that the cycling performance and rate capacity of lithium ion batteries can be great enhanced when the MoS_2 and graphene nanocomposites are employed as the anode materials. Nevertheless, there are just several reports on the application of MoS_2 in the other three domains, especially in the fields of Mg ion batteries. What's more, the practical utilizing of the nanostructured MoS_2 materials is still a lasting issue and a huge challenge to our humankind.

References

1. Yue, G., Lin, J.Y., Tai, S.Y.: A catalytic composite film of MoS_2/graphene flake as a counter electrode for Pt-free dye-sensitized solar cells. Electrochim. Acta **85**, 162–168 (2012)
2. Zhang, X., Luster, B., Church A., et al.: Carbon nanotube-MoS_2 composites as solid lubricants. ACS Appl. Mater. Interfaces **1**(3), 735–739 (2009)
3. Wang, S., Jiang, X., Zheng, H., et al.: Solvothermal synthesis of MoS_2/Carbon nanotube composites with improved electrochemical performance for lithium ion batteries. Nanosci. Nanotechnol. Lett. **4**(4), 378–383 (2012)
4. Li, H., Li, W.J., Ma, L., et al.: Electrochemical lithiation/delithiation performances of 3D flowerlike MoS_2 powders prepared by ionic liquid assisted hydrothermal route. J. Alloy. Compd. **471**(1–2), 442–447 (2009)
5. Feng, C.Q., Ma, J., Li, H., et al.: Synthesis of molybdenum disulfide (MoS_2) for lithium ion battery applications. Mater. Res. Bull. **44**(9), 1811–1815 (2009)
6. Ding, S.J., Zhang, D.Y., Chen, J.S., et al.: Facile synthesis of hierarchical MoS_2 microspheres composed of few-layered nanosheets and their lithium storage properties. Nanoscale **4**(1), 95–98 (2012)
7. Wang, M., Li, G., Xu, H., et al.: Enhanced lithium storage performances of hierarchical hollow MoS_2 nanoparticles assembled from nanosheets. ACS Appl. Mater. Interfaces **5**(3), 1003–1008 (2013)
8. Tenne, R., Margulis, L., Genut, M., et al.: Polyhedral and cylindrical structures of tungsten disulfide. Nature **360**(6403), 444–446 (1992)
9. Feldman, Y., Wasserman, E., Srolovitz, D.J., et al.: High-rate, gas-phase growth of Mos_2 nested inorganic fullerenes and nanotubes. Science **267**(5195), 222–225 (1995)
10. Feldman, Y., Frey, G.L., Homyonfer, M., et al.: Bulk synthesis of inorganic fullerene-like MS(2) (M = Mo, W) from the respective trioxides and the reaction mechanism. J. Am. Chem. Soc. **118**(23), 5362–5367 (1996)
11. Zelenski, C.M., Dorhout, P.K.: Template synthesis of near-monodisperse microscale nanofibers and nanotubules of MoS_2. J. Am. Chem. Soc. **120**(4), 734–742 (1998)
12. Nath, M., Govindaraj, A., Rao, C.N.R.: Simple synthesis of MoS_2 and WS_2 nanotubes. Adv. Mater. **13**(4), 283 (2001)
13. Bonneau, P.R., Jarvis, R.F., Kaner, R.B.: Rapid solid-state synthesis of materials from molybdenum-disulfide to refractories. Nature **349**(6309), 510–512 (1991)

14. Mdleleni, M.M., Hyeon, T., Suslick, K.S.: Sonochemical synthesis of nanostructured molybdenum sulfide. J. Am. Chem. Soc. **120**(24), 6189–6190 (1998)
15. Lee, H., Kanai, M., Kawai, T.: Preparation of transition metal chalcogenide thin films by pulsed laser ablation. Thin Solid Films **277**(1–2), 98–100 (1996)
16. Vollath, D., Szabo, D.V.: Synthesis of nanocrystalline MoS_2 and WS_2 in a microwave plasma. Mater. Lett. **35**(3–4), 236–244 (1998)
17. Smith Ronan, J., King Paul, J., Lotya, M., et al.: Large-scale exfoliation of inorganic layered compounds in aqueous surfactant solutions. Adv. Mater. **23**(34), 3944 (2011)
18. Yao, Y., Lin, Z., Li, Z., et al.: Large-scale production of two-dimensional nanosheets. J. Mater. Chem. **22**(27), 13494–13499 (2012)
19. Lee, Y.H., Zhang, X.Q., Zhang, W., et al.: Synthesis of large-area MoS_2 atomic layers with chemical vapor deposition. Adv. Mater. **24**(17), 2320–2325 (2012)
20. Zeng, Z., Yin, Z., Huang, X., et al.: Single-layer semiconducting nanosheets: high-yield preparation and device fabrication. Angew. Chem.-Int. Ed. **50**(47), 11093–11097 (2011)
21. Castellanos-Gomez, A., Barkelid, M., Goossens, A.M., et al.: Laser-thinning of MoS_2: on demand generation of a single-layer semiconductor. Nano Lett. **12**(6), 3187–3192 (2012)
22. Chen, X.H., Fan, R.: Low-temperature hydrothermal synthesis of transition metal dichalcogenides. Chem. Mater. **13**(3), 802–805 (2001)
23. Chen, J., Kuriyama, N., Yuan, H., et al.: Electrochemical hydrogen storage in MoS_2 nanotubes. J. Am. Chem. Soc. **123**(47), 11813–11814 (2001)
24. Ding, S.J., Chen, J.S., Lou, X.W.: Glucose-assisted growth of MoS_2 nanosheets on CNT backbone for improved lithium storage properties. Chem.-A Eur. J. **17**(47), 13142–13145 (2011)
25. Chhowalla, M., Amaratunga, G.: Thin films of fullerene-like MoS_2 nanoparticles with ultra-low friction and wear. Nature **407**(6801), 164–167 (2000)
26. Zhang, J., Soon Jia, M., Loh Kian, P., et al.: Magnetic molybdenum disulfide nanosheet films. Nano Lett. **7**(8), 2370–2376 (2007)
27. Divigalpitiya, W.M.R., Frindt, R.F., Morrison, S.R.: Inclusion systems of organic-molecules in restacked single-layer molybdenum-disulfide. Science **246**(4928), 369–371 (1989)
28. Sun, M.Y., Adjaye, J., Nelson, A.E.: Theoretical investigations of the structures and properties of molybdenum-based sulfide catalysts. Appl. Catal. A-Gen. **263**(2), 131–143 (2004)
29. Park, S.K., Yu, S.H., Woo, S., et al.: A simple L-cysteine-assisted method for the growth of MoS_2 nanosheets on carbon nanotubes for high-performance lithium ion batteries. Dalton Trans. **42**(7), 2399–2405 (2013)
30. Dresselhaus, M.S., Thomas, I.L.: Alternative energy technologies. Nature **414**(6861), 332–337 (2001)
31. Sen, U.K., Mitra, S.: High-rate and high-energy-density lithium-ion battery anode containing 2D MoS_2 nanowall and cellulose binder. ACS Appl. Mater. Interfaces **5**(4), 1240–1247 (2013)
32. Yang, L.C., Wang, S.N., Mao, J.J., et al.: Hierarchical MoS_2/polyaniline nanowires with excellent electrochemical performance for lithium-ion batteries. Adv. Mater. **25**(8), 1180–1184 (2013)
33. Bruce Peter G., Scrosati, B., Tarascon, J.M.: Nanomaterials for rechargeable lithium batteries. Angew. chem.-Int. ed. **47**(16), 2930–2946 (2008)
34. Winter, M., Brodd, R.J.: What are batteries, fuel cells, and supercapacitors? Chem. Rev. **104**(10), 4245–4269 (2004)
35. Zhou, X.S., Wan, L.J., Guo, Y.G.: Synthesis of MoS_2 nanosheet-graphene nanosheet hybrid materials for stable lithium storage. Chem. Commun. **49**(18), 1838–1840 (2013)
36. Whittingham, M.S.: Lithium batteries and cathode materials. Chem. Rev. **104**(10), 4271–4301 (2004)
37. Liu, H., Su, D.W., Zhou, R.F., et al.: Highly ordered mesoporous MoS_2 with expanded spacing of the (002) crystal plane for ultrafast lithium ion storage. Adv. Energy Mater. **2**(8), 970–975 (2012)

38. Sathish, M., Tomai, T., Honma, I.: Graphene anchored with Fe3O4 nanoparticles as anode for enhanced Li-ion storage. J. Power Sources **217**, 85–91 (2012)
39. Chen, S., Wang, Y., Ahn, H., et al.: Microwave hydrothermal synthesis of high performance tin-graphene nanocomposites for lithium ion batteries. J. Power Sources **216,** 22–27 (2012)
40. Park, S.K., Yu, S.H., Woo, S., et al.: A facile and green strategy for the synthesis of MoS_2 nanospheres with excellent Li-ion storage properties. Cryst. Eng. Comm. **14**(24), 8323–8325 (2012)
41. Hwang, H., Kim, H., Cho, J.: MoS_2 nanoplates consisting of disordered graphene-like layers for high rate lithium battery anode materials. Nano Lett. **11**(11), 4826–4830 (2011)
42. Xiao, J., Choi, D.W., Cosimbescu, L., et al.: Exfoliated MoS_2 nanocomposite as an anode material for lithium ion batteries. Chem. Mater. **22**(16), 4522–4524 (2010)
43. Wang, Z., Chen, T., Chen, W.X., et al.: CTAB-assisted synthesis of single-layer MoS_2-graphene composites as anode materials of Li-ion batteries. J Mater. Chem. A **1**(6), 2202–2210 (2013)
44. Zhang, C.F., Wu, H.B., Guo, Z.P., et al.: Facile synthesis of carbon-coated MoS_2 nanorods with enhanced lithium storage properties. Electrochem. Commun. **20**, 7–10 (2012)
45. Du, G.D., Guo, Z.P., Wang, S.Q., et al.: Superior stability and high capacity of restacked molybdenum disulfide as anode material for lithium ion batteries. Chem. Commun. **46**(7), 1106–1108 (2010)
46. Wang, Q., Li, J.H.: Facilitated lithium storage in MoS_2 overlayers supported on coaxial carbon nanotubes. J. Phys. Chem. C **111**(4), 1675–1682 (2007)
47. Nogueira, A., Znaiguia, R., Uzio, D., et al.: Curved nanostructures of unsupported and Al_2O_3-supported MoS_2 catalysts: synthesis and HDS catalytic properties. Appl. Catal. A-Gen. **429**, 92–105 (2012)
48. Rapoport, L., Bilik, Y., Feldman, Y., et al.: Hollow nanoparticles of WS2 as potential solid-state lubricants. Nature **387**(6635), 791–793 (1997)
49. Sun, M.Y., Adjaye, J., Nelson, A.E.: Theoretical investigations of the structures and properties of molybdenum-based sulfide catalysts. Appl. Catal. A-Gen. **263**(2), 131–143 (2004)
50. Bindumadhavan, K., Srivastava, S.K., Mahanty, S.: MoS_2-MWCNT hybrids as a superior anode in lithium-ion batteries. Chem. Commun. **49**(18), 1823–1825 (2013)
51. Guo, Y.G., Hu, J.S., Wan, L.J.: Nanostructured materials for electrochemical energy conversion and storage devices. Adv. Mater. **20**(15), 2878–2887 (2008)
52. Li, H., Wang, Z., Chen, L., et al.: Research on advanced materials for Li-ion batteries. Adv. Mater. **21**(45), 4593–4607 (2009)
53. Chen, J., Cheng, F.: Combination of lightweight elements and nanostructured materials for batteries. Acc. Chem. Res. **42**(6), 713–723 (2009)
54. Chang, K., Chen, W.X., Ma, L., et al.: Graphene-like MoS_2/amorphous carbon composites with high capacity and excellent stability as anode materials for lithium ion batteries. J. Mater. Chem. **21**(17), 6251–6257 (2011)
55. Chang, K., Chen, W.X.: L-cysteine-assisted synthesis of layered MoS_2/graphene composites with excellent electrochemical performances for lithium ion batteries. ACS Nano **5**(6), 4720–4728 (2011)
56. Das, S.K., Mallavajula, R., Jayaprakash, N., et al.: Self-assembled MoS_2-carbon nanostructures: influence of nanostructuring and carbon on lithium battery performance. J. Mater. Chem. **22**(26), 12988–12992 (2012)
57. Chang, K., Chen, W.X.: Single-layer MoS_2/graphene dispersed in amorphous carbon: towards high electrochemical performances in rechargeable lithium ion batteries. J. Mater. Chem. **21**(43), 17175–17184 (2011)
58. Chang, K., Chen, W.X.: In situ synthesis of MoS_2/graphene nanosheet composites with extraordinarily high electrochemical performance for lithium ion batteries. Chem. Commun. **47**(14), 4252–4254 (2011)
59. Aurbach, D., Lu, Z., Schechter, A., et al.: Prototype systems for rechargeable magnesium batteries. Nature **407**(6805), 724–727 (2000)

60. Peng, B., Chen, J.: Functional materials with high-efficiency energy storage and conversion for batteries and fuel cells. Coord. Chem. Rev. **253**(23–24), 2805–2813 (2009)
61. Aurbach, D., Suresh Gurukar, S., Levi, E., et al.: Progress in rechargeable magnesium battery technology. Adv. Mater. **19**(23), 4260 (2007)
62. Levi, E., Gofer, Y., Aurbach, D.: On the way to rechargeable Mg batteries: the challenge of new cathode materials. Chem. Mater. **22**(3), 860–868 (2010)
63. Novak, P., Imhof, R., Haas, O.: Magnesium insertion electrodes for rechargeable nonaqueous batteries - a competitive alternative to lithium? Electrochim. Acta **45**(1–2), 351–367 (1999)
64. Liang, Y.L., Feng, R.J., Yang, S.Q., et al.: Rechargeable Mg batteries with graphene-like MoS_2 cathode and ultrasmall Mg nanoparticle anode. Adv. Mater. **23**(5), 640 (2011)
65. Tao, Z.L., Xu, L.N., Gou, X.L., et al.: TiS2 nanotubes as the cathode materials of Mg-ion batteries. Chem. Commun. **18**, 2080–2081 (2004)
66. Nuli, Y., Yang, J., Li, Y., et al.: Mesoporous magnesium manganese silicate as cathode materials for rechargeable magnesium batteries. Chem. Commun. **46**(21), 3794–3796 (2010)
67. Li, X.L., Li, Y.D.: MoS_2 nanostructures: synthesis and electrochemical Mg2 + intercalation. J Phys. Chem. B **108**(37), 13893–13900 (2004)
68. Liu, C.J., Tai, S.Y., Chou, S.W., et al.: Facile synthesis of MoS_2/graphene nanocomposite with high catalytic activity toward triiodide reduction in dye-sensitized solar cells. J. Mater. Chem. **22**(39), 21057–21064 (2012)
69. Oregan, B., Gratzel, M.: A low-cost, high-efficiency solar-cell based on dye-sensitized colloidal TIO2 films. Nature **353**(6346), 737–740 (1991)
70. Gratzel, M.: Photoelectrochemical cells. Nature **414**(6861), 338–344 (2001)
71. Graetzel, M.: Recent advances in sensitized mesoscopic solar cells. Acc. Chem. Res. **42**(11SI), 1788–1798 (2009)
72. Lin, J.Y., Chan, C.Y., Chou, S.W.: Electrophoretic deposition of transparent MoS_2-graphene nanosheet composite films as counter electrodes in dye-sensitized solar cells. Chem. Commun. **49**(14), 1440–1442 (2013)
73. Olsen, E., Hagen, G., Lindquist, S.E.: Dissolution of platinum in methoxy propionitrile containing LiI/I-2. Sol. Energy Mater. Sol. Cells **63**(3), 267–273 (2000)
74. Kay, A., Gratzel, M.: Low cost photovoltaic modules based on dye sensitized nanocrystalline titanium dioxide and carbon powder. Sol. Energy Mater. Sol. Cells **44**(1), 99–117 (1996)
75. Imoto, K., Takahashi, K., Yamaguchi, T., et al.: High-performance carbon counter electrode for dye-sensitized solar cells. Sol. Energy Mater. Sol. Cells **79**(4), 459–469 (2003)
76. Murakami Takurou, N., Graetzel, M.: Counter electrodes for DSC: application of functional materials as catalysts. Inorg. Chim. ACTA. **361**(3), 572–580 (2008)
77. Banks, C.E., Davies, T.J., Wildgoose, G.G., et al.: Electrocatalysis at graphite and carbon nanotube modified electrodes: edge-plane sites and tube ends are the reactive sites. Chem. Commun. **7**, 829–841 (2005)
78. Xiao, Y., Lin, J.Y., Tai, S.Y., et al.: Pulse electropolymerization of high performance PEDOT/MWCNT counter electrodes for Pt-free dye-sensitized solar cells. J. Mater. Chem. **22**(37), 19919–19925 (2012)
79. Wang, M., Anghel Alina, M., Marsan, B., et al.: CoS supersedes Pt as efficient electrocatalyst for triiodide reduction in dye-sensitized solar cells. J. Am. Chem. Soc. **131**(44), 15976 (2009)
80. Lin, J.Y., Liao, J.H., Chou, S.W.: Cathodic electrodeposition of highly porous cobalt sulfide counter electrodes for dye-sensitized solar cells. Electrochim. Acta **56**(24), 8818–8826 (2011)
81. Sun, H., Qin, D., Huang, S., et al.: Dye-sensitized solar cells with NiS counter electrodes electrodeposited by a potential reversal technique. Energy Environ. Sci. **4**(8), 2630–2637 (2011)
82. Jiang, Q.W., Li, G.R., Gao, X.P.: Highly ordered TiN nanotube arrays as counter electrodes for dye-sensitized solar cells. Chem. Commun. **44**, 6720–6722 (2009)

83. Li, G.R., Wang, F., Jiang, Q.W., et al.: Carbon nanotubes with titanium nitride as a low-cost counter-electrode material for dye-sensitized solar cells. Angew. Chem.-Int. Ed. **49**(21), 3653–3656 (2010)
84. Jang, J.S., Ham, D.J., Ramasamy, E., et al.: Platinum-free tungsten carbides as an efficient counter electrode for dye sensitized solar cells. Chem. Commun. **46**(45), 8600–8602 (2010)
85. Wu, M., Lin, X., Hagfeldt, A., et al.: Low-cost molybdenum carbide and tungsten carbide counter electrodes for dye-sensitized solar cells. Angew. Chem.-Int. Ed. **50**(15), 3520–3524 (2011)
86. Wu, M., Wang, Y., Lin, X., et al.: Economical and effective sulfide catalysts for dye-sensitized solar cells as counter electrodes. Phys. Chem. Chem. Phys. **13**(43), 19298–19301 (2011)
87. Zhou, W., Yin, Z., Du, Y., et al.: Synthesis of few-layer MoS_2 nanosheet-coated TiO2 nanobelt heterostructures for enhanced photocatalytic activities. Small **9**(1), 140–147 (2013)
88. Mai, N., Tran Phong, D., Pramana Stevin, S., et al.: In situ photo-assisted deposition of MoS_2 electrocatalyst onto zinc cadmium sulphide nanoparticle surfaces to construct an efficient photocatalyst for hydrogen generation. Nanoscale **5**(4), 1479–1482 (2013)
89. Andrew Frame, F., Osterloh Frank, E.: CdSe-MoS_2: a quantum size-confined photocatalyst for hydrogen evolution from water under visible light. J. Phys. Chem. C **114**(23), 10628–10633 (2010)
90. Hinnemann, B., Moses, P.G., Bonde, J., et al.: Biomimetic hydrogen evolution: MoS_2 nanoparticles as catalyst for hydrogen evolution. J. Am. Chem. Soc. **127**(15), 5308–5309 (2005)
91. Tran Phong, D., Pramana Stevin, S., Kale Vinayak, S., et al.: Novel assembly of an MoS_2 electrocatalyst onto a silicon nanowire array electrode to construct a photocathode composed of elements abundant on the Earth for hydrogen generation. Chem.-A Eur. J. **18**(44), 13994–13999 (2012)
92. Fujishima, A., Honda, K.: Electrochemical photolysis of water at a semiconductor electrode. Nature **238**(5358), 37 (1972)
93. Min, S., Lu, G.: Sites for high efficient photocatalytic hydrogen evolution on a limited-layered MoS_2 co-catalyst confined on graphene sheets-the role of graphene. J. Phys. Chem. C. **116**(48), 25415–25424 (2012)
94. Walter Michael, G., Warren Emily, L., Mckone James, R., et al.: Solar water splitting cells. Chem. Rev. **110**(11), 6446–6473 (2010)
95. Chen, G., Li, D., Li, F., et al.: Ball-milling combined calcination synthesis of MoS_2/CdS photocatalysts for high photocatalytic H-2 evolution activity under visible light irradiation. Appl. Catal. A-Gen. **443**, 138–144 (2012)
96. Xu, Z., Wu, G., Yan, H., et al.: Photocatalytic H-2 evolution on MoS_2/CdS catalysts under visible light irradiation. J. Phys. Chem. C **114**(4), 1963–1968 (2010)
97. Zhang, W., Wang, Y., Wang, Z., et al.: Highly efficient and noble metal-free NiS/CdS photocatalysts for H-2 evolution from lactic acid sacrificial solution under visible light. Chem. Commun. **46**(40), 7631–7633 (2010)
98. Liu, H., Zhang, K., Jing, D., et al.: SrS/CdS composite powder as a novel photocatalyst for hydrogen production under visible light irradiation. Int. J. Hydrogen Energy **35**(13SI), 7080–7086 (2010)
99. Zhang, J., Yu, J., Zhang, Y., et al.: Visible light photocatalytic H-2-production activity of CuS/ZnS porous nanosheets based on photoinduced interfacial charge transfer. Nano Lett. **11**(11), 4774–4779 (2011)
100. Xu, Z., Han, J., Ma, G., et al.: Photocatalytic H-2 evolution on CdS loaded with WS2 as co-catalyst under visible light irradiation. J. Phys. Chem. C **115**(24), 12202–12208 (2011)
101. Xu, Z., Yan, H., Wu, G., et al.: Enhancement of photocatalytic H-2 evolution on CdS by loading MOS_2 as co-catalyst under visible light irradiation. J. Am. Chem. Soc. **130**(23), 7176 (2008)
102. Xiang, Q., Yu, J., Jaroniec, M.: Synergetic effect of MoS_2 and graphene as co-catalysts for enhanced photocatalytic H-2 production activity of TiO2 nanoparticles. J. Am. Chem. Soc. **134**(15), 6575–6578 (2012)

Chapter 10
Valley Polarization in Transition-Metal Dichalcogenides by Optical Pumping

Junfeng Dai and Xiaodong Cui

Abstract Degenerate valleys of energy bands well separated in momentum space constitute a discrete degrees of freedom for low-energy electrons. This has led to the emergence of valleytronics, a conceptual electronics based on manipulating valley index. Transition-metal dichalcogenides monolayers have been theoretically proposed as a good platform to realize manipulation of valley degrees of freedom through nonzero Berry curvatures at high symmetric points in Brillouin zone. Very recently several groups reported the selective occupation of the degenerate but inequivalent valleys by circularly polarized optical pumping in transition-metal dichalcogenides monolayers. These experimental evidences reveal the viability of optical valley control in group-VI transition-metal dichalcogenides, and form the basis for integrated valleytronics and spintronics application.

10.1 Introduction

Valleytronics is a more recent development that relies on the fact that the conduction/valence bands of some materials have two or more extremes at equal energies but at different positions in momentum space. The inequivalent valleys constitute a multiple index for low-energy carriers. Due to the large valley separation in the momentum space, the valley index is expected to be robust against scattering by smooth deformations and long wavelength phonons. So the valley index, like carrier and spin, can be used as potential information carrier in valley-based electronic and optoelectronic devices.

J. Dai (✉)
Department of Physics, South University of Science and Technology of China, Shenzhen 518055, China
e-mail: dai.jf@sustc.edu.cn

X. Cui
Physics Department, The University of Hong Kong, Pokfulam Road, Hong Kong, China

Z. M. Wang (ed.), *MoS*$_2$, Lecture Notes in Nanoscale Science and Technology 21, DOI: 10.1007/978-3-319-02850-7_10,

To make a valleytronic device, it is necessary to control the occupation of electrons in these valleys, thereby producing a valley polarization. In many conventional semiconductors, such as Si and AlAs, electrons occupy pockets near or at the equivalent X points of the Brillouin zone. The mechanisms that break the symmetry of the crystal potential can be used to lift the degeneracy of valley. So the piezoelectric and magnetic control of valley degrees of freedom was recently achieved on experiment [1–5]. Different from the silicon-based electronics, the two-dimensional graphene crystal exhibits fascinating electrical and optical properties due to its unusual band structure. In the band structure of graphene, the conduction and valence bands conically contact at a point called Dirac point. There are two inequivalent Dirac points in the Brillouin zone, related by time-reversal symmetry. Because of the large separation in moment space, the intervalley scattering will be suppressed in pure graphene samples. So the independent and degenerated valley degrees of freedom are considered as a kind of robust information carriers against temperature perturbation [6–12]. To distinguish the degenerated and inequivalent valleys, theorists proposed a biased bilayer structure to generate a nonzero and contrasting Berry curvatures and orbital magnetic moments in K and K′ valley. These nonzero and contrasting Berry curvatures and orbital magnetic moments offer us potential tools to manipulate valley degrees of freedom by electric and optical ways [6, 13–17].

Transition-metal dichalcogenides monolayers represent a class of semiconductor materials similar to bilayer graphene, where the conduction and valence-band edges are located at the corners (K points) of the 2D hexagonal Brillouin zone [18–20]. Because of the intrinsic structure inversion asymmetry, the inequivalent valleys constituting a binary index for low-energy carriers owns nonzero but contrasting Berry curvatures and orbital magnetic moments at the symmetry points of Brillouin zone [21]. Consequently, it leads to the contrasting optical selection rules at K and K′ valley. Moreover, inversion symmetry breaking together with strong spin–orbit coupling will result in spin and valley coupling, which allows complementary interplay the spin and valley degrees of freedom. Furthermore, because of the large direct bandgap in the range of visible frequency, this class of materials is ideal candidate for valley and spin-based optoelectronic devices applications through optical and electric means.

10.2 Valley Physics

The family of dichalcogenides MX_2 (M = Mo, W; X = S, Se) has a structure of X–M–X covalently bonded hexagonal quasi-2D network stacked by weak Van der Waals forces. Within each layer of MX_2, the M and X atoms form 2D hexagonal lattices, with the M atom being coordinated by the six neighboring X atoms in a trigonal prismatic geometry (Fig. 10.1b). Taking the M atom as the inversion center, an X atom will be pointed to an empty location. The inversion symmetry is explicitly broken to a MX_2 monolayer with the D_{3h}^1 crystal symmetry. In the bulk

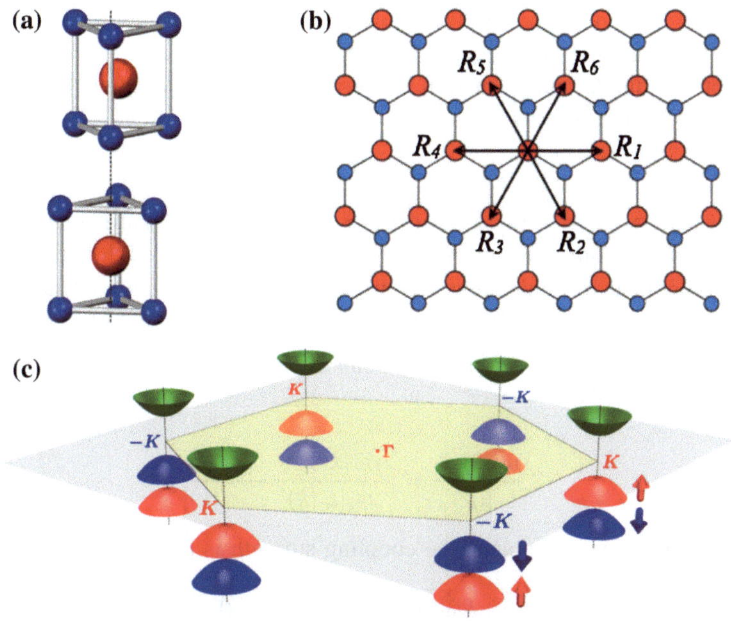

Fig. 10.1 **a** The unit cell of bulk 2H-MoS_2, which has the inversion center located in the middle plane. It contains two unit cells of MoS_2 monolayers, which lacks an inversion center. **b** Top view of the MoS_2 monolayer. R_i are the vectors connecting nearest Mo atoms. **c** Schematic drawing of the band structure at the band edges located at the K points. (Reprinted with permission from [21]. Copyright (2012) by the APS)

form, MX_2 films exhibit 2H stacking order: the neighboring layers are 180° in-plane rotation of each other, where the metal atom of a given layer sitting exactly on top of the chalcogenide atom of the adjacent layer (Fig. 10.1a). Therefore, an even–odd variation in the structural symmetry of ultrathin films appears: inversion symmetry is present in films with even number of layers, but absent in odd number of layers.

The conduction band-edge state is constructed by d_{z^2} orbitals and valance-band is by the hybridization of $d_{x^2-y^2}$ and d_{xy} orbitals. The basis functions at the band edges have the form:

$$|\phi_c> \; = |d_{z^2}> , |\phi_\upsilon^\tau> \; = \frac{1}{\sqrt{2}}(|d_{x^2-y^2}> + i\tau|d_{xy}>)$$

where the subscript $c(\upsilon)$ denotes the conduction (valance) band, and $\tau = \pm 1$ is the valley index. To first order in k, the two-band $k \cdot p$ Hamiltonian can be expressed as:

$$\hat{H}_0 = \mathrm{at}(\tau k_x \hat{\sigma}_x + k_y \hat{\sigma}_y) + \frac{\Delta}{2}\hat{\sigma}_z$$

where $\hat{\sigma}$, a, t, and Δ are the Pauli matrices, the lattice constant, the effective hopping integral, and the energy gap, respectively.

Under the dipole approximation, the k-resolved oscillator strength of $\sigma_{\pm}$ circular polarization given by

$$\wp_{\pm}(k) = \wp_x(k) \pm i\wp_y(k)$$

where $\wp_a(k) \equiv m_0 < u_c(k)|\frac{1}{\hbar}\frac{\partial \hat{H}}{\partial k_a}|u_v(k) >$ is the interband matrix element of the canonical momentum operator, m_0 is the free electron mass [22], and $u_{c(v)}$ is the Bloch component of conduction (valence) bands. In the simplest situation of a two-band model, the projection of the orbital magnetic moment m(k) contributed by the intercellular current circulation along the light propagation direction ($\hat{z}$) can be expressed as [23]

$$-2\frac{m(k)\cdot\hat{z}}{\mu_B} = \frac{|\wp_+(k)|^2 - |\wp_-(k)|^2}{m_e[\varepsilon_c(k) - \varepsilon_v(k)]}$$

For transition near K points, the coupling strength is [10, 21]

$$|\wp_{\pm}(k)|^2 = \frac{m_0^2 a^2 t^2}{h^2}(1 \pm \tau\frac{\Delta'}{\sqrt{\Delta'^2 + 4a^2t^2k^2}})^2$$

where $\Delta' \equiv \Delta - \tau s_z \lambda$ denotes the spin-dependent band gap. Since $\Delta' >> atk$, the interband transitions in the vicinity of the K (K′) point couple exclusively to right (left)-handed circularly polarized light $\sigma+$ ($\sigma-$) (Fig. 10.2). The direct-gap transition at the two degenerate valleys, together with this valley-contrasting selection rule, suggest that one can optically generate and detect valley polarizations in this class of materials.

Moreover, the monolayer transition-metal dichalcogenides have a very strong spin–orbit coupling (SOC) originating from the d orbitals of the heavy metal atoms. Considering the inversion symmetry breaking and SOC, conduction band (CB) minimum located at the K and K′ points is spin degenerate, and valance-band edge state splits to spin up and spin down bands. Time-reversal symmetry requires that the spin splitting at different valleys must be opposite as shown in Fig. 10.1c.

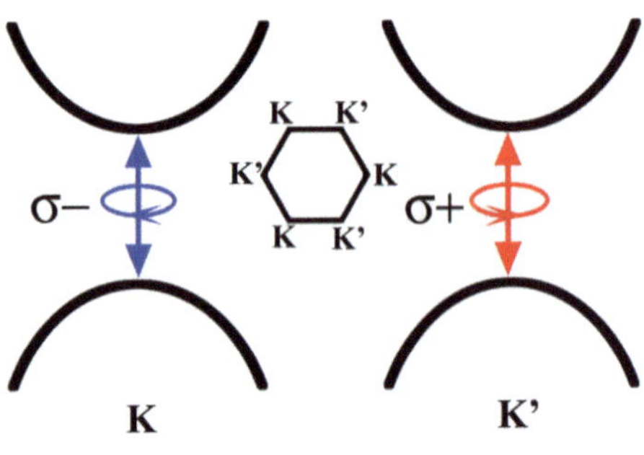

Fig. 10.2 Schematic of proposed valley-dependent selection rules at K and K′ points in crystal momentum space: *left* (*right*)-handed circularly polarized light $\sigma\pm$ ($\sigma-$) only couples to the band-edge transition at K (K′) points for the sake of angular moment conservation and time-reversal symmetry

Therefore, the spin–valley coupling at the valence-band edges suppresses spin and valley relaxation, as flip of each index alone is forbidden by the valley-contrasting spin splitting. As a result, a change in valence carrier spin state and hence valley in k-space will be far less likely in transition-metal dichalcogenides monolayers.

Furthermore, lots of valley-dependent phenomena such as valley polarization, valley Hall effect, will be accompanied by electron spin due to strong spin–valley coupling [21]. These degrees of freedom will form the basis for developing integrated valleytronic and spintronic application in the future.

10.3 Sample Preparation and Characterization

Monolayer MoS_2, WS_2, and WSe_2 could be obtained by mechanical exfoliation from natural or synthetic single crystals with low-track clean room tapes. Ultrathin flakes are first screened with visual colors arising from light interference by a bright field optical microscope with a tungsten lumination [24]. For MoS_2, two characteristic Raman modes: the in-plane vibrational E^1_{2g} mode and the out-of-plane vibrational A_{1g} mode were employed to identify the specific thickness: $\Delta\omega = 19\ cm^{-1}$ to monolayers and $\Delta\omega = 21\ cm^{-1}$ to bilayers, respectively [25]. In the case of WS_2, the frequency difference, $\Delta\omega = 65.5$, 68.3, and 69.2 cm^{-1} was labeled to monolayer, bilayer, and trilayer, respectively. As to WSe_2, two main peaks around 250 cm^{-1} were observed in various thicknesses from monolayer to bulk. However, there is no obvious trend to be observed on both modes. The layer dependence of photoluminescence intensity and energy position are better tools to identify WS_2 and WSe_2 monolayers [26]. In addition, the AFM technique was also used to identify the thickness of samples experimentally.

10.4 Structural Inversion Symmetry Determined by SHG

A key requirement toward nonzero orbital magnetic moments and Berry curvatures is the inversion asymmetry. To check the symmetry properties of ultrathin layers, one could exploit nonlinear optic techniques, e.g., second harmonic generation (SHG). In nonlinear optics, the optical response of SHG is dominated by the second-order nonlinear optical susceptibilities $\chi^{(2)}$, which is determined by crystal structure and can occur only in noncentrosymmetric crystals. So the second harmonic generation (SHG) provides a simple characterizing tool to examine the inversion symmetry in ultrathin film of transition-metal dichalcogenides [27]. Since $\chi^{(2)}$ equals to zero in the presence of structure inversion symmetry, a dramatic even–odd oscillation pattern of SHG intensity is expected to observe with increasing the thickness of samples layer by layer. Experimentally, WS_2 and WSe_2 ultrathin flakes were scanned by a 150 fs pulsed laser beam with a wavelength of 800 nm at normal incidence and the double frequency (400 nm) signal was

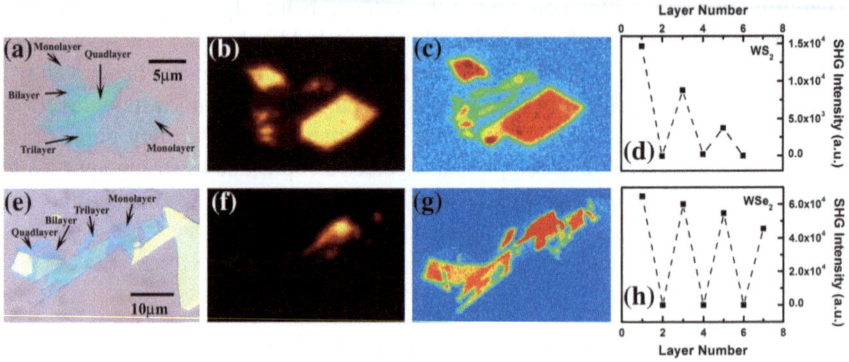

Fig. 10.3 Optical images of WS_2 (**a**) and WSe_2 (**e**) slabs on Si substrates with 300 nm SiO_2 cap-layer; Photoluminescence images at direct-gap transition energy of the corresponding WS_2 (**b**) and WSe_2 (**f**) slabs excited at 2.41 eV. Only monolayers are visible at the present contrast; The corresponding SHG under a 800 nm excitation at normal incidence (150 fs, 80 MHz) on WS_2 (**c**) and WSe_2 (**g**) respectively. The highest intensity labeled in red arises from monolayers. The SHG signal from the silicon/silicon oxide surface is negligible compared with this from multilayers; the relative intensity of SHG as a function of the film thickness in WS_2 (**d**) and WSe_2 (**h**). The SHG shows an even–odd oscillation dependence on the layer number

collected simultaneously. As expected, the SHG signals were not observed and negligible for both WS_2 and WSe_2 flakes with even layer number. However, the strong second harmonic emission arose from flakes with odd number of layers were observed as shown in Fig. 10.3c and g. Notably, the brightest second harmonic emission was observed in both WS_2 and WSe_2 monolayers. The intensity of the second harmonic emission decayed gradually as the layer number increased, which was indicated in Fig. 10.3d and h.

The optical second harmonic of 810-nm pulse in mechanically exfoliated MoS_2 flakes was also reported in Ref. [28]. From the curve of SH power as a function of fundamental power as shown in Fig. 10.4, it indicated the expected quadratic dependence for the SHG process. By measuring power dependence of two orthogonal components of SH on the angle (θ) between the laboratory and the crystalline coordinates, Kumar et al. got the expected $\sin^2 3\theta$ ($\cos^2 3\theta$) relation fitted from experimental data (Fig. 10.4b). Then the magnitude of second-order susceptibility $\chi^{(2)}$ for MoS_2 monolayer was deduced about 10^{-7} m/V [28].

10.5 Polarization-Sensitive Photoluminescence Setup

The polarization-sensitive photoluminescence measurement was carried out with a confocal-like microscopic set up as shown in Fig. 10.5. The linearly polarized excited light, passing through a $1/2\lambda$ waveplate, a beam splitter that assures the polarization of incident beam, and a $1/4\lambda$ waveplate, impinged on the sample

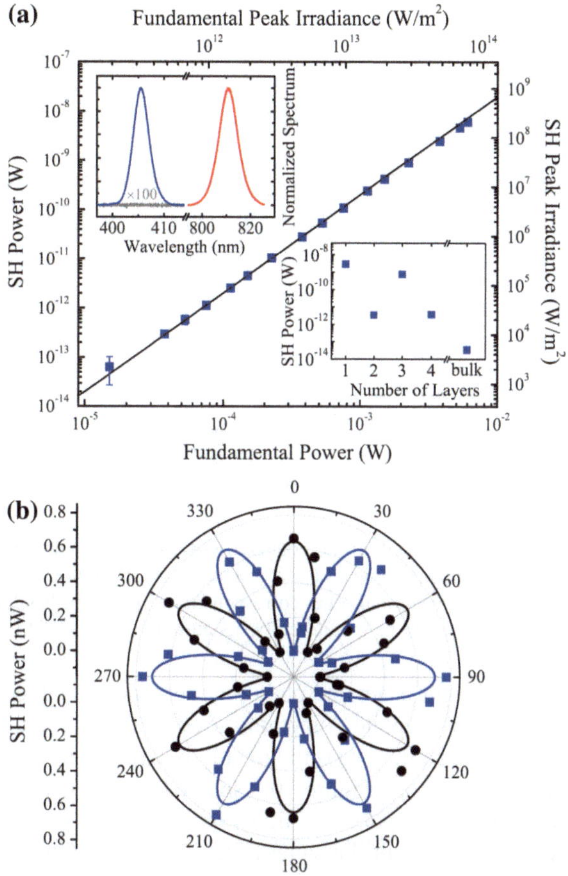

Fig. 10.4 Second harmonic generation from a triangular monolayer MoS_2 flake grown by CVD, as shown in the lower inset of (**a**). The main panel of (**a**) shows the power dependence of second harmonic generation. The *solid line* indicates the expected quadratic dependence. (**b**) Shows angular dependence of the parallel (*blue squares*) and perpendicular (*black circles*) components of the second harmonic, along with the expected dependence (*solid lines*). The upper inset of (**a**) shows a separate measurement of the parallel component with a finer step size near $\theta = 0$. (Reprinted with permission from [28]. Copyright (2013) by the APS)

mounted on the low temperature cryostat. By rotating the $1/2\lambda$ waveplate 45°, the helicity of excited light was switched from left-handed to right-handed circularly polarized light. Then the induced circularly polarized photoluminescence was transferred into two orthogonally polarized beams after passing the same $1/4\lambda$ waveplate. After that, the collimated backscattering light and photoluminescence passed through beam splitter, a beam displacing prism which separated the light beam into two orthogonally polarized beams, a $1/4\lambda$ waveplate as a depolarizer and then was focused to two spots at the entrance slit of the monochromator (focal length = 0.7 m) equipped with CCD (−75 °C). The helicity of the circular polarization could be monitored by its attitude on the CCD image. The CCD images in Fig. 10.5 displayed two branches, upper and lower branch, respectively corresponding to right- and left-handed circularly excited lights.

To do the system (spectrometer, depolarizer) calibration, we first removed the $1/4\lambda$ waveplate in the front of the sample. The linearly polarized light with orthogonal polarization shined on the sample, the corresponding polarized

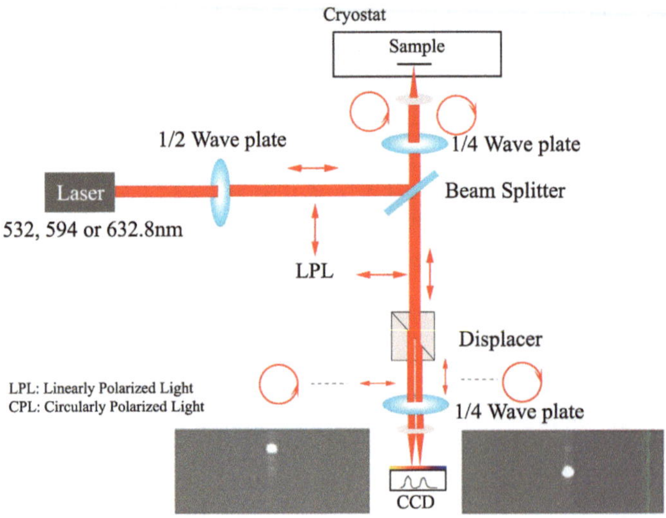

Fig. 10.5 Schematic of the polarization-sensitive photoluminescence measurement. The *lower-left corner* image shows the left-handed circularly polarized light casted on the spectrometer's CCD; the *lower-right* one corresponds to the right-handed. The helicity of the circular polarization could be monitored by its attitude on the CCD image

reflected light beam and photoluminescence was feed on the slit through displacer and $1/4\lambda$ waveplate (function as depolarizer). Rotating the $1/4\lambda$ waveplate, we could balance the two orthogonally polarized photoluminescence and excluded the systematic errors. Figure 10.6 showed the intensity of photoluminescence excited by two linearly polarized lights on orthogonal configuration. By rotating the $1/4\lambda$ waveplate, we got the same intensity of photoluminescence.

The degree of linear polarization was obtained by

$$P = \frac{I(+) - I(-)}{I(+) + I(-)}$$

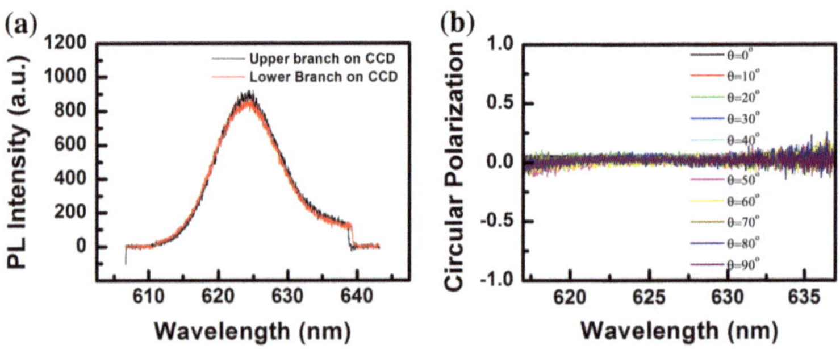

Fig. 10.6 **a** Balanced photoluminescence under orthogonally linearly polarized light excitation. **b** The polarization of photoluminescence is independent of orientation of polarized light

where $I(+)$ and $I(-)$ denoted the intensity of the two orthogonally polarized beams casted in the CCD, corresponding respectively to the PL signal under two orthogonally linearly polarized component of the excited light. So we got that the linear polarization of photoluminescence was equal to zero on this case. By tuning the orientation of incident linearly polarized light, we observed that the photoluminescence polarization was independent of orientation of excited linear light as shown in Fig. 10.6b. So in the current configuration, this setup clearly distinguished the helicity and polarization of the whole spectrum in one CCD exposure with a very high sensitivity.

10.6 Valley Polarization

Based on the analysis of inversion symmetry breaking, the helicity of the luminescence should exactly follow that of the excitation light. In other words, the right-handed circularly polarized excitation generates right-handed luminescence, and the left-handed circularly polarized excitation generates left-handed luminescence. To characterize the circular component in the luminescence spectra, we define the degree of circular polarization

$$P = \frac{I(\sigma+) - I(\sigma-)}{I(\sigma+) + I(\sigma-)}$$

where $I(\sigma\pm)$ is the intensity of the left (right)-handed circular component of photoluminescence. For perfectly circularly polarized light, $P = 1$ ($\sigma+$) or -1 ($\sigma-$).

The circularly polarized luminescence spectra of a pristine MoS_2 monolayer (peak around 1.9 eV) with right- and left-handed circularly polarized excitation (HeNe laser, 1.96 eV) at a near-resonant condition at $T = 10$ K, was studied. The luminescence spectra demonstrated a symmetric polarization for excitation with opposite helicities as shown in Fig. 10.7: $P \approx 0.32$ under $\sigma+$ excitation and $P \approx -0.32$ under $\sigma-$ excitation for the most representative MoS_2 monolayers. This behavior is fully expected under the mechanism of the valley-dependent optical selection rule.

We switched to measure the valley polarization of WS_2 using 594 nm solid-state laser under the near-resonant excitation. The same symmetric polarization for excitation with opposite helicities was observed: $P \approx 40 \sim -55$ % under $\sigma+$ ($-$) excitation for the most representative WS_2 monolayer as shown in Fig. 10.8. The valley polarization kept constant in the whole range of photoluminescence.

Meanwhile, the similar phenomena in MoS_2 monolayers were observed by other groups. Mak et al. reported nearly 100 % valley polarization on resonance excitation with A exciton at 14 K [29], and Cao et al. reported that the degree of circular polarization of the photoluminescence spectra was about 50 % at 83 K [30].

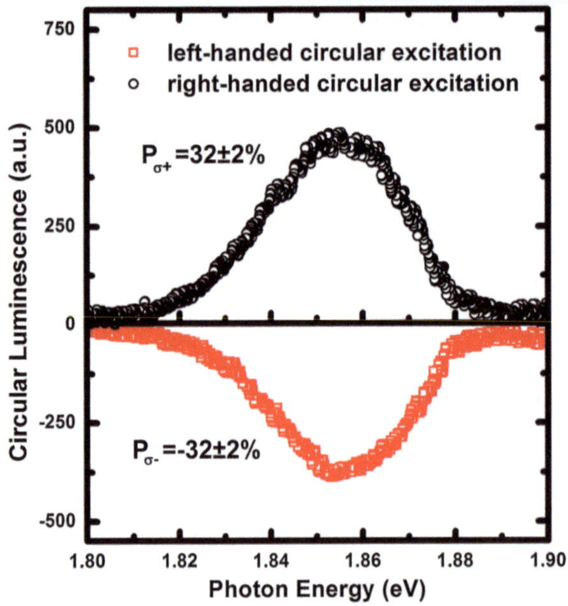

Fig. 10.7 Polarization resolved luminescence spectra under circularly polarized excitation from a HeNe laser at 1.96 eV and 10 K. Circular polarizations of $P = 32 \pm 2$ % and -32 ± 2 % are observed along out-of-plane direction with right- and left-handed circular excitation, respectively

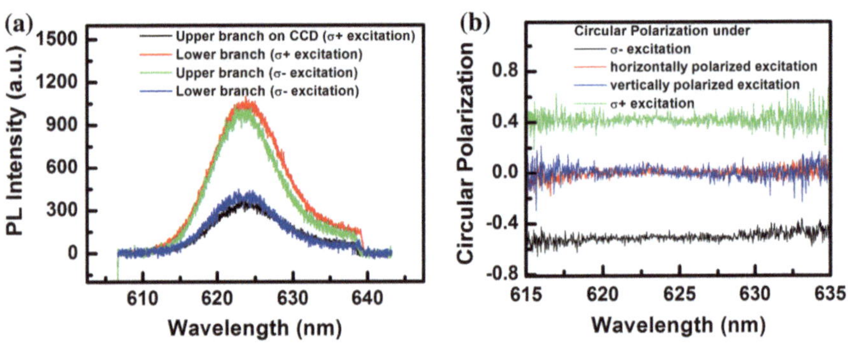

Fig. 10.8 **a** The two branches on CCD of luminescence spectra under circularly polarized excitation from a Solid-state laser at 2 eV and 10 K. **b** The calculated circular polarizations of out-of-plane direction with right- and left-handed circular excitation in the range of photoluminescence: $P \approx 40$ % and $P \approx -55$ % for two cases. The polarization represented by the *central lines* is for linearly polarized light excitation

10.7 Valley Coherence

In valleytronics, valley quantum coherence, namely the coherence between the two well-separated band extrema in momentum space, is a more important and meaningful concept so to achieve coherent manipulation.

Exciton formation mechanisms contain the Coulomb interaction with other carriers and coupling to phonons to transfer the binding energy [31]. Because of the large valley separation in momentum space and the long-range nature of Coulomb interaction, intravalley scattering will dominate. It is independent of the valley index and preserves both the valley polarization and coherence. Because exciton formation through phonon-assisted intravalley scattering are valley independent, such processes also preserve the valley polarization and coherence, though the homogenous broadness would reduce somewhat coherence. Based on the analysis of inversion symmetry breaking, intervalley scattering through the Coulomb interaction and phonons can lead to valley depolarization and decoherence. But it will require large momentum transfer.

When the timescale of intervalley scattering is lower than that of the electron–hole recombination, the valley polarization and valley quantum coherence could be generated by means of optical excitation. Jones et al. reported the coherence between valley excitons through linearly polarized luminescence spectra with polarization up to 40 % on monolayer WSe_2. The evidence that the orientation of photoluminescence always coincided with that of any linearly polarized excitation was shown in Fig. 10.9 [32]. But from our experimental results on WS_2 monolayer shown in Fig. 10.6b, it was shown that the photoluminescence polarization was independent of the orientation of linearly polarized excitation. The ability to address valley polarization and valley coherence makes it possible for the quantum manipulation of valley index for future valleytronic devices.

10.8 Magnetic Independence of Valley Polarization

In many semiconductor systems, such as GaAs bulk and heterostructures, the circular polarization of luminescence from circularly polarized excitation originates from electron (hole) spin polarization as a result of the spin-dependent optical selection rule [33–36]. But in monolayer of MoS_2 and other group-VI dichalcogenides, the optical selection rule originates from orbital magnetic moments at K valley, independent of electron spin. This different mechanism can be distinguished by examining the change in the photoluminescence spectra in an in-plane magnetic field. With a nonzero in-plane g-factor, spin polarization will precess about the in-plane magnetic field, and the time average of the photogenerated carrier spin projected along the Z-axis will be found at

$$S_z = \int_0^{\tau} S_{z0} \exp(-t/\tau_s) \cos(g\mu_B Bt/h)dt$$

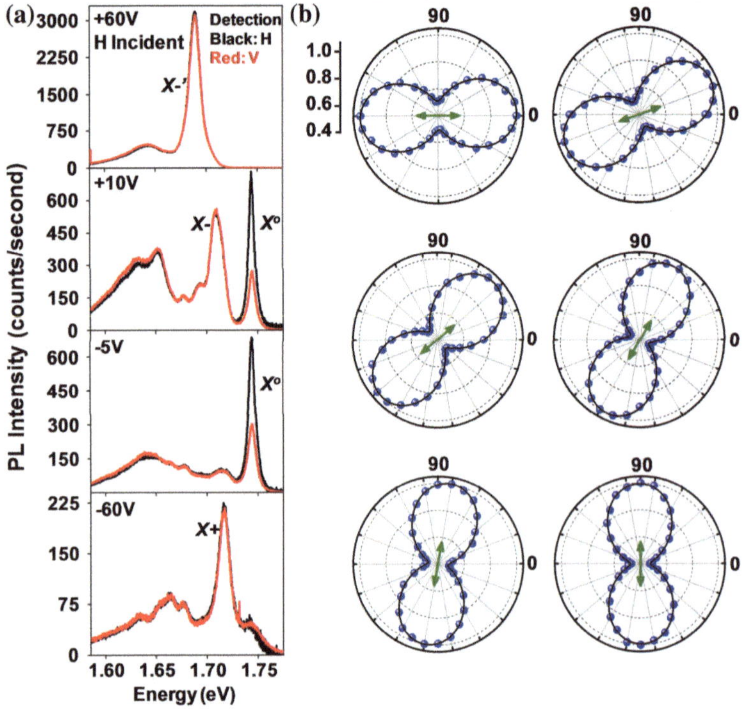

Fig. 10.9 Signature of excitonic valley quantum coherence. **a** Polarization resolved photoluminescence (PL) spectra at selected gate voltages under horizontally polarized excitation. **b** Neutral exciton peak intensity as a function of detection angle for given incident laser polarization (marked by *green arrow*). (Reprinted by permission from Macmillan Publisher Ltd: Nature Nanotechnology [32], copyright (2013))

where S_{Z0} is the initial spin along the Z-axis, τ_S is the spin relaxation time, μ_B is the Bohr magneton, and τ is the lifetime of the photoexcited carriers. Consequently, the polarization of the luminescence under continuous wave excitation should follow

$$P(B) \approx \frac{P(B = 0)}{1 + (g\mu_B B\tau_s/h)^2}$$

where $P(B = 0)$ is the polarization of the luminescence without the magnetic field. This is the well-known Hanle effect. If we assume that the spin relaxation time τ_S is of approximately the same order as the photocarrier lifetime $\tau \approx 10$ ps [37], then the polarization $P(B = 0.65$ T) will drop to a 5 % of $P(B = 0)$.

Experimentally, there was no noticeable difference between the photoluminescence polarization at zero field and in an in-plane field of $B = 0.65$ T. The same magnetic-field independent of circular polarization was reported in Ref. [38]. Sallen et al. observed that the intensity of valley-polarization-sensitive photoluminescence

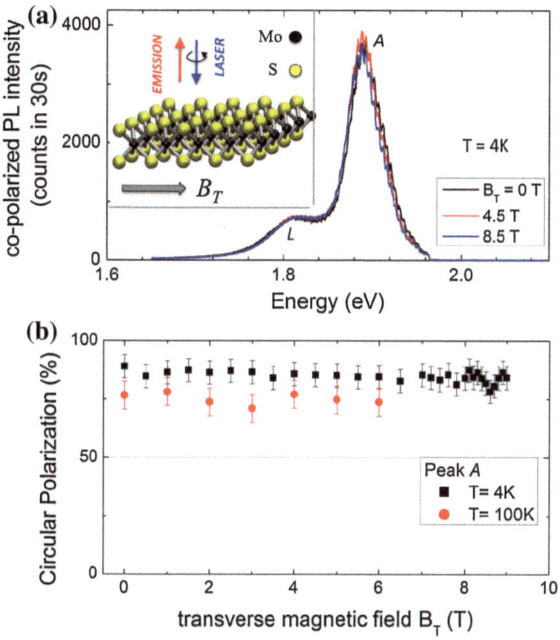

Fig. 10.10 **a** Comparison of PL spectra of a single MoS_2 ML recorded for a transverse magnetic field $B_T = 0$, 4.5, and 8.5 T under otherwise identical experimental conditions using HeNe laser at 1.96 eV. Inset: Experimental geometry. **b** Measured PL circular polarization degree of the PL as a function of the applied transverse magnetic field B_T for A exciton at $T = 4$ K (*squares*) and at $T = 100$ K (*circles*). (Reprinted with permission from [38]. Copyright (2012) by the APS)

shown in Fig. 10.10 was unchanged when an in-plane magnetic field $B = 0$ T, 4.5 T, and 8.5 T at $T = 4$ K was applied. The circular polarization degree of A exciton keep invariant for the magnetic field up to 9 T at 4 K and 6 T at 100 K.

Therefore, spin polarization cannot explain the polarized photoluminescence observed. Instead, this magnetic-field-independent photoluminescence polarization is a fully expected consequence of valley polarization through the valley-dependent selection rule; because the in-plane magnetic field does not couple to the valley index, the valley polarization cannot be rotated by the magnetic field. Even a strong transverse field would not lead to any variable emission polarization.

10.9 Valley Polarization on MoS_2 Bilayers

As to MoS_2 bilayers, they are composed of two structurally identical monolayers stacking with hexagonal symmetry (the sulfur atoms in one layer sit directly upon/below the molybdenum atoms in the other) (shown in Fig. 10.1a) and have D_{6h}^4 symmetry. Inversion symmetry is preserved in the bilayer unit cell with the inversion center located in the middle plane. So different from polarized photoluminescence on MoS_2 monolayers, the valley-dependent selection rule is not allowed in bilayers [10]. Because the MoS_2 sample comes from the natural mineral, the intensity of the luminescence in bilayers is on the same order, but relatively weaker than that of monolayers (Fig. 10.11 Inset). As to the circular polarization of

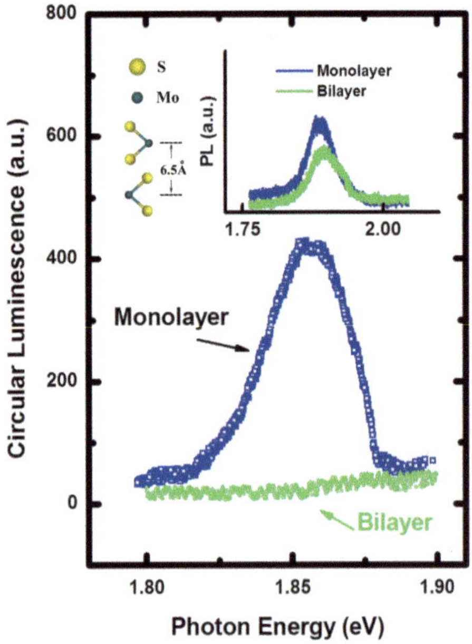

Fig. 10.11 Inset: schematic of a MoS_2 bilayer unit cell; Photoluminescence spectra of monolayer (*blue*) and bilayer (*green*) MoS_2 under the same conditions (base temperature, 10 K; 0.2 mW linearly polarized excitation at 1.96 eV). The intensity of the bilayer luminescence is slightly weaker than that of the monolayer. Main: Circularly polarized components of luminescence spectra from MoS_2 bilayers (*green*) and monolayers (*blue*) under circular excitation at 1.96 eV and 10 K. Negligible circular polarization is observed for the MoS_2 bilayers

luminescence, the striking difference was that it was negligible for MoS_2 bilayers compared with MoS_2 monolayer under the same experimental conditions (Fig. 10.11). If the excitation power is increased above 1×10^5 W cm^{-2}, circular dichroism appeared but weak in bilayers. It is suggested that the heating effects could induce structural anisotropy and consequently break the inversion symmetry in the bilayer. Mak et al. also reported the nearly 20 % valley polarization on MoS_2 bilayers compared with near 100 % polarization on monolayers. They thought that the low valley polarization might originate from slightly broken inversion symmetry due to the substrate charging effect [29].

Since crystal symmetry governs the valley-contrasting properties, controlled inversion symmetry can be used to achieve continuous and reversible manipulation to valley polarization. Analogous to the method used in bilayer graphene [39], the inversion symmetry in MoS_2 bilayers can be broken by applying an electric field perpendicular to the plane, which leads to a potential difference between the two layers. By using polarized photoluminescence measurements, Wu et al. observed that the circular dichroism in MoS_2 bilayers can be continuously tuned from −15 to 15 % with gate voltage (Fig. 10.12a), whereas in MoS_2 monolayers the photoluminescence polarization is gate independent (Fig. 10.12b). The lack of dependence of valley polarization in MoS_2 monolayer on gate voltage also implied that its valley properties were governed by the inherent structure inversion asymmetry [40].

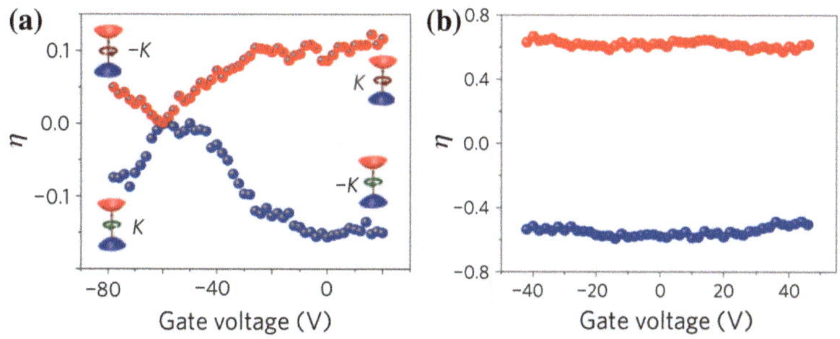

Fig. 10.12 Degree of photoluminescence polarization as a function of gate voltage for bilayer (**a**) and monolayer (**b**) MoS_2. *Red* (*blue*) *dots* denote $\sigma+$ ($\sigma-$) excitation (Reprinted by permission from Macmillan Publisher Ltd: Nature Physics [40], copyright (2013))

10.10 Mechanism of Valley Depolarization

As described in optical selection rules in MoS_2 monolayer (Fig. 10.2), only one valley can be populated under the resonant absorption of circularly polarized light, and circular polarization of emitted photoluminescence is predicted to 100 %. As yet, the valley polarization of transition-metal dichalcogenides reported by several groups is lower than 100 % under resonant circularly polarized pumping both in natural and synthetic samples. It implies that the intervalley scattering plays an important role in the depolarization process.

Figure 10.13 showed our typical temperature dependence of the circular polarization of luminescence from MoS_2 monolayers under circularly polarized excitation. The circular polarization has a flat plateau at ~31 % below 90 K, then dramatically drops with increasing temperature. The small temperature dependence of the polarization at low temperatures implies that the intervalley scattering ($K \leftrightarrow K'$) results mainly from scattering at grain boundaries and atomically sharp deformations where the crystal momentum conservation is relaxed. Because the sample is a natural mining product, abundant impurities and vacancies presumably provide intervalley scattering centers and populate conduction electrons at the energetically degenerate K and K′ valleys.

As the temperature increases above 90 K, the circularly polarized components of the luminescence spectra gradually decrease, which is an indicator of phonons dominating the intervalley scattering at high temperatures. The steady-state photoluminescence polarization is inversely proportional to the valley scattering rate γ_v. Phonons near the K points of the Brillouin zone can directly supply the momentum change for intervalley scattering. In such a situation, the valley scattering rate is proportional to the population of these phonons:

$$\gamma_v \propto \exp(-E_K/k_B T)$$

where E_K is the phonon energy near the K points and k_B is the Boltzman constant. The solid curve shown in Fig. 10.13 is a fit assuming an $\exp(-E_K/k_B T)$ dependence,

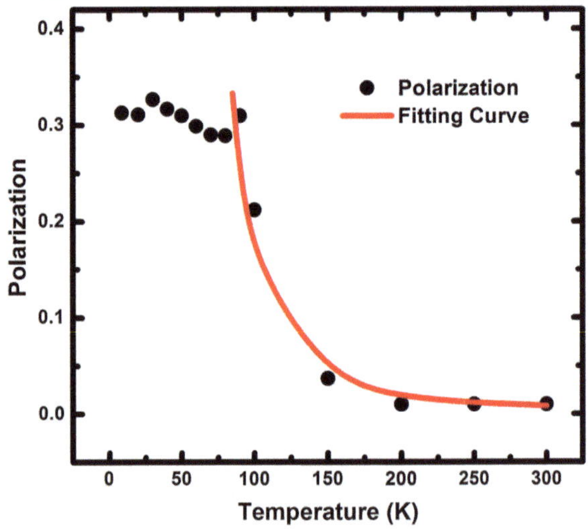

Fig. 10.13 Degree of circular polarization *P* as a function of temperature. The fitting curve (*red*), assuming an intervalley scattering proportional to the phonon population, gives a phonon energy of ~240 cm^{-1}

from which we extract $E_K \approx 240\ cm^{-1}$, consistent with the acoustic phonon energy near the K point reported in the bulk and monolayers [41, 42].

Moreover, if the excitation light was switched to a higher energy at 2.33 eV under the polarization-sensitive photoluminescence measurement, no polarization can be observed in the luminescence spectra. The similar experimental results were also reported in Ref. [29]. The quick depolarization at off resonance excitation could arise from two channels. First, the model of valley-dependent selection rules is discussed only in the vicinity of the K (K′) point [10, 21], whereas only metal *d* orbits are involved. The band properties alter at the various crystal moment spaces. As the optical transition occurs far away from the K points, the contrasting optical selection rules are not guaranteed. The other channel is the depolarization through emission of acoustic phonon at K points during the hot carrier relaxation process, as the temperature-dependent depolarization process discussed above.

Kioseoglou et al. systematically studied the degree of circular polarization of the photoluminescence (peak around 1.9 eV) from MoS_2 monolayers under circularly polarized photoexcitation with different energy, which ranges from 1.98 to 2.08 eV (Fig. 10.14) [43]. The circular polarization of the photoluminescence is about 40 % for excitation near the bandgap and has a power-law decrease as the excitation energy increases. They identified in-plane longitudinal acoustic (LA) phonon-assisted intervalley scattering as the primary spin relaxation mechanism, which decreases the measured valley polarization [29, 43]. From the phonon-dispersion curve for MoS_2 monolayer [44], the longitudinal acoustic with energy 30 meV at the K points of the Brillouin zone serves as the lowest energy in-plane phonons. Therefore, when pumping energies exceed E_A (A exciton) +2LA (acoustic phonon), the intervalley scattering becomes accessible and the depolarization process will start.

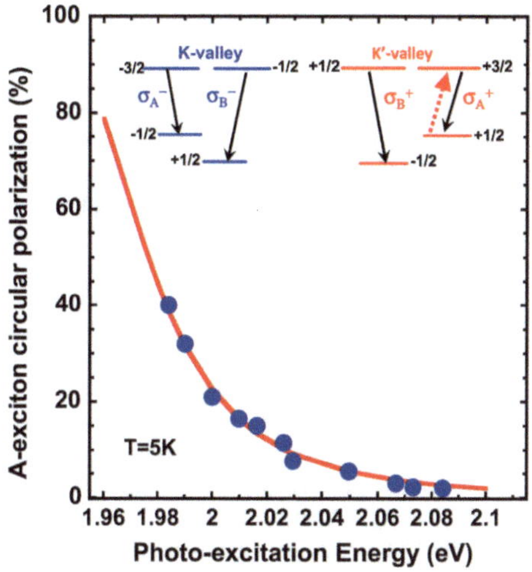

Fig. 10.14 Circular polarization of the A exciton as a function of the photoexcitation energy E_{pump}. The *solid line* represents a fit to the data (*circles*) with $P_{circ} = 1/[1 + \Upsilon(E_{pump} - E^*)^{\delta}]$. Inset: dipole allowed optical transitions based on the quantum selection rules. Reprinted with permission from Applied Physics Letters 101(22): 221907–221904. Copyright 2012, American Institute of Physics

To obtain a complete understanding and microscopic pictures of valley depolarization in this class of materials, time-resolved polarization-sensitive measurements should be used to investigate the radiative lifetime and the dynamics of the depolarization. As yet the report at monolayer level is lacking. Also, the transport measurement may be an alternative approach to study the mechanism due to intervalley scattering [45].

10.11 Conclusions

The circularly polarized luminescence from MoS_2 and WS_2 monolayers under circularly polarized excitation in the case of near-resonant condition was reported by several groups. Theoretically, it originates from the contrasting selection rules for optical transitions in the vicinity of K and K′ points. The main depolarization mechanism experimentally is discussed, which comes from the nonresonant absorption and intervalley scattering. Meanwhile, the valley coherence was observed from the identical orientation of polarized photoluminescence with that of linearly polarized excitation. Moreover, the control of valley polarization on MoS_2 bilayers was achieved by electric field that breaks the inversion symmetry. These provide us a viable tool to generate, detect, and manipulate valley polarization in group-VI transition-metal dichalcogenides monolayers and bilayers, which may form the basis for developing integrated valleytronic and spintronic applications.

Acknowledgments The author wishes to thank Dr. W. Yao, Dr. H.Z. Lu, and Dr. H.L. Zeng for valuable discussions and support. The author also thanks Mr. B.R. Zhu, Mr. L. Xie, Mr. R.C. He, and Miss X. Chen for help.

References

1. Shkolnikov, Y.P., et al.: Valley splitting of AlAs two-dimensional electrons in a perpendicular magnetic field. Phys. Rev. Lett. **89**(22), 226805 (2002)
2. Gunawan, O., et al.: Valley susceptibility of an interacting two-dimensional electron system. Phys. Rev. Lett. **97**(18), 186404 (2006)
3. Zhu, Z., et al.: Field-induced polarization of dirac valleys in bismuth. Nat. Phys. **8**(1), 89–94 (2012)
4. Gunawan, O., et al.: Quantized conductance in an AlAs two-dimensional electron system quantum point contact. Phys. Rev. B **74**(15), 155436 (2006)
5. Lay, T.S., et al.: High-quality two-dimensional electron system confined in an AlAs quantum well. Appl. Phys. Lett. **62**(24), 3120–3122 (1993)
6. Xiao, D., Yao, W., Niu, Q.: Valley-contrasting physics in graphene: magnetic moment and topological transport. Phys. Rev. Lett. **99**(23), 236809 (2007)
7. Rycerz, A., Tworzydlo, J., Beenakker, C.W.J.: Valley filter and valley valve in graphene. Nat. Phys. **3**(3), 172–175 (2007)
8. Tombros, N., et al.: Quantized conductance of a suspended graphene nanoconstriction. Nat. Phys. **7**(9), 697–700 (2011)
9. Gunlycke, D., White, C.T.: Graphene valley filter using a line defect. Phys. Rev. Lett. **106**(13), 136806 (2011)
10. Yao, W., Xiao, D., Niu, Q.: Valley-dependent optoelectronics from inversion symmetry breaking. Phys. Rev. B **77**(23), 235406 (2008)
11. Zhang, F., et al.: Spontaneous quantum hall states in chirally stacked few-layer graphene systems. Phys. Rev. Lett. **106**(15), 156801 (2011)
12. Akhmerov, A.R., Beenakker, C.W.J.: Detection of valley polarization in graphene by a superconducting contact. Phys. Rev. Lett. **98**(15), 157003 (2007)
13. Gwo, S., Shih, C.K.: Site-selective imaging in scanning tunneling microscopy of graphite: the nature of site asymmetry. Phys. Rev. B **47**(19), 13059–13062 (1993)
14. Ohta, T., et al.: Controlling the electronic structure of bilayer graphene. Science **313**(5789), 951–954 (2006)
15. McCann, E., Fal'ko, V.I.: landau-level degeneracy and quantum hall effect in a graphite bilayer. Phys. Rev. Lett. **96**(8), 086805 (2006)
16. Castro, E.V., et al.: Biased bilayer graphene: semiconductor with a gap tunable by the electric field effect. Phys. Rev. Lett. **99**(21), 216802 (2007)
17. Min, H., et al.: Ab initio theory of gate induced gaps in graphene bilayers. Phys. Rev. B **75**(15), 155115 (2007)
18. Li, T., Galli, G.: Electronic properties of MoS_2 nanoparticles. J. Phys. Chem. C **111**(44), 16192–16196 (2007)
19. Lebègue, S., Eriksson, O.: Electronic structure of two-dimensional crystals from ab initio theory. Phys. Rev. B **79**(11), 115409 (2009)
20. Zhu, Z.Y., Cheng, Y.C., Schwingenschlögl, U.: Giant spin-orbit-induced spin splitting in two-dimensional transition-metal dichalcogenide semiconductors. Phys. Rev. B **84**(15), 153402 (2011)
21. Xiao, D., et al.: Coupled spin and valley physics in monolayers of MoS_2 and other group-VI dichalcogenides. Phys. Rev. Lett. **108**(19), 196802 (2012)
22. Souza, I., Vanderbilt, D.: Dichroic f-sum rule and the orbital magnetization of crystals. Phys. Rev. B **77**(5), 054438 (2008)
23. Yu, P.Y., Cardona, M.: Fundamentals of Semiconductors: Physics and Materials Properties, 3rd ed., Berlin: Springer (2001)
24. Castellanos-Gomez, A., Agrait, N., Rubio-Bollinger, G: Optical identification of atomically thin dichalcogenide crystals. Appl. Phys. Lett. **96**(21), 213116 (2010)
25. Lee, C., et al.: Anomalous lattice vibrations of single- and few-layer MoS_2. ACS Nano **4**(5), 2695–2700 (2010)

26. Zeng, H., et al.: Optical signature of symmetry variations and spin-valley coupling in atomically thin tungsten dichalcogenides. Sci. Rep. **3**, 1608 (2013)
27. Shen, Y.R.: Principles of Nonlinear Optics. Wiley, New York (1984)
28. Kumar, N., et al.: Second harmonic microscopy of monolayer MoS_2. Phys. Rev. B **87**(16), 161403 (2013)
29. Mak, K.F., et al.: Control of valley polarization in monolayer MoS_2 by optical helicity. Nat. Nano. **7**(8), 494–498 (2012)
30. Cao, T., et al.: Valley-selective circular dichroism of monolayer molybdenum disulphide. Nat. Commun. **3**, 887 (2012)
31. Gibbs, H.M., Khitrova, G., Koch, S.W.: Exciton-polariton light-semiconductor coupling effects. Nat. Photon. **5**(5), 273 (2011)
32. Jones, A.M., Yu, H., Ghimire, N., Wu, S., Aivazian, G., Ross, J.S., Zhao, B., Yan, J., Mandrus, D., Xiao, D., Yao, W., Xu, X.: Optical generation of excitonic valley coherence in monolayer WSe_2. arXiv **1303**, 5318 (2013)
33. Parsons, R.R.: Band-to-band optical pumping in solids and polarized photoluminescence. Phys. Rev. Lett. **23**(20), 1152–1154 (1969)
34. Dyakonov, M.I.: Introduction to spin physics in semiconductors. Physica E **35**(2), 246–250 (2006)
35. Yang, C.L., et al.: Determination of the sign of g factors for conduction electrons using time-resolved Kerr rotation. Appl. Phys. Lett. **96**(15), 152109 (2010)
36. Dai, J., et al.: Magnetoelectric photocurrent generated by direct interband transitions in InGaAs/InAlAs two-dimensional electron gas. Phys. Rev. Lett. **104**(24), 246601 (2010)
37. Korn, T., et al.: Low-temperature photocarrier dynamics in monolayer MoS_2. Appl. Phys. Lett. **99**(10), 102109 (2011)
38. Sallen, G., et al.: Robust optical emission polarization in MoS_2 monolayers through selective valley excitation. Phys. Rev. B **86**(8), 081301 (2012)
39. Zhang, Y., et al.: Direct observation of a widely tunable bandgap in bilayer graphene. Nature **459**(7248), 820–823 (2009)
40. Wu, S., et al.: Electrical tuning of valley magnetic moment through symmetry control in bilayer MoS_2. Nat. Phys. **9**(3), 149–153 (2013)
41. Wakabayashi, N., Smith, H.G., Nicklow, R.M.: Lattice dynamics of hexagonal MoS_2 studied by neutron scattering. Phys. Rev. B **12**(2), 659–663 (1975)
42. Jiménez Sandoval, S., et al.: Raman study and lattice dynamics of single molecular layers of MoS_2. Phys. Rev. B **44**(8)3955–3962 (1991)
43. Kioseoglou, G., et al.: Valley polarization and intervalley scattering in monolayer MoS_2. Appl. Phys. Lett. **101**(22), 221907 (2012)
44. Molina-Sánchez, A., Wirtz, L.: Phonons in single-layer and few-layer MoS_2 and WS_2. Phys. Rev. B **84**(15), 155413 (2011)
45. Lu, H.-Z., et al.: Intervalley scattering and localization behaviors of spin-valley coupled dirac fermions. Phys. Rev. Lett. **110**(1), 016806 (2013)

Index

Z. M. Wang (ed.), *MoS_2*, Lecture Notes in Nanoscale Science and Technology 21,
DOI: 10.1007/978-3-319-02850-7,

MIX
Papier aus verantwortungsvollen Quellen
Paper from responsible sources
FSC® C105338

If you have any concerns about our products,
you can contact us on
ProductSafety@springernature.com

In case Publisher is established outside the EU,
the EU authorized representative is:
Springer Nature Customer Service Center GmbH
Europaplatz 3, 69115 Heidelberg, Germany

Printed by Libri Plureos GmbH
in Hamburg, Germany